Watson and DNA

WATSON AND DNA

Making a Scientific Revolution

VICTOR K. MCELHENY

A Merloyd Lawrence Book
BASIC BOOKS
A Member of the Perseus Books Group
New York

Copyright © 2004 by Victor K. McElheny

Hardback first published in 2003 by Perseus Publishing
Paperback first published in 2004 by Basic Books

Library of Congress Control Number: 2002114461
ISBN 0–7382–0341–6 (hc)   ISBN 0-7382-0866-3 (pbk)

Books published by Basic Books are available at special discounts for bulk purchases in the U.S. by corporations, institutions, and other organizations. For more information, please contact the Special Markets Department at the Perseus Books Group, 11 Cambridge Center, Cambridge, MA 02142, or call (800) 255-1514 or (617) 252-5298, or e-mail specialmarkets@perseusbooks.com.

Text design by Lisa Kreinbrink
Set in 11.5-point AGaramond by the Perseus Books Group

First printing, January 2003
1 2 3 4 5 6 7 8 9 10—06 05 04 03

For Ruth

CONTENTS

PREFACE

This account describes the improbable career of James Dewey Watson, a central figure in the revolution that has transformed biology again and again in the 50 years since he and Francis Crick discovered that DNA is a double helix, thus shedding bright new light on how characteristics are transmitted from generation to generation.

With the complete sequence of human DNA virtually assembled by the Human Genome Project that Watson launched, and innumerable celebrations of the fiftieth anniversary of the double helix, 2003 is an apt moment for further study of the DNA revolution. To be sure, the new biology's full effect on human life lies in the future. But it has already begun to change both agriculture and medicine and to alter profoundly our ideas of how living species evolve and are related to each other. The exploitation of the new genetic knowledge seems certain to transform prevention and treatment of diseases from crudity to precision.

The drumfire of discoveries by a few generations of biologists adds up to a huge challenge of intellectual history. One way of sorting out some of the complex intellectual genealogies is to follow the professional life of Jim Watson, who may have influenced the thinking of biologists more than any other scientist during this half century.

That influence is clearly a surprise. Watson could so easily have been brushed aside as a crazy kid and an arrogant pest, whose great discovery was a fluke based on others' data. But after much agonizing over what he would do for an encore, he settled on a career of getting things going. He became an intellectual manager on a vast scale—without showing the fatherly instincts of a Niels Bohr.

His conversion into a scientific impresario, picking talent and topics with brazen opportunism, building Cold Spring Harbor Laboratory into one of the world's greatest centers of biology, was the equivalent of Einstein taking over and running the Institute for Advanced Study. None of this was sweetheart stuff. Jim grabbed young scientists at their time of doing really new things, drove them for years of what felt like a war, and then sent them off to academic security. Very few duds ever got anywhere near him.

Jim was no angel; neither was Toscanini. People were influenced by Watson because of his combination of love of science with a fantastic ability to absorb and transmit information and analysis. People just wanted to talk to him, even if it meant getting their heads torn off.

Watson's actual story, 50 years of mornings after, is far subtler and more profound than would be guessed from either his canonical textbook of the new field of molecular biology or his brazen tale of how he and Francis Crick found the double helix.

What follows does not focus on Watson's family, or private life, but rather on his personality as he worked with an unusually confident and bold group of people who were determined to do great things in science. Fashioned without interviews with Watson or his family, or access to his papers, this distillation of the testimony of witnesses arose from my 40 years of following and reporting about the revolution in biology, from four years of working at Cold Spring Harbor Laboratory, and from the remarkably frank judgments of some 50 scientists who worked with him and generously granted interviews.

Acknowledgments

I wish to thank the subject of this book, James Dewey Watson, for invaluable help over many years in steering me to scientists, including Mark Ptashne and Walter Gilbert, who were doing some of the most interesting work in biology—which is probably the most exciting field of science in the last 50 years—and for the lessons I learned in four years of working for him on a high-wire project in environmental risk assessment. I wish to thank him and his wife, Elizabeth, for the friendship they have shown to me and my wife. Watson did not participate in this project. Over the last several years, he has been occupied with the production of several books of his own, and a major public-television series that is to be broadcast in 2003, when the fiftieth anniversary of the discovery of the DNA double helix occurs.

Crucial to this book has been the willingness of some 50 scientists, many of whom I have worked with as a science reporter, to be interviewed in the last three years about their experiences with and judgments of Watson, and for their conviction that the time had come for a summary account of Watson's amazing career. A list of those interviewed is printed in the back of this book. About a dozen of those interviewed agreed to go over the manuscript for errors and omissions, including Nancy Hopkins and Phillip Sharp of MIT, Paul Doty and Tom Maniatis of Harvard, Norton Zinder of Rockefeller University, Mark Ptashne of the Memorial Sloan-Kettering Cancer Center, and David Botstein of Stanford. They have helped me avoid many small and large mistakes. The book has also been examined by several journalists, including Harold Evans of New York, Jonathan Weiner of Pennsylvania, and Matt Ridley of Northumberlandshire in England. I must also thank the many others who also agreed to do interviews, if the time for researching and writing this account had permitted.

Also crucial has been the education provided over decades by biologists working at the frontier of their fields, including those I interviewed in their laboratories in Europe in the 1960s, such as Sydney Brenner, Francis Crick, Max Perutz, John Kendrew, and Fred Sanger at the Medical Research Council laboratory in Cambridge, England, and François Jacob, André Lwoff, and François Gros of the Pasteur Institute in Paris.

The knowledge of developments in biology since the 1960s that I have acquired is due in no small measure to the encouragement of editors such as Philip Abelson of *Science*, Thomas Winship and Ian Menzies of the *Boston Globe*, and Walter Sullivan, Henry Lieberman, John Wilford, and John Lee of the *New York Times*. Over the years I have been immeasurably aided by conversations with fellow journalists, such as successive groups of Knight Fellows at MIT, and with such colleagues at MIT as Philip Khoury, Carl Kaysen, Kenneth Keniston, Merritt Roe Smith, Nancy Hopkins, Phillip Sharp, Gobind Khorana, and Salvador Luria.

Even in a biography heavily reliant on personal impressions, including my own, archives are indispensable. I wish to thank the archivists at MIT for their work in assembling the great collection of documents and oral histories of the recombinant DNA controversy in the 1970s. The archivists at California Institute of Technology made it easy to study the collected papers of Max Delbrück and George Beadle. Archivists at Harvard University and Cold Spring Harbor Laboratory were indispensable in selecting suitable images for this book.

Without belief in the enterprise of a one-volume study of Watson and DNA, the project would have been impossible. I must thank most particularly my editor, Merloyd Lawrence, of Perseus Books, for insisting that this book be written and providing detailed reactions to the narrative. Seamlessly embedded in the text and footnotes is the skill of my copy editor, Kate Scott of Falmouth, Massachusetts, who has insisted on consistency and clarity. I must also thank my agent Jill Kneerim of the Hill and Barlow Agency, particularly for help in drafting the proposal. Of immense help in funding the travel and interview-transcription and other research essential to the project has been a grant from the Alfred P. Sloan Foundation of New York. I wish to thank both Doron Weber for securing this support and Arthur L. Singer, Jr., who worked for many years at the foundation, for his consistent guidance.

My greatest debt is to my wife, Ruth, who has supported this project from the day that I convinced her that I should undertake it, and provided limitless research and editing skills.

Cambridge, Massachusetts, 15 September 2002

PROLOGUE:
19 OCTOBER 1962

The scientist spoke passionately about spring in Cambridge, England, in 1953, about the unattainable girls he called popsies, about tennis, and about the race to be first to discover the structure of the gene. It was the morning of 19 October 1962, and, sitting in the front row of a Harvard lecture hall, I left my dated note-page blank. The students in the Natural Sciences 5 class had expected to continue hearing about the revolution in biology from serene, white-haired George Wald, already one of the smoothest and best-known teachers in America. The subject of the course was biology, not a biology of accumulating facts about living creatures but a biology whose focus was protein and nucleic acid molecules carrying out the work of the cell under the direction of genes—a biology concerned with the transfer of information from where it was stored to where it was used. Today, instead of Wald, a Martian string bean with wispy hair burst into the well of the lecture theater. A teaching assistant told us that 34-year-old James Watson had heard from Stockholm the previous day that he'd won the Nobel Prize.

The students cheered as if for a home team. Then, nervously brushing back his wisps of hair, Watson thrust on them the special, frenzied romance he wanted his life to be. For an hour he recited, in a hysterical, flippant stage whisper, the adolescent tale that would become, a few years later, his sensational book *The Double Helix*. Whenever Watson was about to say something he thought funny, his eyes would open as wide as they could and he would suck in his breath through his teeth, and break into an anticipatory smile that was more like a grimace. His account was a torrent of emotions, false starts, and surprises.

There was no sense of a man at a stage of life different from that of his youthful, puzzled, amazed listeners. Maybe he had never grown up, or never had had the stuffing knocked out of him by his childhood peers. The self-editing inculcated in most of us was absent. It was a bewildering performance, oscillating between intense self-deprecation and burning arrogance.

The students had very likely never encountered a person as passionate about his subject as Jim Watson. He was incandescent, all right, while somehow not repellent. All of a sudden, the DNA revolution could be seen as a wild inside joke—bizarre and romantic—and he was letting us in on it.

To intelligent young people he stood for the promise of the new molecular biology: they could attack abundant problems at the most fundamental levels of life. In his world, the smart ones were the young, not "the old gang." To be sure, I thought this wild man was in pain, and thought he needed protection, and a loving relationship with a woman, but he also was witty and racy, ready to be surprised and stimulated by new facts—and he was serious about nature.

At the time he spoke to George Wald's students, Watson was navigating in a storm of new ideas and experiments. Tensely cooperating and competing at the same time, scientists in labs at Harvard, Paris, and Cambridge, England, were racing to describe exactly how DNA drove the business of living cells: How did DNA duplicate itself, and how did it regulate the cell's bewilderingly complicated second-by-second workings? The competing labs were focused on deciphering the code in which DNA language was written and in discovering how various types of a related substance, RNA, got the DNA message out to the cell's protein factories. They also wanted to know how genes in cells that are specialized for particular tasks are shut off, as they must be; when a few of them do not, the result is lethal, an uncontrollably growing cancer cell.

As a young science reporter in the heady days after the Russians launched Sputnik in October 1957, I was predisposed to be excited by Jim Watson, even though he was living a story so rocky that it has become over the decades a kind of third rail of biography. He spoke to the students with such lighthearted and brutal frankness that one wondered if he mightn't be the most indiscreet scientist in the 300-year history of modern science. To the brash youths he was recruiting at Harvard and elsewhere to sweep aside the old biology, he offered no fatherly sympathy, but demanded that they be interesting—doing the next significant thing in science. Otherwise he would simply walk away. He mumbled in class, lecturing to the blackboard. He peppered his talks with seemingly irrelevant and sometimes malicious gossip about colleagues and competitors. Even when he talked with coworkers, he tended to look at his shoes. He frequently burst out in unpredictable anger. He once said, "'Nice' is what you do when you have nothing else to offer!" His countless acts of generosity were equally unpredictable, however calculated they may have been.

Jim Watson wanted to be relevant, useful, helpful, influential, and, above all, never bored. He was an uncompromising rationalist, a believer in facts, and totally committed to the doctrine that to solve problems one must break them into manageable pieces. His style was to bring that day's discoveries straight into the classroom, complete with anxiety and striving and gossip. This style was already attracting many of the best young people, including Harvard undergraduates, into the world of DNA science.

Aspiring young biologists fervently wanted Jim's attention in spite of the fact that they felt anxiety leap the moment he entered the room. While writing the canonical opening advertisement for the field, *Molecular Biology of the Gene*, in 1965, Jim become an intellectual impresario at Harvard, building a laboratory that many regarded as one of the most exciting in the world. He was learning the lessons he would later use in his phenomenal 25-year rebuilding of Cold Spring Harbor Laboratory, on Long Island, and in getting the Human Genome Project rolling. Watson was teaching thousands of new biologists how to think about their field when it was exploding like a supernova.

By the time of the Nobel announcement, Watson had largely left direct performance of scientific research. The day before, when journalists descended on his laboratory, the photographers demanded that the new Nobel Prize winner put on a white coat, something he never did. After resisting for a while, he complied, and gave a shy but triumphant smile.

The students listening that morning, though uncertain what Watson would say next, knew they were hearing the principal revolutionary of the brand-new world of DNA. With Francis Crick he had uncovered its double-helical structure, and like Crick he had striven to make additional significant discoveries—although he sensed that never again in his life would he make a discovery so important to science.

The purpose of George Wald's class, Natural Sciences 5, was to introduce freshmen and sophomores to the racing advances taking place in biology. Only nine years had passed since the discovery that the genes were embodied in a double helix of DNA, but scientists had confirmed the Watson-Crick model and gone a long way in unraveling the basic steps for genetically controlling the machinery of the living cell. The discoveries were turning the operations of life from the ineffable to the knowable. Life's workings were proving amenable to observation and experiment. In the approach often denounced as "reductionist," biology now focused on the smallest and simplest components of life.

DNA was the subject for which the well-known South African biologist Sydney Brenner proposed the motto "Think small, talk big." He added, "Genes are, in fact, molecules, and they use molecular machinery to transform the messages they contain into other molecules."[1] In 2002, Brenner shared in the Nobel Prize in Medicine.

Drawing heavily from physics and chemistry, these precise studies seemed likely to have vast relevance to increased understanding of the causes of cancer or hereditary diseases. Biology was beginning to seem less remote, and more practical, than the genetic-manipulation fantasies of a "brave new world." Amid unceasing talk in the early 1960s about total destruction from nuclear bombs, here was a science of peace.[2]

On that day, however, in the brutal concrete environment of Harvard's Allston Burr Lecture Hall, Watson was showing his tendency to trash himself worse than anyone planning a hatchet job, and by the same means ward off hagiographers. The "ice-blue eyes" of a ferociously clear, if intuitive, intellect were not immediately apparent.[3]

An audience in those Freud-conscious days of the early 1960s, when nurture—environment—was held to be more important than nature—heredity—might have guessed that Watson's childhood was bizarre and unhappy. But it wasn't. His problem was that he was a genius.

1

Books and Birds: "Growing Up" in Chicago

*I never even tried to be an adolescent. . . . I never went to
teenage parties. I never tried to talk like a teenager. That
probably made people dislike me. . . . I didn't fit in. I didn't
want to fit in. I basically passed from being a child to an adult.*
James D. Watson, *Washington Post*, 12 September 1989

"NO MONEY BUT LOTS OF BOOKS"

In Jim Watson's case, genius did not imply a high score on a test. He combined exceptional curiosity with exceptional willfulness and a special kind of self-confidence that made him refuse to waste time on subjects that weren't important. His vast capacity to absorb information and pull it together was allied with an insistence on arranging that information, intuitively and impulsively, into new patterns. People wondered and continued to wonder where the energy and urgency and plethora of ideas came from.

If there are rules about raising a genius, Jim Watson's parents appear to have grasped them instinctively. In a way that seems to have bolstered their son's confidence and love of learning, they faithfully followed a set of rules that a magazine writer amusingly stated in 2001: 1. Avoid the word "genius." 2. Find as many learning experiences as you can manage. 3. Don't overstructure the child's life. 4. Don't expect your child to be popular with age-mates. 5. Don't hold your child back in school in hopes of "normal" development.[1]

A child who possesses special gifts—where did they come from?—such as a very unusual type of intellect, must have stability, rules, and encouragement. Jim's young life, which he recalled time after time in speeches and interviews over several decades, appears to have been controlled and normal, if pinched for lack of money. In Jim's world, love meant marriage.

It was a life filled with curiosity, but not, as with many scientists, triggered by a teacher who motivated him. In 2000, Watson recalled, "I think I was interested before I had a teacher."[2]

This curiosity was what kept him going, along with the pleasure of reading facts and interpreting them. "Even before high school, I wanted to master the basic laws of nature," he said. Darwin fascinated him.[3]

Life at the Watson home was very different from the one imagined many years later by the great French biologist André Lwoff. Exasperated by what he saw as Watson's merciless gossip about friends, rivals, and, above all, himself, Lwoff complained of the just-married 40-year-old genius, "All things considered, it seems as though Jim's heart has not been nurtured and touched long enough by a loving and beloved person." But this civilized, even sympathetic comment does not reflect what Jim recalls.[4]

It probably helped Jim that his parents were not too old. James Dewey Watson, Sr., a lapsed Episcopalian and loyal Democrat with well-off Republican relatives, was turning 31 when his son was born on 6 April 1928. Jim's mother, Jean Mitchell Watson, the Catholic daughter of an Irish mother and a Scottish father, a tailor, was 27. Right through the Depression of the 1930s, they both held jobs and paid their bills, didn't operate a car, and clung tightly to the middle class. James Watson worked as a bill collector, and his wife did secretarial and administrative work. They made their way despite health problems. His mother's heart had been weakened by a childhood streptococcal infection and she died at the age of 56.[5] By the time Jim was born his father had already begun an eventually lethal 40-year habit of smoking two packs of Camels a day.[6]

It surely helped their son that both James and Jean Watson loved learning and respected it above business as a career. Their son recalled, "My family had no money but lots of books."[7] Most of the books in the modest Watson home at 7922 South Luella Avenue, on Chicago's South Side—which Watson later delighted in saying was closer to the steel mills of Gary, Indiana, further south, than to the University of Chicago[8]— were on philosophy. Jim avoided the philosophy books, but he devoured the volumes that were full of facts of any kind, especially if those facts were scientific. By the time Jim was seven, he preferred books to toys as Christmas presents.[9] A cherished gift was a book on the migration of birds that an uncle gave him the Christmas he was seven, presaging many years of constant bird-watching with his father.[10]

It also helped that Jim was not an only child. He was close to his sister Elizabeth, called Betty, who was two years younger than him, and remained so throughout her life. She married a U.S. intelligence official in Japan in the fall of 1953. Earlier that year, at the end of a lengthy stay in Europe, Betty typed the manuscript of the initial announcement of the Watson-Crick structure for DNA (deoxyribonucleic acid).[11] Lwoff, angrily reviewing Jim's famously indiscreet *The Double Helix*, in 1968, wrote of her as the "principal object of her brother's attachment and potential for affection."[12]

A further steadying influence was his maternal grandmother, Elizabeth Mitchell, nicknamed Nana, who lived nearby until 1933, when he was five, and then moved into the Watsons' heavily mortgaged "bungalow" on Chicago's South Side. Each day, Nana welcomed Jim and his sister home from school and made them supper when their mother was not yet home from work.[13] Jean Watson took care of her mother for more than 20 years, indeed until three years before her own premature death from heart disease. She had lost her father when he was killed in an accident caused by a runaway horse on her seventh birthday, Christmas Eve, 1907.

After lack of money forced Jean to drop out of the University of Chicago, she took a job at the LaSalle Extension University. There she met James Watson, Sr., who had become trapped in the role of the school's bill collector, and she married him in 1925. Jean Watson was a born organizer. At various times she worked for the Red Cross. In her basement, working for $12 a night, she helped get out the vote for the Democratic machine of Chicago,[14] and served as a secretary in the admissions and housing offices of the University of Chicago.[15]

Both Jim and Betty Watson left high school early to attend the university, which had an unusual policy of taking students at around 15. Their parents made sure they applied for scholarships. They lived at home, a three-cent, 30-minute streetcar ride north to the university campus.[16]

Jim remained close to his parents. After graduating from college in 1947, when he was 19, and going on to graduate school at Indiana University down in Bloomington, he visited them frequently in Chicago. By the time Betty graduated in 1949, the Watsons had bought a home near the Indiana Dunes State Park, where he and his father had often gone bird-watching, and Jim visited them there. At last the Watsons had a car, although, sadly, Jean Watson had to use it frequently for exhausting drives to visit her mother at a nursing home.[17] When away from home, Watson wrote to his mother frequently. Indeed, those letters provided the chronology for *The Double Helix*.[18]

In both speeches and press interviews, Watson described his parents with a light touch, but respectfully. He recalled his mother as "very supportive" of the family. She made Watson wear rubbers over his shoes, although Watson "hated it."[19] It was a sore point that Jim long remembered. In 1974 he said, "As a boy in Chicago, I used to equate good manners with wearing rubbers over your shoes—something your parents got mad about when you forgot, but somehow sissy and not at all connected with the world of science that I wanted to be part of." His main worry then was "the limitations of my brain."[20] Watson remembered his mother as having "good taste but no money." She felt that if you had money, you weren't supposed to show it off. In arguments with her son about the sources of human personality, she upheld the importance of inheritance, of nature, whereas Jim, an uncompromising liberal, upheld the role of environment, of nurture.

When the South African biologist Sydney Brenner visited the Watson home in 1954, he was impressed by Jean Watson, whom he recalled as "the organizing person . . . the brains . . . the mainstay" of the household, the "breadwinner [who] held it together." According to Brenner, Jim's father, by contrast, surrounded himself with books and yearned to be a philosopher.

His father's job as a bill collector was "miserable," Jim recalled. He would have been happier as a schoolteacher. He "worshipped persons of reason and took particular pleasure in reading the thoughts of the great philosophers." On his visit to the Watson home, Brenner observed that "Jim's father was a kind of failure in a sense that he was never able to do the things that in fact he was unable to do, if I could put it that way."[21] For the decade he survived his wife, Jim's father lived with his son.

Watson never saw his parents mistreat anybody. Always on the side of the underdog, they believed in social justice in a time when it was simple to recognize the enemies: Hitler and those who opposed Franklin Roosevelt. Watson remembered Roosevelt as his "first real hero. . . . Then, if you had a family car, you could afford to be a Republican, but if you had been knocked down by the Depression, common sense made you a Democrat."[22] The Watsons thought: "Unions were good. The *Chicago Tribune* was bad. Roosevelt was good. Churchill was good."[23]

Jim's parents believed in knowledge as liberation from illness, poverty, and superstition. They were convinced that truth came from observation and experiment, not "revelations,"[24] and that finding out about life and its place in the universe was "a glorious endeavor that should be undertaken for its own sake."[25]

When he was in Horace Mann elementary school in Chicago, Jim sneaked up one day to his teacher's desk to see what his I.Q. was. "Pretty low," around 120, and yet he recorded later that he read 400 to 500 words a minute. Nonetheless, Jim did not think of himself as a prodigy. "I never felt myself smart as a young kid."[26] He thought of himself as an underdog who must work harder.[27]

He was small, unathletic, bookish, and "excessively shy," and looked to adults for friendship. In school, he recalled being "so unpopular."[28] His unpopularity led to bullying that he still resented decades later. "Dumb kids bored me," he remembered. "Sometimes they beat me up. They knew I wasn't one of them."[29] Early in the 1980s he told a young scientist that he had been "getting back" ever since.[30]

He was unpopular also with his father's well-off Republican relatives, such as an uncle who lectured at the Art Institute of Chicago and another uncle who taught physics at Yale. Prickly and undiplomatic even in childhood, he couldn't stop talking about Roosevelt. "Markedly bright and never accustomed to hide the fact," as one later writer put it, he thought he shouldn't worry about being like people he didn't respect.[31] One should not curry favor from fools.[32] Why pay lip service to forms of behavior that, at best, generate more hypocrisy? As he remarked in 1974, politeness and conventions remained for "minor minds unwilling to take chances."[33]

Although he was relentlessly optimistic, there were occasional cranky, morose comments. In 1975, speaking at a school attended by his two nieces, Watson hoped for a day when science would be indispensable to the education of all, not "limited to those compulsively bright kids, whose enthusiasm for perpetual learning makes them congenitally unable to put up with the polite banter that characterizes the successful lawyer or banker or businessman."[34]

Jim looked to books for a "way of seeing the world I wanted to enter."[35] When he was 10, he recalled, he stayed up nights reading the *World Almanac*.[36] Every Friday, Watson and his father would walk about a mile from their house on Luella Avenue, just south of Seventy-ninth Street, to the library on Seventy-third to browse through the stacks and borrow books.[37] His best friends seemed to be books. He recalled later, "I got through a great deal of my childhood reading novels because the environment was boring. If a child on the South Side of Chicago learned much of anything, he learned it from books, not from peers. Peers aren't generally very interesting when you're young. It's books that give you a sense of the outside world."[38]

BIRD-WATCHING QUIZ KID

Soon after 12-year-old Jim was confirmed in the Catholic church, he crossed a divide. Like many boys at the approach of adolescence, he shifted from his mother's orbit to his father's. He stopped going to Mass and instead went bird-watching with his father. Although Nana surely was disappointed by Jim's decision to stop going to Sunday Mass at Our Lady of Peace Church, Jim did not recall her speaking out about it.

Jim's attention turned to a new issue: the scientific nature of life, which was at that time still "fairly mysterious." He later recalled, "I had always had a desire to know what life was, following in the footsteps of my father. He couldn't stand religion. My mother was nominally a Catholic, and until I was twelve, I went to a Catholic church and was confirmed. Then I came to the conclusion that the church was just a bunch of fascists that supported Franco. I stopped going on Sunday mornings and watched the birds with my father instead. The Catholic Church at that time had a pretty dismal world view."[39]

Around this time, he found a chance to shine in a highly public forum.

In June of 1940, *The Quiz Kids,* a half-hour radio show conceived by a neighbor of the Watsons', Louis G. Cowan, went on the air late Friday evenings as a summer substitute. Cowan had thought of it as a junior edition of the highly popular and witty *Information Please.* Sponsored by Alka-Seltzer and broadcast from Studio E in Chicago's Merchandise Mart, *The Quiz Kids* was so successful that in September it moved to 8 P.M. on Wednesdays, where it ran for two years before settling into Sunday slots until it went off the air in 1953.

Jim, the late-night reader of the *World Almanac,* was one of the stream of "geniuses," some of them only six years old, who were drafted to compete in answering brain twisters posed by the down-to-earth quizmaster, Joe Kelly, who often played the dunce. The topics ranged from mathematics and natural history to the Bible and Shakespeare. Five children appeared on each show, and the three who scored highest stayed on for the next week. Each child received a $100 war bond for appearing. Watson used his to buy a pair of binoculars.[40]

The show became such a recognized part of popular culture, that in later years journalists delighted in mentioning that Jim had been a contestant. He told one journalist, "The only reason I was on was that the producer of the program was literally our next-door neighbor. I was bright enough so that I knew a lot of facts." Jim lasted three broadcasts, but was tripped up by questions about Shakespeare and the Old Testament. He recalled once, "They had a Jewish girl and asked a lot of questions on the Old Testament." He may have been referring to Ruth Duskin, then seven, who became a regular. Jim later consoled himself with the thought that his father "would have been angry with me" if he had known the answers to the religious questions. Competing on the show helped his ego: "I knew I wasn't hopeless."[41] His sister Betty was less charitable. She thought Jim had lost out because he was too straightforward for the producers, who wanted audience-grabbing remarks from their little geniuses.

At about this time, Jim began attending the famous Laboratory School of the University of Chicago, a progressive high school that was the brainchild of the philosopher John Dewey, whose ideas on education and other subjects entranced Jim's father. But even there, he remained a loner:

> I didn't fit in too well with, say, kids my own age. Books were my way of
> seeing the world I wanted to enter. Birds were a way for me to get into science. Partly it was just trying to see rare birds. But with time, I wanted to
> understand how they migrated, a problem we still don't quite understand,
> in terms of, how can the brain really know where it is and take these long
> hops across oceans?[42]

During these years, Watson, like many other leading biologists of his generation, was pulled toward science by reading Sinclair Lewis's *Arrowsmith*, which is an archetypal narrative of the place of basic science in treating disease. In 1968, when his own book, *The Double Helix,* was beginning to grab similar attention among young people aiming to go into science, Watson told reporters, "Basically the only book I ever read that told me about it was *Arrowsmith*."[43]

Lewis's Pulitzer Prize–winning 1925 novel was almost uncannily full of echoes of Jim's future. The leading character, Martin Arrowsmith, is pulled between two poles of life science—the practice of medicine using what is known now, and the discovery of new facts about life processes that open the way to the medicine of the future. The idealistic Arrowsmith gets started as a physician, often working for fools, poseurs, and money-grubbers, but eventually he takes a scientist's vows and enters the religion of research. After the explosion by Darwin and many others of the notion of a divine plan, science is what is left of philosophy.

Paul de Kruif, scientific consultant to Lewis, wrote in *Microbe Hunters* that science puts a sword in the hand of those fighting disease. In finding the microorganism that causes tuberculosis, de Kruif writes, the great microbiologist Robert Koch began "to change the whole business of doctors from a foolish hocus-pocus with pills and leeches into an intelligent fight where science instead of superstition was the weapon."[44] It was just this belief in science as the path to conquering disease that led to the founding of the Rockefeller Institute for Medical Research in New York in 1901—almost simultaneously with the rebirth of Mendelian genetics—which had been forgotten for decades—around 1900.[45]

In *Arrowsmith,* the representative of the demanding religion of science is a professor named Max Gottlieb. A few of his students do not just want "to kill patients"—they want to escape the commonly unavoidable helplessness of doctors in those days and search for knowledge that would cure disease. They want "to work with bugs and make mistakes . . . to wait and doubt." Arrowsmith becomes one of these, enchanted by the bubbling and steaming devices for microbiological research, working in the lab by himself at night, and winning Gottlieb's praise for "craftsmanship," which he calls the artistic impulse that is crucial for "the beautiful dullness of long labors." When Gottlieb challenges Arrowsmith over his research, Arrowsmith angrily retorts that he has done the work over and over. "I can't help what the dogma is . . . I only know what I observe!" Gottlieb is delighted. But circumstances force Arrowsmith away from the lab into clinical medicine. Not until years later can he join Gottlieb at a research center in New York (based on the actual Rockefeller Institute). Arriving, he learns from Gottlieb one of the harshest lessons for a beginning scientist: independence. "You are not to help me. You are to do your own work." The first question is, "What do you want to do?"

Gottlieb derides "the doctors who want to snatch our science before it is tested and rush around hoping they heal people, and spoiling all the clues with their footsteps." A real scientist is "the only real revolutionary . . . because he alone knows how little he knows." He adds, "He must be heartless. He lives in a cold, clear light."

When Arrowsmith's wife, Leora, complains to him that he has little sympathy for the shy, the lonely, or the stupid, he says, "I'm too much absorbed in my work, or in doping stuff out, to waste time on morons!"

Leora accompanies her husband to a plague-stricken island in the West Indies, where Arrowsmith faces an agonizing choice between the claims of science for precise knowledge and those of people suffering from the dreadful epidemic who cry out for treatment, even if untested. Exploring his cure for plague, he is unable to withhold the therapy from a control group to get clear statistics on whether his treatment is superior to others. Eventually Leora contracts the plague that Martin has gone to the island to study, and dies alone while her husband is away doing field work. Science could be tragic as well as thrilling.[46]

AVERY AND SCHRÖDINGER

Just a few months before 15-year-old Jim entered the University of Chicago in the fall of 1943, a physicist in exile from Austria, Erwin Schrödinger, was giving a series of lectures in Dublin. In one lecture he told his audience that the gene, the unit of heredity in living things, "must reflect," as Watson wrote later, "the precise arrangement of atoms within the molecules of heredity located on chromosomes." The message was, "To truly understand life, we must pursue genetics at the molecular level."[47]

Many years later, the molecular biologist Gunther S. Stent put it this way: "In retrospect, the most important point made by Schrödinger was that the gene is to be thought of as an *information carrier*. And the only reasonable way in which genes could be imagined to carry their hereditary information is by embodying a succession of a small number of different repeating elements, or symbols, whose exact pattern of succession represents an encoded genetic message."[48]

As an example of this principle in action, Schrödinger cited the Morse telegraphic code of combinations of dots and dashes to represent the letters of words.

Schrödinger drew on research done 10 years earlier by three scientists; one of the three was Max Delbrück, a physicist who had changed his focus to biology and later fled Hitler's Germany. A few years later, Delbrück became one of Watson's scientific parents. In a way, Delbrück was the hero of *What Is Life?*, the little book made from Schrödinger's Dublin lectures. In it Schrödinger heralded a day when living processes could be studied by physicists.

About the same time as Schrödinger's lectures, the American geneticist Oswald T. Avery and two colleagues at the Rockefeller Institute were making an epochal discovery:[49] After years of doubt-filled research, they found that they could not escape a puzzling and startling conclusion. In their studies of pneumococci, bacteria that cause pneumonia, they found that the units of heredity did not reside in complex proteins but instead in a seemingly boring chemical called DNA, which was found only in the nuclei of cells. They said that their evidence "supports the belief that a nucleic acid of the deoxyribose type is the fundamental unit of the transforming principle of *Pneumococcus* Type III." They had explained the transformation puzzle first found by the English pathologist Frederick Griffith in 1928. Watson once described this puzzle: "You could change the morphology of bacteria by adding a sort of filtrate of one bacteria to another," but no one knew what the filtrate was.[50]

Avery's finding inspired Erwin Chargaff, also an émigré, who became one of the leaders in DNA research: "This discovery," he wrote in 1976, "almost abruptly, appeared to foreshadow a chemistry of heredity." It showed where to look for "the first text of a new language," a code that probably depended on the sequence of nucleotides in DNA. Chargaff resolved to enter the field, and so did the British physicist Maurice Wilkins.[51]

Watson was to write later, "Given the fact that DNA was known to occur in the chromosomes of all cells, Avery's experiments strongly suggested that future experiments would show that all genes were composed of DNA."[52] Yet the insight caused little stir. Did bacteria have genetics like higher organisms? "At that time," Watson reflected in 1981, "bacteria were thought to be different from other cells. They didn't have a nucleus. People thought that maybe bacteria just grew and divided in some very amorphous fashion, and that they didn't have genetics." Even though Max Delbrück and his Italian colleague, Salvador Luria—also a refugee physicist-turned-biologist—had found, also in 1943, what looked like the genuine mutations of higher cells in bacteria, it was not until 1946 that Joshua Lederberg discovered that bacteria could mate and recombine genetic material, just like higher cells. Until then, "the relationship of bacteria to other forms of life was very unclear."[53] Besides, the structure of DNA seemed uninteresting, with its nucleotide units of ribose sugars combined with phosphates and bases monotonously repeating over and over according to what was called the "tetranucleotide hypothesis." Chargaff found that the four different nucleotides arranged themselves in particular ratios. In widely different species, the amount of adenine always equaled that of thymine, and the amount of cytosine and guanine were also identical. This was true even though the amounts of A=T and G=C varied greatly from species to species. DNA was just beginning to look like a molecule that could code genetic information.

Within five years, Chargaff's dogged biochemistry had overthrown the tetranucleotide hypothesis.[54]

Until Chargaff's findings, biologists had trouble believing that DNA really carried the genetic information of "higher cells." Jim told a reporter in 1994 that Avery's finding "was one of the great discoveries of science, but it didn't at that time get the recognition."[55] Alfred Hershey, who nine years after Avery discovered that DNA was the genetic material of bacterial viruses—himself a very quiet man—thought Avery and his colleagues "were just too modest. They refused to advertise."[56]

LEARNING TO THINK: COLLEGE AT 15

Shortly before Avery submitted his paper to *The Journal of Experimental Medicine* in 1943, Jim became one of the few students to be admitted each year to the University of Chicago after only two years of high school. Precocity, Jim maintained ever afterward, had nothing to do with it. Robert Hutchins, the iconoclast president of the University of Chicago, believed that the last two years of high school were useless, stunting the faculty of getting beyond the lists of facts to the challenge to think. Watson thought later that he was chosen because "I was a keen reader with a much better-than-ordinary memory."[57]

Jim's original goal was to become a naturalist, but he began to change his mind as he found that science was more than listing and describing. His first two years at the university were nothing special, he recalled. He later described himself as "too small, too skinny, too homely, and too young to go out with girls."[58] He did not feel at ease until the third year.[59] In the fall of 1945 Jim came upon Schrödinger's *What Is Life?* in the university library. At almost the same time, across the Atlantic, the book was inspiring an English physicist named Francis Crick, who was not put off by what he regarded as Schrödinger's ignorance of chemistry. Compelling and imaginative, the book suggested to Crick that "biological problems could be *thought* about—in physical terms." Exciting discoveries appeared to lie not far off.[60] It was a blow at vitalism, the idea that life somehow transcended physics and chemistry. Even at the beginning of his argument, Schrödinger asserted that "the obvious inability of present-day physics to account for [the events of life] is no reason at all for doubting that they can be accounted for" by physical science. As Watson recalled later, Schrödinger "very elegantly propounded the belief that genes were the key components of living cells and that, to understand what life is, we must know how genes act."[61]

Schrödinger's picture of coded messages that had to lie in the chromosomes—"Every complete set of chromosomes contains the code" — enchanted Jim. Schrödinger argued that "the essence of life had to be information, which was faithfully copied every time a chromosome divided." The genes, each of which consisted of, at most, a few million atoms, achieve a permanence that is "almost absolute" in transmitting traits to succeeding generations. Humans are creatures whose being is "entirely based" on such marvelous interplay. The only greater marvel is that human beings "possess the power of acquiring considerable knowledge about it." But how, Watson asked himself, could a cell carry information? Bound by the perceptions of those days, Schrödinger guessed that the key was a protein. Despite Avery's experiments, Watson recalled, "No one talked much of DNA then."[62]

Toward the end of the book he admired so much, Jim detected a mystical tone that put him off. "I'd been taught to avoid anything which was in any way mystical. . . . You had to have observations and experiments and just avoid truth by revelation."[63] Still, the Schrödinger book weakened his idea of becoming a naturalist: "Suddenly birds seemed objectives for outdoor fun, not for serious science."[64]

Not every aspiring scientist who read Schrödinger's little book was swept off his feet. The most notable skeptic was Sydney Brenner, who confessed in 1968, "I read it and absolutely did not understand what he was driving at."[65] To Brenner, the book represented a "romantic dream" of Delbrück's that new laws of physics would come out of studying life. He scoffed at this. At Cambridge University in England, where Brenner worked, the faith was that life would turn out to be "some fancy kind of chemistry." When biologists knocked at the door of the cell, "someone would reply, 'There's nobody here but us machines.' " Delbrück's romantic dream was smashed in 1953, Brenner said, by Watson and Crick.

Most of Watson's grades at the University of Chicago were B's. "I don't think whenever you get a B and your ambition is to be an academic, you feel very good." In philosophy, his grade was somewhere between a B and C. "I hated it, every aspect of it." Then and fifty years later.[66] But he was an unusual B student. His professor in embryology and vertebrate zoology, Paul Weiss, said in 1973 that Watson "was (or appeared to be) completely indifferent to anything that went on in the class; he never took any notes— and yet at the end of the course he came [out] on top of the class."[67] (Despite this encomium, the same professor, just a few years later, sought to cut off Jim's fellowship when he had the temerity to shift from Copenhagen to Cambridge, England, without either permission or qualifications to study the structure of big molecules, including DNA, with X-rays.)

The situation looked very different to Jim. He took an unusual stance toward his Chicago courses. The final exams bore little relation to what he heard in the lectures—so it was useless to take notes. Instead, he concentrated intently on whether the lecturer's remarks added up. He found that "it was the ideas, not the facts that led to good grades in Robert Hutchins's college."[68] He began to appreciate that the task was to learn "what laws [of nature] really make sense and what are silly and shouldn't be there."[69]

He was learning what he recalled as three principles. The first was to go directly to the great sources and not bother with the interpretations. "Go to the original source things whenever you can." The second was the importance of theory—figuring out how to put the facts together in a rational scheme. "You always wanted to know not what happened, but why it happened. So you wanted an explanation in terms of causality." The third was to concentrate "on learning how to think, as opposed to improving memorization skills." The message he was getting at the university was, "Don't think about unimportant things."[70]

When he started the university at age 15, "the students were bright and I was scared," he recalled, but "by the time of my senior year, I wasn't scared."[71] Not only was he learning what was important, but also "what needs to be done" now—not what was needed in the past.

In his mind, he continued to treat authority with little reverence. "I guess I never felt part of an establishment. Conventional wisdom is often wrong." As he looked back on these years, Watson guessed he was "a fighter" who wasn't satisfied with "people who avoid the truth."[72] At a university far younger than Harvard, "You were never held back by manners, and crap was best called crap."[73] This lack of politeness did not help his relations with classmates—or later with the genteel people at Harvard. Jim did not see himself as a shrinking violet, even though, as he recalled, he thought of himself as still "too short to see the need to move outside the security of my family home."[74] He had to charge though "a hard intellectual obstacle course" of great books. Fellow students pulled no punches. "An illogical argument inevitably resulted in verbal slaps that seldom elicited the sympathy of others."[75] He recalled, "They knock you down, and then you had to get up, and no one was gonna pick you up. But it's good to be in a place where they knock you down." Watson was being taught to be different "if later I was to succeed."[76]

Near the end of his four years at the university, as he was considering where to go to graduate school, Jim sat in on the course in physiological genetics taught by Sewall Wright, one of the leading geneticists in the United States.[77] From Schrödinger's book Jim had learned that the key to life and heredity was the gene, which carried specific instructions and somehow was duplicated with perfect fidelity from generation to generation. The Schrödinger book inspired him to attend Wright's lectures. There, he heard Wright mention the work of Oswald Avery—as he had begun doing since 1944, when Avery's paper was published and drew modest interest. Jim came away with three questions: "What is the gene? . . . How is the gene copied? . . . How does the gene function?"[78] He needed to go to graduate school and learn genetics.

2

TARGET, THE GENE: BLOOMINGTON AND "PARADISE"

When you get into science, you realize the people above you are not gods; that they're very human and often they'll say things which you'll regard as just wrong. That's why you want to get close to great people.
JAMES D. WATSON, *TALK OF THE NATION*, 2 JUNE 2000

"HIGH-POWER MINDS"

Contingency always fascinated Jim Watson. How easily things could have turned out differently! How narrowly he missed reaching a dead end! In his last year at Chicago, as he was finishing his degree in zoology and applying to graduate schools, chance played a big role. His plans were all so improbable. As a relic of his old aspirations to be a naturalist and a curator of birds at a top museum, he mentioned ornithology as an interest in his applications. He had little formal training in genetics and yet genetics was what he intended to study. The California Institute of Technology (Caltech), in Pasadena, was strong in genetics, but they turned him down—was it the C in physics, or distrust of Robert Hutchins's unconventional policies at the University of Chicago? Harvard, weak in genetics at the time, accepted him but offered no financial assistance.[1]

Fortunately, his adviser had seen to it that he also applied to Indiana University. Watson did want to move into genetics, and Indiana was strong in genetics. On its faculty were such luminaries as Herman Muller, the pioneer of using radiation to induce mutations in fruit flies, and two others who worked with microorganisms: Tracy Sonneborn, who did genetics with paramecia, and Salvador Luria, an Italian émigré physician and physicist turned biologist who studied viruses that prey on bacteria. These tiny packages of genes are called bacteriophages—phages for short—and their genetics could be described statistically. Jim wrote many years later that Salva Luria and Max Delbrück had "changed the face of genetics by making the bacteria the obvious organisms for research on the nature of the gene. Experiments could be done in a day instead of weeks."[2]

Muller, who had come to Indiana the year before, in Jim's opinion "was perhaps the only American geneticist with intellectual credentials superior to those of Sewall Wright." Jim heard that the younger geneticists, Sonneborn and Luria, "were both very clever and I might want to do my thesis work with them."[3]

Dean Fernandus Payne wrote Jim that he was ready to take a chance on him, but added sharply that if his heart was still set on ornithology, he should try someplace else, like Cornell. After this, Jim visited Bloomington with his father, and the university awarded him one of half a dozen available fellowships; Jim's was worth $900 a year.[4]

As a zoology student starting in on lectures at Bloomington, Jim was at first attracted to Muller's *Drosophila* fruit flies, but even though he got an A in Muller's course, "Mutation and the Gene," he decided that *Drosophila's* "better days were over." Although students adored Sonneborn and were afraid of Luria's apparent arrogance toward those he believed were wrong, Jim was more interested in Luria's bacteriophages as a tool of genetics than in paramecia. What was more, Salva collaborated with Delbrück. Jim later recalled Luria as "a rather quiet individual [who did not] have a sense of theater."[5]

Jim was obsessed by competition. How lucky it was that Caltech had turned him down! "There I would have felt myself inferior to many of my peers with much better backgrounds in physics, math and chemistry. But at Indiana, I was the only incoming Ph.D. student who already had started to think about genes." Chicago's "hard intellectual obstacle course," its insistence on "the big picture as opposed to the details that go nowhere," was paying off.[6]

Now tall and thin and awkward, Jim usually wore tennis shoes as he moved about the campus. In the corridors, he would walk past fellow students with a far-away look in his eyes, disdaining talk. Sonneborn invited him to join Friday evening seminars at his house. There he irritated some graduate students by his way of steering the conversation where he wanted it to go, and annoyed all by "his habit of opening a book to read when speakers are dull and unintelligent."[7] Shy or not, he boldly, singlemindedly went up to Muller, Sonneborn, or Luria after seminars for discussions. Once again, his preferred contacts were with his elders.

At Bloomington, Jim's "adolescent fantasies" were being fulfilled among the "high-power minds" he was meeting for the first time. He exulted, "The truly bright did not live like our relatives or nearby neighbors and wasted little time worrying about how they looked, or the polish on their cars, or whether their lawns were overrun with crab grass." He was learning that he could burst through age barriers—and not destroy his career by setting his elders straight. Could he be in the same league as leading American biologists?[8] Having become "part of the highest form of human achievement," science, he felt "light-years away from the uninformed prejudices of the poor or the callous self-satisfaction of the rich."[9]

In Watson's paper for Sonneborn's class, which he wrote on the genetics of *Chlamydomonas,* he was not shy about writing critically of the work of a scientist named Moewus. "Some of the statements reported as facts were merely wishful thinking. It is hard to imagine how all of [the] work was ever done." Later, Moewus admitted that the experiments were unrepeatable.[10]

Indiana, Jim reflected later, probably was "the best place in the world" for him. There, "people were trying to move into the future. . . . When you go to graduate school, you've got to become future-oriented." Jim aspired "to make the next big step in a major scientific thing."[11] No one knew how to do it, but he could reject the professors who clearly weren't getting anywhere. Half a century later, in an on-line exchange, he advised a student not to work in an area where there are "too many facts."[12]

At Bloomington, the significance of what he had heard in Wright's course at Chicago began to hit Jim "with a vengeance." As units of heredity, genes had to replicate. Luria's lectures pulled him "into the center of the gene-replication dilemma," which was that no one knew how they did it.[13] It was clear, Luria said in the 1970s, "that we were thinking about nothing but the gene."[14] At the time, Watson reflected later, one good idea was that genes probably provided the information for making proteins, but biologists still had only a foggy notion of what a protein was. A second good idea was that it might be DNA that carried the genes, and that viruses could yield answers about that sooner than larger organisms could.[15] But in those days, scientists still lacked a clear idea of what a virus was. Despite Avery's finding, published in 1944, that DNA carried the genes in *Pneumococcus,* many scientists found it hard to believe that genes for such complex things as proteins could be anywhere but in equally complex proteins. Watson took a course with one of the disbelievers, Felix Haurowitz, and got an A.[16]

Like Max Delbrück, his copioneer on bacteriophages, "Lu," as Salva Luria signed his letters, was "not afraid to say what . . . was bad science."[17] Together with Delbrück, Luria had "changed the face of genetics" by showing that the bacteria their viruses preyed on were "the obvious organisms for research on the nature of the gene. Experiments could be done in a day instead of weeks." He also showed that the bacteriophage viruses could form mutants "every bit as stable as those found in bacteria." Delightfully to Jim, Luria was impolite, unlike many professors who were "too gentlemanly to unmask the trivial."[18] He was attracted by the simplicity of Luria's bacteriophage system. After only a few days in Luria's course on viruses, and even though he did not know Luria at all, and was just a zoology student among 40 in the class, Jim asked to start working in Luria's lab in the spring term of 1948.[19] Although Salva found Jim at that time "even more odd than later," he accepted him promptly. He was "tremendously intelligent, with this mixture of self-assurance and uncertainty of himself that very often bright kids have."[20]

Getting into a good lab was vital, as Watson appreciated later, and it was best "if you work for a young person who's later going to be important." The ideas are likely to be new and the professor will not be surrounded by too many students.[21] "I think you're unlikely to make an impact unless you get into a really important lab at a young age because you're unlikely to know what problem to work on," Watson said in 1992.[22]

Luria set him to work on bacteriophages that had been made inactive by X-rays; Luria himself had been studying phages inactivated by less energetic ultraviolet rays, and their reactivation in a process he called "multiplicity reactivation." Jim's work on phages treated with X-rays, he saw later, was "a routine extension of Luria's prior work,"[23] and so he did not need to be clever in planning the next day's experiments. In Salva's opinion, Jim "did a rather simple problem for his thesis but very beautifully." Although Jim looked disheveled all the time, his notebooks were more perfect than the notebooks of anyone else Luria ever saw. His odd student was "a mess—except in things that mattered." Luria rejected any idea that he had "programmed" Watson. He thought his biggest contribution to Jim was "to give him a pleasant environment."[24]

One of Jim's lab mates was Renato Dulbecco from Italy, later a Nobel Prize winner for his pioneering discoveries on viruses that induce cancer. Dulbecco's family had not yet joined him, and he and Watson occasionally dined together at the Indiana Union. Jim tried out an idea on Renato, based on a theory of Luria's, that the T2 bacterial virus had 25 genes. Why not calculate a crude weight of a gene by using electron-microscope pictures that roughly indicated a total weight of the virus? Dulbecco wasn't interested, perhaps because he was skeptical of Luria's theory, or for a more general reason: "Despite Avery, McCarty, and MacLeod, we were not at all sure" that DNA was the only component of phages that "carried genetic specificity."[25]

"FUN AND GAMES" WITH MAX

In the spring of 1948, not long after Jim started working in Salva's lab, Max Delbrück stopped off in Bloomington for a day and met this excited student, who thought of him as "a legendary figure" because his ideas were so prominent in Schrödinger's *What Is Life?*[26] Jim was pleasantly surprised. He had expected a balding, overweight German. Instead, at 42, the thin and crew-cut Max appeared youthful in body and spirit, and he talked straight. "He did not beat around the bush and the intent of his words was always clear."[27]

Delbrück was a veteran of the great physicist Niels Bohr's Institute for Theoretical Physics in Copenhagen. He talked of Bohr's hope that some principle of complementarity, like that needed for understanding quantum mechanics in physics, would explain biology. One day in August 1932, Max had rushed over from the Copenhagen train station to hear Bohr lecture on applying the physicist's concept of complementarity in biology. Delbrück was so struck that he made it his mission to bring physics into biology. Delbrück, perhaps Bohr's greatest contribution to biology, fervently hoped, but in vain, that biology would yield new laws of physics.[28] Luria didn't share this hope. "I was always skeptical. . . . I always thought this was lots of baloney."[29] By the late 1930s, Delbrück had hit on tiny bacteriophages—which infected bacteria and were less than one ten thousandth of a millimeter in length—as the target of his research. These ultramicroscopic phages could reproduce. Within half an hour of being invaded, a phage-infected bacterium would burst and release 100 or more progeny phages. So now Delbrück could state the "central" problem, in the words of Gunther Stent: "Just how does the parental phage particle manage to produce its crop of a hundred progeny in that half-hour?" Delbrück, Luria, and the taciturn biochemist Alfred D. Hershey of Washington University in St. Louis became the nucleus of a small band of phage workers infused with "the desire to solve the mystery of the nature of the gene."[30] Delbrück was a relentless simplifier, and hence was constantly irritated that nature was so prodigal in adding functions at each age of evolution. The bacteriophage called T4 had 150 genes, many of them used only in emergencies. "Nature," he said, "provided more than was needed."[31]

Jim was inspired. "In the presence of Delbrück, I hoped that I might someday participate just a little in some great revelation."[32]

According to Luria, Delbrück did not so much pick good people as attract them by his intelligence and the excitement of spending a day with him at the blackboard, writing and erasing, "chewing on a problem." He insisted that "science had to be fun."[33]

One of Max's students at Caltech was Robert Sinsheimer, who wrote that the sheer force of Delbrück's personality influenced students and colleagues alike. "Bright and rigorously logical, he imposed a quantitative intellectual discipline" on a largely qualitative and unfocused field. Delbrück tackled "problems that could be approached quantitatively, analyzed abstractly, and preferably studied with simple equipment." As a combination of pater familias and Herr Professor, he created "an extraordinary cohesion" among phage workers, insisting that they concentrate on only a few types of phage—so that their results would be comparable. To Sinsheimer, Max could be "mercilessly caustic" in scientific debate—a stock response was "I don't believe a word of it"—but he was "never mean-spirited." By blunt and open criticism, by suspicion of "convoluted and arcane argument," by exposing ambiguities and uncertainties in a speaker's chain of reasoning, he persistently demanded clear concepts and logical presentation. The best policy for giving presentations, Max held, was, "Assume [listeners] are totally ignorant but infinitely intelligent." It was "sink or swim" for his students, combined with parties and pranks and excursions to the California desert for hiking and climbing.[34] The French Nobel Prize winner François Jacob wrote that Delbrück's "rigor, his frankness, his way of going to the heart of a problem were combined with his surprising youthfulness, of mind as of body." Jacob wrote:

> He loved jokes and gambling. But the game that for him mattered above all else was science: an open, direct science without secrets or any affectation of mystery, where the efforts of all were joined in mutual complement. He cared little about who was first to bring off a particular experiment. The essential thing was that it had been done. Only coherence and relevance mattered: the coherence of theories and of conceptions; the relevance of facts.

Delbrück was a confirmed reductionist. He believed, as Jacob wrote, that the biologist had to choose a "system" that defined "the experimenter's freedom to maneuver, the nature of the questions he is free to ask, and even, often, the type of answer he can obtain." There were so many questions. "To make them accessible to experiment, the questions had to be changed, broken down into parts."[35] The bacterial viruses were the smallest packages of genes known. Decades later, as the Human Genome Project was revving up, the biologist Ron Davis of Stanford applied the doctrine to the genetics of plants: "It was Max Delbrück who started the concept that you . . . can't work on a whole bunch of different organisms. You have to work on one and only one."[36]

Delbrück expected that he would make progress on the gene faster by staying clear of chemistry and the details of the exact nature of the gene. Chemists would resolve the precise role of DNA while biologists focused on how genes are duplicated. Luria, explaining the small influence on them in those days of Avery's great discovery that DNA carries the genes, admitted that he and Delbrück had a partly unconscious distaste for biochemistry and biochemists. "I don't think we attached great importance to whether the gene was a protein or a nucleic acid."[37]

Delbrück's "fun and games" style was not lost on the young Watson. Max adopted a policy of "calculated bizarreness" and exhibited what Watson thought was "an extraordinary mixture of arrogance and decency." Delbrück "despised all forms of pretense and had . . . little use for tortoise-like minds" buried in the past. So giving a talk before him "was for most scientists a frightening experience. He actually wanted to learn if you had a 'take-home message.' Unclear talks could provoke his scorn, as did unreadable scientific papers." Delbrück never hesitated "to interrupt when the message was unclear or patently erroneous." Even if the interruptions were not meant personally, biologist Norton Zinder recalled many years later, "Max *destroyed* young presenters with his interruptions. Even the hardiest could scarcely recover."[38]

Under Delbrück's influence, the decades-old laboratories at Cold Spring Harbor on the North Shore of Long Island were, after the war, becoming "*the* site to which the best and the brightest of the gene-dominated scientific world came to meet for the exchange of ideas." Their protégés could attend a growing number of summer courses, including the one on bacteriophages that Delbrück began in 1945 as "a lifeboat" for scientists who were in danger of going stale.[39] Watson remembered "Max's unchallenged scientific collectivism." The idea was: "What mattered most . . . were the phage facts and ideas, not the individuals who brought them forth." And so Delbrück's followers could "take real joy in the discoveries of others."[40]

Soon, at Cold Spring Harbor, Jim was to experience Max's personality for much more than a day, and to begin building a picture of what science was supposed to be. Salva Luria took Jim and Renato Dulbecco with him to Cold Spring Harbor for the summer of 1948, where Luria and Delbrück had begun their summer phage experiments together in 1941. As refugees from wartime enemies, Italy and Germany, they were not pulled into war research. They and their small band of disciples, when not together for summers and small meetings, telephoned and wrote each other constantly.

Since 1933, Cold Spring Harbor had been the site of its increasingly famous annual Symposium on Quantitative Biology, which was designed to focus on an urgent problem in fundamental biology. Year-round research at Cold Spring Harbor was divided between the Biology Laboratory, supported by wealthy local people organized into the Long Island Biological Association, and a Genetics Unit of the Carnegie Institution of Washington. Established by Andrew Carnegie in 1902, the institution operates labs in several fields at various locations around the United States and abroad, among them the Mount Wilson and Palomar observatories in California. The Carnegie Institution took over the Eugenics Records Office, which had already been set up at Cold Spring Harbor, but had to close it in the late 1930s because its work was judged to be racist and unscientific.[41] The scientific director of the laboratories was Milislav Demerec, whom Watson credited with bringing the labs into modern biology. Many years later, Sydney Brenner, who was recruited for a research visit to the United States by Demerec, recalled the relentless economy Demerec practiced. If Demerec came into a lab and spotted a gas burner going he would shut it off. But Brenner praised the initiative of "Old Demerec" in "shifting off *Drosophila* into the bacterial genetics which was very new then."[42]

At Cold Spring Harbor, salt water and its smells lapped against the laboratory's grounds. From this "oasis of calm and peace," Jacob later wrote, the "excessiveness" of New York seemed far away.[43] The grounds were strung out along Bungtown Road, which ran from south to north a short distance from the harbor's southwestern shore. This crudely paved road, much of it enclosed by jungly growth that flourished in the humid air, had gotten its name from the local nineteenth-century whaling industry. Blubber was rendered into oil that was stored in barrels stopped by bungs. Walking up and down this rural road was excellent for talking science. White frame houses from the village's whaling days were dotted amid the trees and lawns bordering the harbor. Institutional buildings included a beautiful white frame laboratory built in the 1890s, a stuccoed concrete dining hall and dormitory called Blackford, built shortly after 1900, and a research building and a library, both in Italianate style.

The lawns could be used for volleyball or evening softball matches. The softballs frequently overshot the playing area into the cornfield where the geneticist Barbara McClintock's experiments led to her theory of "movable genetic elements," for which she received the Nobel Prize many years later. One could lounge on the lawns, look out across the water to the village of Cold Spring Harbor, and talk about the most interesting problems.

Toward the northern end of Bungtown Road a sand spit belonging to the laboratory almost closed the inner harbor off from the outer. One could go swimming and sailing and clamming—or play tennis. One could canoe across the harbor to the village "in pursuit of . . . ice cream sundaes or clams on the half shell."[44] Between experiments, scientists could talk almost incessantly about the gene. In this beautiful place, in an informal and intimate atmosphere, good and bad science could be sorted out in "honest ways." To Jim, "There was only one question, 'What is the gene?' It was paradise."[45]

He was entering "a privileged inner circle of scientists" at the summit of modern biology, "before he had done anything to deserve it." He could test himself against his elders and complete his liberation from traditional biology, even though he felt he was "more an observer than a real player." The scientists were all on a first-name basis, and there was no "personal penalty" for disagreeing with Max. Jim was excited to find that "high-level science could be more than long days in the lab and much mental sweat." It could include "outdoor fun and silly moments."[46] He recalled later: "When you get into science, you realize the people above you are not gods; that they're very human and often they'll say things which you'll regard as just wrong. That's why you want to get close to great people 'cause . . . in a sense you discover they're not so great and you don't feel very great and so . . . maybe you'll find an idea before they do."[47]

Jim was mesmerized by Max, who was becoming his intellectual father and remained so for a few years. He was taken with "Max's firm yet soft way of speaking," and delighted in the company of Delbrück's wife, Manny, a lifelong favorite.

As much as possible I tried to be near him—when he was eating in Blackford, or writing equations on the Blackford Hall Fireplace Room blackboard, or hitting tennis balls so much harder than I could, or swimming off the sand spit raft. Then, he was about to turn 42, and at 20 I was almost young enough to be his son. Others, observing our similar tall, thin shapes and my never subtle attempts to mimic Max's behavior, jokingly began to call me *son of Max*.[48]

Late that summer the phage group assembled at Cold Spring Harbor for experiments and the annual phage course. Among those taking the course were Gunther Stent and Seymour Benzer, who both became famous biologists. To Watson, the gathering "was not for amateurs." Half a century later, when the number of scientists thinking about DNA every day may have numbered a quarter of a million, Watson wrote that at that time, "logic, never emotion or commercial considerations, set the tone."[49]

X-RAY SURVIVAL

Soon after that summer, when Jim was back at Indiana working on his thesis, Dulbecco delivered a blow that forced Luria's lab to repeat the work of the last year and a half. He discovered photoreactivation in phages. Apparently multiplicity reactivation would not, by itself, explain the genetic organization of phages. Now, Watson's thesis on phages bombarded with X-rays was "much less likely to yield anything very valuable."[50] Somewhat discouraged, Luria wrote Delbrück that "the whole picture of X-ray phage is very foggy."[51] But the complexity of chemically induced indirect effects of the X-rays kept Jim from worrying "whether they would be very significant." He realized that the time for doing a thesis "was primarily a time for learning until, eventually, I could stand on my own feet as opposed to those of Luria."[52]

That fall, Jim, Salva, and Renato drove up to Chicago to take part in one of the frequent phage meetings staged by the famous Hungarian-born atomic physicist Leo Szilard, who in the aftermath of the explosion of the atomic bomb had shifted his interests from chain reactions to the gene. The portly and ebullient Szilard "crushed" Jim by telling him he had to learn to speak clearly. "Leo would start interrupting me as soon as I started to speak." Perhaps Szilard wouldn't have been so fierce if Jim had been saying nothing of importance.[53]

Jim plowed onward with his research on effects of chemicals on phage reactivation. His mentors arranged for him to speak at meetings in Bloomington and Oak Ridge, Tennessee. In Bloomington, Elie Wollman of the Pasteur Institute in Paris summarized his colleague André Lwoff's significant discovery of a new virus survival trick. Certain phages could penetrate a bacterium and not, as was usual, make the cell turn out hundreds of phage copies. Instead, under some kind of genetic restraint, the viral genes would behave as part of the bacterial genome in a phenomenon called "lysogeny." Delbrück, as the pope of the phage "church," had dismissed lysogeny as "heresy." But now Lwoff had definite proof of the phenomenon's existence. Also at Bloomington, Alfred Hershey described the discovery in bacteriophages of the "crossing over" of genetic material, which biologists had hitherto thought was possible only in sexually reproducing organisms. Much later, with the advantage of hindsight, Gunther Stent, who attended the meeting, thought that if Hershey's experiments had been properly understood, his epochal discovery three years later—that virus genes are carried by DNA alone—could have occurred three years earlier. Hershey's discovery in 1952 would greatly intensify Watson's focus on DNA.[54]

At Luria's suggestion, Jim spent the summer of 1949 at Caltech, to which Dulbecco would soon move. The phage group gathered at Caltech instead of Cold Spring Harbor because Manny Delbrück was expecting a baby, and couldn't travel. Max and Salva spent most of their time writing, and Jim did what he would remember as "token weekday experiments," but there were phage seminars several times a week.[55] Max told Jim that he was lucky that his thesis was "boring." Otherwise he would have been stuck with following it up, "instead of having time to think and learn."[56] He could have been "trapped into a rat race where people wanted you to solve everything immediately." Many years later, asked about his thesis work, Watson said he had been reviewing his notebooks, "but they failed to reveal the exact ups and downs of my thinking about the direct and indirect effects of X-rays as revealed by phage survival curves."[57]

Soon, as Watson was passing his preliminary exams, Luria's and Del-brück's thoughts turned to where their protégé should go after receiving his doctoral degree. Luria thought he should go to Europe. Europe, with its slower pace, was better for developing innovative scientific ideas. And further, Jim needed to immerse himself in what Luria could not bear to learn: biochemistry. In Copenhagen there was an excellent biochemist, Herman Kalckar, who had attended a phage course at Cold Spring Harbor. A friend of Max's, Kalckar was interested in the synthesis of nucleic acids. "DNA seemed to be the thing because of Avery." Better still, in the fall of 1949 Kalckar attended a phage meeting in Chicago and agreed to take Watson (and Stent) for postdoctoral fellowships the next fall. An annual fellowship stipend of $3,000 was secured. Salva thought Kalckar would arrange for Jim "to learn the X-ray techniques of the Braggs [in England] and maybe apply it to study the molecular structure of DNA." Max had a more mystical reason for sending Jim to Copenhagen. Niels Bohr, who had influenced Max profoundly, was still there. "Therefore I should go and pick up some of the spirit of Copenhagen," Jim recalled. "I did. Most of the spirit was given on Friday night."[58]

In his final year in Bloomington, Jim's confidence grew. He thought of himself as a genuine member of the phage group, with which he had spent the summers of 1948 and 1949. He was convinced that he knew the factual details of the last 10 years of phage research better than Salva or Max. In the Szilard meetings up in Chicago, he spoke up when he didn't "understand an argument or an experiment." Now "effectively on my own," he dreamed up several new ideas on how X-rays inactivate phages. They proved to be off base, but he was having fun.[59]

The struggle in the spring of 1950 was "to write up minor results" for his thesis. He thought of it as "torture." He wrote the thesis in a month, but Salva "did not like it" and took it home for rewriting. "Not surprisingly," Watson recalled, "the thesis was accepted without fuss at my Ph.D. exam in late May." He later reflected that he got his degree fast, "not because I was really that bright, but because there was very much less to learn."[60]

By then, Jim knew "more than subconsciously . . . that even the most elegant phage experiments were unlikely to reveal the gene at the crucial molecular level. Somehow, I had to move closer toward the chemistry of DNA."[61]

In 1950, after a summer month with Max at Caltech, where Linus Pauling, the colossus of the chemical bond, was king, Jim went east for a final six weeks at Cold Spring Harbor before taking the boat to Europe. At Cold Spring Harbor, "practical jokes dominated the mood." One evening he, Manny Delbrück, and two others let the air out of friends' tires when they were over at Neptune's Cove in the village. The payback was buckets of water thrown over their beds.[62] Late in August, the first

phage conference—another stage in the growth of the phage "church"—
drew some 30 scientists to Cold Spring Harbor. A big topic was following radioactive phosphorus from one generation of phage to the next.
Summarizing the conference, as he would for many years, was Alfred
Hershey, who had just moved to Cold Spring Harbor from St. Louis.
(He later shared a Nobel Prize with Luria and Delbrück.) Watson recalled that when Hershey summarized the meeting, he often spoke more
words in an hour "than he had spoken to outsiders over the past year."[63]

"LONG WINTER OF RAIN AND DARKNESS"

Soon after Jim reached Copenhagen and Kalckar's lab, he became afraid
that he had made a bad move—he found that he was bored by biochemistry. He and Gunther Stent were more phage-conscious than Kalckar
had expected. Kalckar's habits seemed "vague." Worse, Kalckar wasn't really interested in DNA but only in "very small things," mainly the role of
enzymes, those catalyst-like worker proteins, in putting DNA together.[64]
To Jim, "there was no way [Kalckar's] experiments on how DNA precur-
sors are made could help determine the structure of the gene." Seeing little chance of getting guidance from Kalckar, Watson spent most of his
time across town in the laboratory of Ole Maaloe, working on how radioactive phosphorus was handed down from parents to progeny. He
lived in a pension filled with theoretical physicists visiting Bohr's laboratory. He thought the Danish girls "very pretty" but he knew few of
them.[65] Still, he wrote to Max that "as a city, Copenhagen is very nice."[66]

But as the days shortened and "the long winter of rain and darkness"[67]
deepened, Herman Kalckar's marriage dissolved, and the atmosphere in
his lab became "just plain depressing." Jim wrote Max, "I find it difficult
to describe the very morbid feeling which pervaded Herman's lab during
this interval."[68] It was a relief to get permission—and travel expenses—
from the U.S. fellowship headquarters to accompany Kalckar and his
new girlfriend south to the sun, to the Zoological Station in Naples.

It was one of the shrines of classical biology, but during his weeks
there Jim found "really nothing to do. . . . It was hard to be interested in
it."[69] He pushed to finish a paper he and Maaloe had written and corre-
sponded about it with Max, who was planning to submit the paper to
the *Proceedings of the National Academy of Sciences*. Max, rejecting what
Watson remembered as his "turgid style," rewrote both the introduction
and the discussion of the results.[70] At one point in the draft, Watson and
Maaloe had written that they were confident that by means of radioactive phosphorus "the genetic material can be labeled." Significantly, Max
changed "genetic material" to "virus *particle*." He was not yet ready to re-
strict the genes of viruses to viral DNA.[71]

"WHY NOT ME?"

At Naples, contingency influenced Jim's career in a big way. He happened to attend a conference on macromolecules, both proteins and nucleic acids. He was interested in the topic because he didn't see how biochemistry and genetics would get together and go beyond trying to figure out how genes were made without determining the structure of a gene. He knew by then that "the essence of the gene was a molecule. Genetic experiments were never going to say anything deep about the molecule. I had to study the molecule."[72] By another accident, Maurice Wilkins from King's College London substituted for his boss, John T. Randall, as a speaker. Several years earlier, Wilkins had become friends with the brash Francis Crick, who, like Wilkins, had left physics for biology.

At the conference, as Jim listened, Wilkins started by stating that "the study of crystalline nucleoproteins in living cells may help one to approach more closely the problem of gene structure." Near the end, Wilkins showed an X-ray diffraction photograph of DNA fibers that he had taken. Suddenly Watson was aware that "there existed someone who actually was trying to solve the structure of DNA, which seemed a likely candidate for the gene."[73] Instead of the murky pictures made earlier, "these pictures were very good. . . . There was a well-defined structure. . . . This time there was a marvelous structure that someone could find." Responding strongly to visual evidence, as he would do so often in the future, he wondered who could study this structure. "Why not me?"[74]

On an excursion to the Greek temples at Paestum not far from Naples, Jim tried to talk to Wilkins about coming to work with him in London. But Wilkins shied away. Wilkins recalled that he found "Jim Watson very interested in DNA, but I couldn't make out from what point of view. He was a bit of a puzzle to me. I didn't quite know what to make of him."[75]

Heading back to Copenhagen, Jim stopped off in Geneva and there heard of Linus Pauling's great achievement of finding large stretches of the amino acid subunits of complex proteins arranged in what he called the alpha helix.[76] Helices were in the air. To continue his work Jim needed a new place to go the next year: Maaloe was off to Caltech for a year, and Jim was probably beneath the notice of the godlike Pauling, a titan at Caltech. There was one place left: Cavendish Laboratory in Cambridge, England, which was the world's leading center for using X-rays to probe the structure of macromolecules and where, as Jim would soon find out, Crick was working. By a stroke of luck, Salvador Luria met the biophysicist John Kendrew of the Cavendish and arranged for Watson to go there.[77]

3

STUMBLING ON GOLD: TWO SMART ALECKS IN CAMBRIDGE

*I know many people, at least when I was young, who thought I
was quite unbearable.*

JAMES D. WATSON, STOCKHOLM, 10 DECEMBER 1962

"A NAGGING YET PRODUCTIVE SYMBIOSIS"
By the fall of 1951, when Watson started talking with Francis Crick, who
was as iconoclastic and determined as he was, he was already convinced
that the clamorous marketplace of science was no place for secrets. Jim
and Francis believed that you must be fiercely competitive, but in order
to go fast, you must share information in the expectation of learning
things rather than withhold it from fear of theft. Over the next year and
a half, they served each other as teachers and devil's advocates—impo-
litely, relentlessly. For them science was inherently interactive. Professional competence in isolation was not enough to get important results.
They were sure that science has, inevitably, many elements of guessing
and playfulness, even farce—one could be wrong so often and so much
of the time; there were so many things staring you in the face that you
and often your rivals didn't see; a beautiful insight could solve a jigsaw
puzzle even when the best solid evidence had been ignored.

Crick first heard about Watson's arrival in Cambridge from his wife,
Odile, who told him one day as he returned from work that Max Perutz,
of the molecular biology unit at the Cavendish Laboratory, had brought
Jim around to the house to meet her husband. Odile told Francis, "Max
was here with a young American he wanted you to meet and—you know
what—he had no hair!" She meant, according to Crick, "that Jim had a
crew cut, then a novelty in Cambridge. As time went on, Jim's hair got

longer and longer, as he tried to take on the local coloration, though he never got so far as to sport the long hair that men wore in the sixties."[1]

Within half an hour of their meeting, the shy, tall 23-year-old biologist and the talkative 36-year-old physicist with the unruly eyebrows[2] talked of "guessing" the structure of DNA.[3] Independently, each was sure that DNA, as the site of genes, was the most important problem around.[4] Having become frustrated with experiments on the genetics of bacteriophages, Jim "wanted to learn more about the actual structures of the molecules which the geneticists talked about so passionately."[5] Protein structure, which was what John Kendrew had brought Jim to Cambridge to work on, was out the window. For a time, Crick laid aside his thesis on the blood protein, hemoglobin, to work on DNA.

Crick, the son of a shoe-factory owner in Northampton, England,[6] was a physicist turned biologist, a man of exceptional boldness and insight. After World War II, R. V. Jones, the head of Britain's wartime scientific intelligence effort, had fun talking with Crick, who spent the war in scientific intelligence at the Admiralty. They discussed how genetic information must in fact be organized at an atomic scale, as Max Delbrück had appreciated so long ago, in 1935. Otherwise it couldn't fit into the head of a sperm. Crick so "relieved the dullness of our post-war situation," Jones recalled, that he wanted Crick to succeed him. Crick was not uninterested. Early in 1947, he wrote Jones with characteristic directness, suggesting how he might lobby to save a unified scientific intelligence office from an uncomprehending review committee: "One point needs bringing out most strongly: It's no use just reorganizing with just the same old gang. We must have someone more lively to head the thing." After the war, the British military was on the run from such liveliness. So Crick went back to science, but not to physics.[7] To him, the choice was between neurobiology and "the division between the living and the non-living." He chose the latter.[8] Thirty years later, he would go into neurobiology.

Jim and Francis, an odd pair, shared what Crick later described as a "religious" aim: "to try to show that areas apparently too mysterious to be explained by physics and chemistry, could be so explained."[9] But the "large and genial" Crick was "confident, ebullient, articulate," while the "thin and angular" Watson, a loner, was "diffident, his words . . . brief." Nonetheless, Jim held to his convictions in spite of his collaborator's "forceful arguments."[10] As they looked back years later, Jim and Francis both felt that the other was the first person he had encountered who thought as he did. As Watson put it, "Before then I had been with lots of bright people and I couldn't agree with any of them."[11] Now, in Francis Crick, Jim had met someone who wasn't wasting time on secondary issues. "With Francis to talk to, my fate was sealed."[12] The following spring he would write Max Delbrück that Crick was "always the person to whom the best of us go if we wish to talk out a half-baked theory, since he is always ready to be both interested and yet devastating in his ruthless logic."[13] Theirs was "a nagging yet productive symbiosis in which neither could do without the special abilities of the other."[14]

Crick's logical mind worked best when he could bring theory to bear on a factual problem. He later remarked, "There has to be a theory or logical structure or I just don't remember it."[15] He always would look first for a simple solution, Jim observed, "even if the problem [was] very, very difficult."[16] The almost-daily conversations with Watson delighted Crick, who had few people to communicate with. Watson had "met all these people," leaders in biology. For Francis, Jim was "the first person in the outside world."[17] They shared, according to Crick, "a certain youthful arrogance, a ruthlessness. . . .[A]n impatience with sloppy thinking came naturally to both of us."[18] They admired each other: "I've always felt that Francis is much more clever than I," Watson said. "His brain works much faster."[19] Late in 1951, Watson wrote Delbrück that Crick was "no doubt the brightest person I have ever known and the nearest approach to Pauling. . . . He never stops talking or thinking."[20]

Although they lacked training for working on DNA structure, they also had few preconceptions, and they possessed much energy and persistence. As two historians, Franklin Portugal and Jack Cohen, put it: "It was precisely their lack of preconceived notions that gave Watson and Crick an advantage in flexibility in considering many possibilities that might have been rejected in a more systematic and deliberate approach."[21] They were confident that "success means disagreeing with others on fundamental things." Real scientific ability requires the "agnostic attitude" that rejects conventional wisdom.

The Watson of the DNA conversations was, in the eyes of the Nobel Prize–winning biologist Peter Medawar, a precocious genius with a style that could have led English schools to steer him toward literary studies. In fact, just then, British universities were producing a score of outstanding molecular biologists. Over all of them Jim "had one towering advantage," Medawar wrote. "In addition to being extremely clever he had something important to be clever *about*."[22] Discovering the nature of the gene "was the most important objective in biology." It was a good working hypothesis that "the gene was what provided the information for life."[23] It was clear that "the main challenge of biology was to understand gene replication and the way in which genes control protein synthesis." Such problems, Jim was convinced, "could be logically attacked only when the structure of the gene became known."[24] He thought of himself as the only one in Cambridge who "lived solely to understand how DNA functioned as the gene, and who had first-hand practical experience in using bacterial viruses to get close to the self-replication of the gene."[25] In that spirit of confidence, he was willing to "learn enough facts so I could talk with [Francis]."[26]

Behind these brash "bad boys of biology," who refused to play by the rules, lay half a century of genetics, beginning with the rediscovery in 1900 of Gregor Mendel's laws of heredity, which he had worked out in the 1860s. With the help of tiny, fast-reproducing organisms like *Drosophila* fruit flies, genes had been mapped as if they were strung out in a line. X-rays had been used to produce artificial mutations in abundance. It appeared to have been established that for each of the worker proteins called enzymes there was a single gene. Experiments on the transformation of microbes from nonvirulent to disease-bearing had led to Oswald Avery's puzzling discovery that DNA was the site of hereditary characteristics. With swiftly multiplying microbes—and their phages—genetic experiments could be carried out overnight.

One new instrument after another allowed biochemists to measure living processes with growing accuracy. One of these was X-ray crystallography, a technique that had at one time been limited to studying the structure of simple molecules like salt, but that could now be turned on complex molecules like proteins. Ultracentrifuges allowed biological molecules to be separated into bands in a test tube, and the techniques of chromatography and electrophoresis, which involved spreading molecules out on sheets of paper and gel, could determine the presence of even tiny amounts of the chemicals in living things. The electron microscope was beginning to reveal the shapes of tiny viruses. Radioisotopes, a byproduct of the Manhattan Project to develop an atom bomb in World War II, allowed biochemists to trace where labeled molecules went.

Alongside such discoveries in basic science ran more than a century of medical advances that gripped popular imagination: not merely anesthesia or asepsis in the operating room, but X-rays for diagnosis, the identification of the major disease-carrying microorganisms leading to vaccines against them, the discovery of the role of vitamins in nutrition, the harnessing of animal insulin to fight diabetes in humans, and, under the pressure of World War II, the development of antibiotics.

The Cavendish Laboratory was entering this realm of applying physics to biology. In the first decades of the twentieth century, Ernest Rutherford, called by Watson "the greatest experimental physicist ever to live,"[27] had led the Cavendish to glory in atomic physics. But by 1951, Rutherford was "a distant memory." After World War II, Britain's Medical Research Council—the government's principal agency for supporting the life sciences—bankrolled several groups in biophysics. Overseeing the new biophysics unit at the Cavendish was Lawrence Bragg, the codiscoverer of X-ray crystallography, who with his father had won the Nobel Prize for physics in 1915. Lawrence Bragg and his colleagues, Watson thought, aimed "to transform biology from a morass of seemingly limitless and often boring facts." Two researchers at the Cavendish had started to use X-rays to attack the detailed structure of big proteins: Max Perutz was studying hemoglobin and John Kendrew, the related but simpler myoglobin. Hugh Huxley probed the workings of muscle tissue. The Cambridge group was on its way to being, as Jim later wrote, "the most productive center for biology in the history of science."[28] And they had several other pioneers in life science nearby, including several later Nobel Prize winners: the biochemist Alexander Todd, who deciphered the linkages of the sugar-phosphate backbone of DNA; the chemist Fred Sanger, who unraveled the sequence of amino-acid building blocks in insulin and later invented a method of sequencing DNA; and Rodney Porter, a student of the immune system. The elements of physics, chemistry, and biology that were needed to attack the problem of DNA lay close together. Watson later recalled, "This highly diverse group of Cambridge academics focused on biology in a vastly more inspired way than any I had ever experienced at any American university."[29]

In the next two years, while Jim met the galaxy of talent at Cambridge, the pace was leisurely. He later referred to himself as "slightly underemployed."[30] He and Francis received little mail and few visitors; there were few seminars to attend—and no late nights. Decades later Crick recalled nostalgically, "Well, we didn't rush in to work in the morning, let me put it that way."[31] Jim talked with Francis in their dark office, which had been assigned to them so that others wouldn't be distracted by their incessant talk. They talked at lunch together at the Eagle pub and afterward walked along the banks of the Cam River or punted up to Grantchester; they talked at tea, and at dinner cooked by Odile. At tea, Jim could meet such "Cambridge characters" as the cosmologists Hermann Bondi, Thomas Gold, and Fred Hoyle.[32] According to Crick, his conversations with Watson were mostly "complicated intellectual discussions concerning points in crystallography and biochemistry. The major motive was to understand."[33] In this atmosphere, Watson's ideas of how to live would be "totally transformed," leaving him "emotionally centered halfway between the East Coast and England, and certainly never capable of the superficially unmannered life of California."[34]

As he got going on DNA structure, Jim began paying attention to important new insights about how and where DNA works in living cells. In Brussels, Jean Brachet and his colleagues had found a correlation between the amount of RNA in a cell and the amount of protein synthesis going on. The outer regions, or cytoplasm, of the busy cells had a great many virus-sized particles, later called ribosomes, containing RNA and protein. These particles were not in the nucleus, where the DNA was, a clear sign that DNA didn't take part directly in making proteins. In Boston, Paul Zamecnik's lab at Massachusetts General Hospital had pinpointed the synthesis of protein to the ribosomes. But how they worked was unknown.[35]

While Jim and Francis fired each other up, the immense figure of Linus Pauling loomed over their conversations. Pauling, the author of the classic text *The Nature of the Chemical Bond,* had an uncanny ability to use—or ignore—the X-ray information about biological molecules, such as proteins, to deduce their detailed structure. Only a few months before, Pauling found that great stretches of the strings of amino acids in proteins were coiled in what he called the alpha helix and others were arranged in what he called beta sheets. Both of these structures obeyed what scientists have called "Pauling Principles," according to which the shapes of proteins made of chains of amino acids are tightly restricted.[36] Pauling's great coup caused severe heartache at the Cavendish.[37] Pauling had grasped the correct orientation of the bond between amino acids in protein chains; as Perutz and Kendrew had not. It was small consolation that Perutz immediately performed a clever experiment strongly confirming what Pauling had found. Linus must not be allowed to grab the brass ring again.

As they thought about DNA, Watson and Crick were wondering urgently, "Can we think like Linus?" They were "highly motivated to succeed" and also certain that Pauling would turn his mind to DNA, as he soon did. From Pauling's success in finding the alpha helix, Crick drew the lesson that one must not "place too much reliance on any single piece of evidence. It might turn out to be misleading."[38]

Pauling had taught Crick and Watson something far more important than the possibility that DNA might be a helix. The real lesson was that "exact and careful model-building could embody constraints that the final answer had in any case to satisfy. Sometimes this could lead to the correct structure, using only a minimum of the direct experimental evidence."[39] Perhaps the Pauling Principles could be applied to DNA.[40] This was as encouraging as Maurice Wilkins's crystallographic images, which pointed to a "well-defined structure." Francis and Jim sensed that "there might be a shortcut to the answer, that things might not be *quite* as complicated as they seemed."[41]

According to Francis, Jim's attitude was still more brash. "No *good* model ever accounted for *all* the facts, since some data were bound to be misleading if not plain wrong. A theory that *did* fit all the data would have been 'carpentered' to do this and would thus be open to suspicion."[42] In Crick's eyes, Jim "just wanted the answer, and whether he got it by sound methods or flashy ones did not bother him one bit. All he wanted was to get it *as quickly as possible.*"[43]

Both Jim and Francis eventually realized that their behavior was at times insufferable, if they didn't realize it at the time. But, in retrospect, they had more than a minimal chance of finding the structure of DNA. They passionately wanted to know the details, they had leisure to consider the problem intensely and then drop it for a while. They had the knack of challenging each other's ideas constantly in "a candid but non-hostile manner."[44] Above all, they picked the right problem and stuck to it. Crick: "Two people are better than one. Otherwise you get in a rut, you get too fond of your own ideas, you get the wrong idea, you can't get out of it."[45] In Crick's opinion, "Solitary thinkers cling to their ideas."[46] Watson didn't "find it easy to think things through by himself." On the other hand, Crick needed help in escaping from slapdash tendencies. "In this area," he said, you didn't "know what ideas you're looking for anyway." He later wrote: "It's true that by blundering about we stumbled on gold, but the fact remains that we were looking for gold."[47]

That fall Jim was worried about his mother's health, as she had had an operation, but he was nonetheless eager to have an active social life. But as in Copenhagen, "lively and pretty girls for one's parties" were scarce in both Cambridge and Oxford, where he visited frequently.[48] He was no suave party goer, to the embarrassment of his sister, Betty, who was in Cambridge for long stretches then. "I wasn't the sort of person you asked to go to a dinner party when I was 22. I would neither amuse them [the other guests] nor put them at ease. The only thing I cared about was the gene—and girls."[49] Nonetheless, Jim found time to promote a romantic interest of his friend Avrion Mitchison. He wrote Delbrück in his neat, tiny handwriting, "I have been engaged in numerous intrigues to bring them together. Unfortunately, when I bring them together he becomes too nervous to say much and so I must do all of the talking."[50] Given Watson's own shyness with women, surely well known to Max and his wife, Manny, this statement is remarkable.

FIASCO

The Paulingesque model-building approach of Crick and Watson was virtually the opposite of the very slow, methodical way that Maurice Wilkins and Rosalind Franklin were working at the Medical Research Council–supported biophysics lab of King's College London, not far from the Strand, in a dark basement three floors beneath the main building's quadrangle. It was a claustrophobic environment that made Franklin anxious.[51] Franklin and Wilkins were not working together, or even communicating much. Franklin was described by the writer Brenda Maddox as "a quick, single-minded young woman with a passion for argument," whereas Wilkins was "an exceedingly reserved man with a hesitant, circumlocutory manner and an aversion to direct eye contact."[52] Jim was certain that Franklin was brighter than he was, but inflexible, "as though she'd been trained to do geometry so wasn't going to use algebra." Despite Watson's lifelong disdain for Franklin (who died in 1958), his unflattering portrait of her in *The Double Helix* (1968) has helped her achieve, posthumously, some of the fame she deserves. A new biography by Maddox shows how profoundly Watson misunderstood Franklin's relations with her well-off Jewish family of bankers and publishers. He apparently knew nothing of her enthusiasms, ranging from climbing mountains to sewing. Like many others, he failed to appreciate the quality of the work she did before her significant role in the discovery of the nature of DNA, from which all biology of the last half century derives. Jim sensed her combativeness but did not grasp the factors that fostered it.[53]

Back in 1950, Wilkins had obtained from Rudolf Signer, a researcher in Bern, Switzerland, samples of DNA that were far purer, far more intact than those he had previously used. With equipment he and colleagues had built for earlier research, Wilkins was able to study DNA's optical properties with greater precision, and he turned some of the purer DNA over to Raymond Gosling, a postdoctoral fellow, for X-ray analysis.[54]

Then, early in 1951, Franklin, a Cambridge-trained physical chemist and an expert in the effect of heat on the carbon molecules of coal, came over to King's, after two happy years in Jacques Mering's government chemistry lab in Paris. Her mission was to speed up the X-ray work, although, as she reflected later, she had never worked with single crystals or with biological substances.[55] She was not aware of the relevance of DNA's structure to its duplication. She did not "live DNA," as Watson put it,[56] and gave it up when she left King's two years later. In recruiting her, John Randall, the head of the lab, described her task so that she thought the X-ray work on DNA had been turned over to her.[57] A "superb experimentalist,"[58] she took off from the base of DNA work already accomplished at King's and, working with the Signer DNA, made rapid progress over some months when Wilkins was away a lot.

She discovered that DNA came in two forms, one containing more water than the other. In September 1951, Wilkins came back from a series of conferences, bearing new samples of DNA from the Columbia University laboratory of Erwin Chargaff. Franklin's pictures of the Signer DNA were exciting and Wilkins began offering interpretations of them, but Franklin, intensely professional, regarded Wilkins as an amateur, an interloper. "How dare you interpret my data for me?" she exclaimed. After this explosion, a truce was patched up: Wilkins left the Signer material to Gosling and Franklin, while he kept the Chargaff samples, which proved much harder to work with.[59] He did obtain X-ray reflections from the wetter, extended form of DNA called B, but he could not get his samples to undergo the transition to the drier, more crystalline form called A. He could not get A at all, and the resolution in Wilkins's pictures was less good than in Franklin's pictures. Wilkins moaned that Franklin had been given not only the best samples but the best cameras—and then she refused to share data. Franklin, realizing she had made a mistake coming to the cold atmosphere of King's, soon began looking for another lab at the University of London.[60]

In the 1990s, the journalist-historian Horace Judson tracked down seven of the eight women who worked in Randall's group at King's (there were 23 men). One of them, Honor Fell, told Judson that she saw no sign of sex discrimination against Franklin, and that both Franklin and Wilkins were "rather difficult people." Franklin seemed unready to join the informal socializing among the women in those days of food rationing. "I suppose," another of the women in the lab, Mary Fraser, wrote from Australia, "we assumed she would fit into the casual role of relaxing amidst the beakers, balances, centrifuges, and petri dishes . . . but she didn't. Rosalind didn't seem to want to mix. Her manner and speech were rather brusque and everyone automatically switched off, clammed up, and obviously never got to know her." Franklin did not join the excursions to the nearby Strand Palace for lunch, Sylvia Jackson reported, nor to a pub in Epping, to which they went by car.[61]

At King's, model building was distrusted. The hope was that slow, detailed analysis of the pictures would yield the structure. Although Wilkins and Franklin shared a general impression of Watson and Crick as butterflies "flipping around with lots of brilliance but not much solidity," the two researchers worked separately. Much later, Wilkins reflected that King's had lost the race for the DNA structure very early, "because we didn't find it possible to work together."[62]

By contrast with scientists at King's, Watson and Crick were excited by DNA models, especially after Wilkins, a friend of Crick's, visited the Cricks in Cambridge soon after Jim's arrival, in October 1951. With Watson also there, Wilkins gave them several bits of information. Surprisingly, Maurice thought that DNA was helical and favored the notion of three helices wound around each other. He had no X-ray pictures better than those of the summer of 1950. Because of Franklin's resistance to Wilkins's entering DNA work, he no longer knew what Franklin and Gosling were finding. According to Watson, "She didn't want to talk to him." Maurice regretted that "he'd given away his problem."[63] But the biophysics people at King's were to discuss their work at a colloquium in November, and Watson was welcome to sit in.[64]

Crick did not go to the seminar, which Franklin was to present. After a recent plunge into gloom, Francis had swiftly soared into euphoria. The gloom came from a sudden quarrel with Lawrence Bragg, who had had a "nice idea"—but it was the same idea Crick had expounded to him six months before! Crick reminded Bragg of this in Max Perutz's office. If Crick's assertion was true, Bragg might have committed a cardinal scientific sin, appropriating another's insight as his own. To the newcomer Watson, the resulting set-to was "a terrible episode . . . a pretty awful sort of encounter."[65] The upshot was that Crick, whose loud voice irritated Bragg, did not have to leave the Cavendish—at least, not immediately. He was expected to finish up his thesis on hemoglobin and go somewhere else. Later, Crick recalled, Bragg admitted that Crick had formulated the "nice idea" better than he had.

Soon after this contretemps, Crick's morale shot up, to the point of overconfidence. In late October 1951, Bragg showed Crick and "quite a senior and very clever Scottish crystallographer," William Cochran, a paper about the X-ray diffraction patterns that helices should produce. Crick and Cochran set to work separately on the problem, and Cochran, by an "elegant" method, and Crick, by devising "an enormously clumsy thing,"[66] reached the same conclusions. "That's what a helix should look like!" Crick exclaimed in delight.[67] In Watson's words, Crick wanted to tell everyone that finally "he actually had an idea that worked." He was preparing to rush to Oxford to talk about it with the leading crystallographer, Dorothy Crowfoot Hodgkin (who in 1964 won the Nobel Prize in Chemistry for working out the structures of such molecules as vitamin B12 and penicillin).[68] The new helical theory increased Crick's and Watson's eagerness to try building a model of DNA.

Attending Rosalind Franklin's seminar in London that November, Jim had the mission of learning new facts that could be used to build a DNA model. He expected that Franklin would talk to her audience of about 15 people about such matters as model building and whether a simple solution to the structure was possible. But Watson was, as Crick recalled, "only a new boy in crystallography." How was he to know the difference between a "unit cell" and an "asymmetric unit"?[69] Rosalind Franklin spoke without "a trace of warmth or frivolity" and delivered "a sermon on caution"[70] that emphasized experimental difficulties. Evidently, she was certain that "the only way to establish the DNA structure was by pure crystallographic approaches."[71]

But she had by no means decisively rejected the idea of a helical structure for DNA, as her notes for the lecture show. Jim did not understand Rosalind's discussion of what happens when the dry, crystalline A form of DNA fiber takes up water and lengthens into the wetter B form. What did that have to do with a DNA model? As usual taking no notes, he missed a crucial set of measurements of the dimensions of the DNA fiber. He did not see these again until February 1953. His mind wandering, he carried away an absurdly low figure for the water content, which clearly implied that the molecule was not held together by bonds between hydrogen atoms.[72]

Afterward Jim joined Francis on the train to Oxford for the visit to Hodgkin and briefed his friend on what he thought he had heard. "We'll build a model!" Crick said, and they excitedly set to work over the next week.[73] Many pieces of information, developed over years by different scientists, lay at hand. DNA had a backbone of linked sugar and phosphate groups. There were the four "bases," adenine, thymine, guanine, and cytosine, which would later be known by the initials A, T, G, and C. These fell into two classes, larger ones called purines (A and G) and smaller ones called pyrimidines (T and C). Clearly, the bases must be in a varied sequence to carry hereditary information. Apparently, the hydrogen atoms in the molecule jumped around, making bonding impossible. Jim and Francis may not yet have read papers by a young Norwegian scientist, Sven Furberg, who was working in London; Furberg placed the bases inside the sugar-phosphate chain and perpendicular to the sugar-phosphate axis.[74]

Rapidly they devised a three-chain model, with the phosphates on the inside. Colleagues, including John Kendrew, insisted that they promptly summon the King's researchers to see it, even though, as Watson recalled, it wasn't "very pretty . . . [not] what you call a 'natural.' "[75] In retrospect he called it "this awful molecule."[76] The London researchers hopped the train up to Cambridge the next morning, "scared to death" that the two clowns at the Cavendish might have got it right.[77] They spotted elementary errors—first and foremost Watson's low figure for water in the molecule. The correct water content made it clear, Crick recalled, "that we were completely on the wrong track."[78] He also reflected, "We didn't know as much as [Franklin] did. She dismissed the whole thing as nonsense and she was quite right."[79] Franklin told them her pictures showed the phosphates on the outside. It was "an embarrassment," Wilkins recalled, "to see these highly intelligent Cambridge chaps turning up with something which obviously was crazily wrong. The whole thing was inside out."[80] Franklin realized, as Watson recalled, that "she was dealing with incompetents."[81] Before the Londoners swept off to the train back to town, Watson and Crick told them they should build models—but they weren't about to.

Impatient with the unhurried approach at King's, Crick and Watson had swung right away for the fences—but the result was a pop fly. They leapt to a model that was "completely incorrect," invited the people from King's to see it, and were told that they had missed obvious points. Any idea of a Cavendish-King's collaboration on model building died.

Watson fiddled with the models for a while, but gave it up because "there were too many bad contacts. And besides, what would one do with a three-chain DNA?" Crick reflected: "In science, when you have an idea and you've been enthusiastic about it, and it's clear it's wrong, you must just push it away and get rid of it." Soon after, Bragg and Randall, as the heads of the Cambridge and London laboratories, had a little talk, and it was decreed that Jim and Francis should leave DNA alone. They did not appeal Bragg's decision. "Since we didn't know what to do, it was easy to stop." They promptly turned over their shop-models, or "jigs," of DNA components to King's.[82]

When Watson looked back on this episode many years later, he was convinced that he and Crick should have solved the structure of the DNA cylinder then. For that matter, Franklin and Pauling should also have found it early, but for the vagaries of fate and personality. Watson delighted in pointing out that the discovery was staring all of them in the face. Ever after, he said, "I always felt very creepy about what would have happened if [Pauling] had done it."[83] A problem was that neither he nor Francis nor Rosalind Franklin knew any chemistry, and so none of them appreciated plain chemical evidence that the bases in DNA were hydrogen-bonded. There were other scraps of evidence they had not put together, such as X-ray pictures showing that there was a distance of 3.4 angstrom units between bases along DNA, and that bases lay at right angles to the long axis of the fiber.

After the war, English scientists found that DNA could be "denatured" by heating, or changing the pH alkaline-to-acid ratio in the surrounding liquid. With hindsight Watson regarded this as "really good evidence" of hydrogen bonding in DNA. Then there was the evidence about the ratios of the four bases in various living organisms. In 1950 and 1951, Erwin Chargaff had published his findings that in several species the amount of adenine equaled that of thymine, and that the same was true of guanine and cytosine, even though the proportions of A=T and G=C varied from species to species.[84] Logically, the next task was to figure out how to make a hydrogen-bonded structure involving the bases. Pauling had worked out ideas for how antibodies and their antigens fit together in the immune system. Almost 50 years later Watson remarked, "You could say, well, you put these hydrogen bonds on those [bases] which are basically equivalent. So it was there! . . . Francis and I really should have put everything together. . . . You could just say we were incompetent."[85]

At almost the moment of the debacle of the model presentation, Jim wrote Max Delbrück bravely, almost brazenly: "We believe the structure of DNA may crack very soon. Time will tell. At present we are quite optimistic. Our method is to completely ignore the X-ray evidence."[86]

Unfazed by nervousness over his draft status or a fight with his Merck fellowship committee over his unauthorized move from Copenhagen to Cambridge (which cost him $1,000), Jim was off to spend Christmas and New Year's at Carradale, the country home of his friends the Mitchisons, which he described as "a Scottish Castle in the Highlands."[87] The failed model? "Christmas was coming and someone invited me to their house in Scotland. So I really didn't think about it very seriously."[88]

CAN'T STOP THINKING ABOUT DNA

Although Jim and Francis were officially off the DNA case, nobody could stop them from thinking about it. And all through 1952, far from marking time, the two learned new DNA facts that gave them fresh insights—and a fresh sense of urgency. Ostensibly, Crick had returned to his thesis on hemoglobin, and Watson, clearly all thumbs with crystallizing myoglobin for Kendrew, was off on a study of tobacco mosaic virus (TMV), a rodlike spiral of identical protein units around a core of RNA. He boned up on the mathematics of helical configurations and on growing, mounting, and X-raying crystals. To Jim TMV was attractive: "It has RNA and no one else is working on it."[89] The RNA in TMV, Jim hoped, might provide a backdoor trail to DNA. By June 1952, formally supervised by the plant virologist Roy Markham, Jim succeeded in getting excellent X-ray pictures of TMV that showed an unmistakable helical pattern, but it was too complex to throw new light on DNA structure. Despite his exuberance, Jim—and Francis—knew that "no more dividends could come quickly though TMV. . . . The way to DNA was not through TMV."[90]

April 1952

In the spring, Jim's obsession with DNA was reinforced by electrifying news from Alfred Hershey at Cold Spring Harbor.[91] Hershey, now in his early forties, had carried out with Martha Chase the famous "Waring blender" experiment. In Watson's words, the experiment proved "that phage DNA—not its protein component—contains its genes."[92] To trace the fate of the virus protein and DNA, Hershey and Chase had labeled the DNA of a virus called T2 with a radioisotope of phosphorus, P32, and the T2 protein with a radioisotope of sulfur, S35. This allowed them to check whether the entire T2 virus went inside the bacterium, or just the protein or just the DNA. Allowing a few minutes for infection, Hershey and Chase then used the Waring blender and a centrifuge to separate heavier cell-wall material from lighter cell fluid. Most of the DNA was in the cell fluid, while the protein stayed behind on the cell walls. Hershey himself was not totally convinced, Watson recalled later, because he didn't consider that the experiment was sufficiently rigorous. All that had been shown was that "most of the [virus] protein doesn't get in [the bacteria]."[93]

Gunther Stent, the virologist and historian of molecular biology, described the result this way: "When a phage particle infects its bacterial host cell, only the DNA of the phage enters the cell; the protein of the phage remains outside, devoid of any further function in the reproductive drama about to ensue within." Now one could begin to think of DNA as responsible not only for the regular operations of a living cell, but also for its own duplication.[94] The results strongly supported the idea of the virus as a protein structure wrapped around DNA, a kind of protein syringe that injected DNA into the bacterium to take over the bacterial chemical machinery and make many copies of itself. The effect on further research, according to Stent, was "immediate and profound. . . . From that time on, all genetic thought was focused on DNA."

Until then, the evidence for genes contained in DNA had only been found in the microbes Avery had studied a decade before. But now, the genes of viruses also were clearly seen to be in DNA. And not just of any viruses, but of the very bacteriophages whose study Delbrück and Luria had pioneered. When Hershey wrote to Watson of this glimpse of DNA's role, Jim immediately was "thrilled," and determined to tell the news at a microbiology meeting at Oxford in April 1952.[95] He was filling in to read a paper of his mentor, Luria. Two weeks before the conference, Luria, amid the McCarthyite frenzy of those days, had been denied a passport because of his leftist views (although his paper had been written out in advance and distributed to those attending the conference). Luria's paper still expressed skepticism that DNA was the seat of the genes. Despite the awkwardness of having to provide a slight emendation of Luria's ideas, Watson, filled with excitement at Hershey's proof of DNA's importance, read out sections of Hershey's letter to an audience of 400. He said, "It is tempting to conclude that the virus protein functions largely as a protective coat for the DNA, and that the perpetuation of genetic specificity is largely or entirely a function of the DNA."[96] At question time, he gave his opinion that the viral DNA was like "a hat inside of a hat box."[97]

The audience, little aware of who Hershey was and almost certainly put off by the gawky American, yawned. Only a delegation from the Pasteur Institute in Paris really listened.[98] André Lwoff and François Jacob—both of whom later shared in the 1965 Nobel Prize in Physiology or Medicine—as well as Seymour Benzer and Gunther Stent were impressed "at the hard simplicity, the dry solidity of [Hershey's] experiment."[99] They "knew that Hershey's experiments were not trivial and that from now on everyone was going to place more emphasis on DNA."[100]

Jacob later wrote of his excitement at the Oxford meeting. This was where he "discovered the feverish atmosphere of colloquia, where what matters at least as much as the public talks are the encounters, contacts, gossip, word of mouth, the chance to learn who is investigating what, of gleaning information, of explaining what one is doing or at least what one wants to leak out and let people believe one is doing."[101]

Jacob, confronting for the first time the brash American culture of science, was stunned by the sight of Watson brandishing the letter from Hershey and reporting its "neat irrefutable experiment." Watson was "an amazing character." Jacob set down this description of Watson:

Tall, gawky, scraggly, he had an inimitable style. Inimitable in his dress: shirttails flying, knees in the air, socks down around his ankles. Inimitable in his bewildered manner, his mannerisms, his eyes always bulging, his mouth always open, he uttered short, choppy sentences punctuated by 'Ah! Ah!' Inimitable also in his way of entering a room, cocking his head like a rooster looking for the finest hen, to locate the most important scientist present and charging over to his side. A surprising mixture of awkwardness and shrewdness. Of childishness in the things of life and maturity in those of science.[102]

May 1952

DNA kept coming up. Two weeks later, in May, Crick and Watson went to London to attend a one-day Royal Society meeting on proteins. There, Wilkins told Watson that he had taken "extremely excellent" X-ray images of DNA and that he still thought DNA was helical. Possibly relieved that Watson was deep in TMV, Wilkins was frank about what was happening at King's: The feud with Franklin had grown more bitter. Admitting that King's had done nothing with the Cambridge jigs, Maurice said he would be happy to give them back. Although Watson and Crick kept pestering him to build models, lest Pauling get DNA first, Wilkins, in Crick's words, "wanted to do it the sound way and not be bullied by us."[103]

For months before the Royal Society meeting, Jim and Francis had been aware that Linus was scheduled to talk there, and they felt "a slight fear" of the masterful scientist.[104] His interest in DNA was rumbling on the horizon. They knew that Pauling had written Wilkins for a copy of his new X-ray photos of DNA, which an American researcher had told him about, and that when Wilkins hesitated, he wrote Randall, who likewise refused to supply the images. Crick remarked that the thought was, "There is a convention . . . in science that when you've done a lot of work and got some experimental data, you . . . should have the first chance of interpreting it."[105] Pauling was interested in DNA, even if, as he often said later, it was just another molecule. Surely, when Linus came over to England he would ask to see the pictures, although it seemed likely to Watson that the King's people would still refuse. But just as the meeting was to start, there was another McCarthyite scandal. As Pauling was preparing to board his flight from New York to London, his passport was seized because of his left-wing sympathies.[106] Derailed by politics from the DNA trail, for the moment, Pauling's interest shifted to the supercoiling of alpha helices in proteins. Close call for Jim and Francis.

It just so happened that on the night of the Royal Society meeting, which she did not attend, Rosalind Franklin began working on the DNA picture that would astound Watson with its clear helical pattern when Wilkins showed it to him nine months later. Two weeks earlier she had photographed a single fiber of the condensed, drier, crystalline, A form, and had concluded (wrongly, it turned out) that it was not a helix. But now the sample had changed irreversibly to the extended, "wetter," B form. She took a picture. The next night, 2 May 1952, she took photo number 51, which she deemed not just good but "very good," and recorded her measurements.[107] Wilkins later referred to it as "the best, and most helical-looking 'B' pattern."[108] But she laid it aside because the A form was giving her the more precise data that she, as a professional crystallographer, preferred. Unhappy at King's, she already was arranging to move to another lab, that of the famous crystallographer J. D. Bernal, at another University of London college, Birkbeck.[109]

They all were getting close to solving the mystery of DNA's structure. Particularly Rosalind. In 1999 Watson stated, "Rosalind had the clue that should have solved it."[110] She had evidence that should have told her that DNA consisted of two chains, with base sequences running opposite each other, that is, in a complementary fashion.

Just like Francis the previous fall, Rosalind was eager to tell Dorothy Hodgkin about her results. She went to Oxford and told Hodgkin her conclusions about the structure, showing that the molecule had three possible "space groups." Two of the space groups, said Franklin, had mirror symmetry, but Hodgkin told her that "there is no mirror symmetry in biology." She advised Franklin to learn more about space groups. Why not have a talk with Jack Dunitz, a Caltech scientist visiting the Hodgkin laboratory, who was present at the interview. They could go into the next room.[111] Franklin felt insulted and left without discussing her third space group. This just happened to be the one that Crick had studied in hemoglobin. Franklin did not pick up the clue, Watson said at a lecture many years later. "If she had used it, I wouldn't be giving this lecture."[112]

On 20 May 1952, Watson wrote at length to his mentor, Max Delbrück, about his plans for the future. Delbrück had arranged a fellowship from the National Foundation for Infantile Paralysis, for work at Caltech. Watson asked to take the first year of this fellowship in Cambridge because in the coming year the particular X-ray equipment he needed for his still-unfinished TMV work would be lacking in Pasadena. Delbrück and Pauling agreed to this. In the fall of 1953, Jim would come to Pasadena to continue working on "the structural side of the virus field" and then, as he did unsuccessfully, "go on to the structure of RNA." It was also clear from the letter that DNA was very much on Jim's mind. Despite Maurice Wilkins's success in taking X-ray pictures, he was still unwilling to try model building à la Pauling. The feud at King's—which Jim and Francis heard of frequently from Maurice—was stalling any "real effort" to find the DNA structure. The previous winter's efforts at model building by Watson and Crick were suspended "for the political reason of not working on the problem of a close friend." He added, "If, however, the King's people persist in doing nothing we shall again try our luck."[113]

June 1952

In June 1952 Watson's mother came from America to visit him and his sister, Betty, who was spending time at Cambridge. Jean Mitchell Watson immediately insisted that her son cut his hair, and he agreed to some shortening.[114] Then they went to Scotland to explore an area where her father's ancestors had lived.[115]

The same month, Rosalind wrote J. D. Bernal at Birkbeck that Randall, the boss of the King's lab, had no objection to her moving to Bernal's lab. In July she held a mock funeral party for the DNA helix.[116]

The comedy of might-have-beens went on. "Although we weren't working on the structure of DNA in the summer of '52," Crick recalled, "we were always thinking about it and turning it over in our minds."[117] In June, Jim mentioned Chargaff's ratios, A=T and G=C, to Francis, but the latter did not start thinking about them at once. He seems to have forgotten what Jim told him. He began wondering about electrostatic forces between bases lying one above the other.

One evening, Crick and John Griffith, a mathematician (he was a nephew of Frederick Griffith, who first observed genetic transformation in microbes), heard a talk by the cosmologist Thomas Gold on "the perfect cosmological principle."[118] Over a drink afterward at the Bun Shop, Crick wondered aloud what the perfect analogous biological principle could be.[119] The obvious answer, Francis told Griffith, was "the self-replication of the gene."[120] So how did it happen? "Since the bases are flat, perhaps it is so that they can stack on top of one another and attract," said Griffith. "Why not work out if adenine attracts adenine, and so on?"[121] Crick asked Griffith how the bases would attract each other. Of course, hydrogen bonding was ruled out because the hydrogens were thought to move about freely. What about the flat sides of a pair of adenines sliding over each other to bond, and so on? Unknown to Crick, Griffith already was working on the attractions between bases. He disagreed with Crick's like-with-like formulation; he favored copying "by the alternative formation of complementary surfaces," just as Pauling and Delbrück had asserted before the war during decades of discussion of a "positive" forming next to a "negative." The young mathematician went away to do calculations in the light of quantum mechanics.

When Crick caught up with Griffith a few days later in the Cavendish tea queue, Griffith told him he didn't favor like-with-like bonding. His calculations pointed toward adenine binding with thymine, and guanine with cytosine—but, as Watson ruefully recalled later, "on top of each other." In Watson's recollection, Crick was getting "lousy chemical advice from a good chemist."[122] Still, Crick began thinking in terms of A making B and B making A—complementary replication. This view fit Chargaff's ratios, with the amount of adenine equaling that of thymine, and the amount of guanine equaling that of cytosine.

July 1952

It happened that in July, Chargaff was coming to the Cavendish on his way to an international congress of biochemists in Paris. John Kendrew arranged for Francis and Jim to meet the censorious Chargaff for coffee after a meal at Kendrew's college. The two bad boys told Chargaff they thought they could crack the DNA structure with model building. Chargaff's evident distaste for Jim's now-long hair and Chicago accent made Jim think Chargaff was dismissing him as a nut.[123] Chargaff didn't take either of the obvious amateurs seriously. They wanted to know about the ratios, and Chargaff amiably reminded them that it was all in the scientific literature. His scornful estimate of the two young men was confirmed when Crick, in midflight about the quantum-mechanical implications of Griffith's results, forgot which base was which. He did not know which bases had NH_2, amino groups. You could always look these up in a book! Chargaff drew the formulas for the two smart alecks. They were so ignorant. He recalled, "I never met two men who knew so little and aspired to so much." They talked a lot about the "pitch" of the bases with respect to the long axis of DNA. After the humiliating interview, Chargaff jotted a note: "two pitchmen in search of a helix." He was not in a hurry to find the DNA structure. Watson and Crick's ambition, and their worry about Pauling's beating them to the structure, left Chargaff cold.[124]

Watson regarded Chargaff as "despicable." Chargaff, he knew, "didn't like me. I didn't like him." For months Jim "just did not want to think about his ratios at all."[125] He was "determined to solve the structure without using Chargaff's data." Crick's reaction was different, "electric." He promptly got Griffith to remind him which base preferred which, and looked up Chargaff's papers at the library.[126]

Pauling's passport had been restored and he attended the Paris biochemistry conference. Linus was the big fish. Having heard, almost certainly from Delbrück, about Francis and Jim's quest for DNA, Pauling was interested in them. As it turned out, Linus and Jim discussed the Hershey-Chase experiments. Pauling clearly was very interested in DNA, remarking that the DNA structure required X-ray studies of its components. Evidently, Pauling thought Watson should "do something about DNA."[127]

Just after the big congress, the mandarin biologist André Lwoff held a meeting at the Abbey of Royaumont near Paris. His impression of Jim stayed with him: "It is evening in the solemn drawing room of the abbey. In the room is a fifteenth-century oak table, on which there is a bust of Henri IV. A young American scientist wearing shorts, has climbed on the table and is squatting beside the king. An unforgettable vision."[128]

Yet another close call. Pauling and Chargaff took the same boat back to America. But Pauling also didn't like Chargaff and so he didn't talk with him and learn the vital ratios. Later, Pauling thought this cost him the double helix. Watson looked back delightedly: "So again I was saved by Chargaff's personality."[129] Yet another close call.

LINUS'S MISTAKE

Later in the year, Watson met the 40-year-old biologist William Hayes, who worked at Hammersmith Hospital in London, at a summer conference in Italy. Jim began bacterial genetic experiments with Hayes, finding him "a very decent fellow, quite modest and unassuming." Hayes was one of the discoverers of sexual reproduction in bacteria. Watson found Hayes's work "very pretty," and he advised Delbrück to invite him to Caltech.[130] Delbrück did so. The trips to London to see Hayes often gave Watson the opportunity to go to dinner with Maurice Wilkins.

Sometime in these months, Watson wrote a short formula on a little piece of paper and tacked it up in his room at Clare College. He was already looking forward to how the gene would direct the making of proteins. He wrote: "DNA makes RNA makes protein."[131]

December 1952

In mid-December 1952, the Biophysics Committee of the Medical Research Council made one of its usual visits to an MRC-supported biophysics lab, in this case, Randall's group at King's. As usual, Randall pulled together reports from each lab, and each committee member, including Max Perutz, received a copy. In the reports, both Wilkins and Franklin summarized their DNA findings, and Franklin included the same unit dimensions of the A form that she had reported at King's the year before.[132]

Watson went off for a skiing holiday in Switzerland with his sister, Betty. In a letter to Delbrück he said, "My skiing holiday was quite amusing despite a lack of talent for this sport." He went on to Milan for three days to talk with the famous geneticist Luca Cavalli-Sforza about bacterial genetics, the topic of his work with Hayes, and then to Brachet's lab in Brussels, where he heard "a very pretty story about Tobacco Mosaic Virus." He still hoped to pinpoint the location of the tiny amount of RNA inside TMV's protein shell.[133]

Wednesday, 28 January 1953, After Lunch

Soon afterward, Pauling reentered the contest. He had finished with supercoiling in proteins and had taken up DNA again. He wrote his son, Peter, who had begun graduate study at Cambridge, and the chemist Alexander Todd that he had worked out a DNA structure that he could believe in. He gave no details. Peter showed the letter to Crick and Watson, and they passed the letter back and forth in frustration. In his heart, Watson thought Pauling must have gotten it right. Crick was not so sure. Just before New Year's, Pauling had written his son that he had just sent off a paper about the DNA structure. He was sending a copy to Bragg (who did not show it to Crick for fear of distracting him from his thesis). Would Peter like a copy? Thinking that Bragg understood even less than he did about the matter, he replied yes. Pauling sent it on 21 January 1953.[134]

A week later, on 28 January, the manuscript arrived. Just after lunch, Peter Pauling, who in those days "had no idea of what a gene was," came into the Watson-Crick office with the manuscript sticking out of his pocket and "a big grin" on his face, and said his father had a three-chain model of DNA. Jim and Francis felt dread. Crick recalled, with understatement, "We were really on tenterhooks." Had their chance to solve DNA evaporated? Had Linus, even without the good X-ray pictures from King's, conquered again, as he had with the alpha helix?[135]

Jim took the document first, and swiftly scanned the summary and introduction, and then the illustrations. Something wasn't right. The phosphate groups were not ionized. Pauling needed them un-ionized if his proposed structure was to hang together. In Pauling's scheme, hydroxyl (OH) groups must be on the phosphates so that hydrogen bonds could pack the phosphates together. But didn't they have to be ionized for DNA to be an acid? Watson thought: "Pauling's either a very, very great chemist and has discovered some new principle, or something is crazy."[136] Linus had made "a stupendous mistake."[137] Crick and Watson "just couldn't believe it. Why did he do this? Why did he say a nucleic acid wasn't an acid?"[138] Jim raced over to Roy Markham's and Alexander Todd's labs to check with the two chemists. The phosphates had to be ionized. DNA was indeed an acid.[139]

Many years later, Watson would say dismissively that although Pauling's great success had been with the carbon atom, the heart of organic chemistry, he otherwise was an inorganic chemist. In Watson's eyes, Pauling was not thinking like a biologist. He never thought of two chains or how the molecule might be copied. Linus's model was "really an ugly bastard . . . bad—contacts which shouldn't exist." And yet "he liked it." The problem, to Watson, was that the godlike Pauling never "talked to anyone."[140] And he didn't go to the library himself—he sent a graduate student—and so missed crucial facts. The two Cambridge adventurers thought they would be "able to conquer the reigning champion. Champions get dethroned."[141]

But Watson and Crick drew little immediate comfort from Pauling's error. It was so obvious that Pauling was sure to see it soon, along with colleagues reading his soon-to-be-published paper, and then go hell-for-leather for the right solution. Pauling had devised "his dreadful structure [but] we were afraid that someone would realize he was off-base. We just wanted to get it. I'd say it was impossible not to think about it."[142] They were still in the game, but they probably had six weeks at most to get the structure. Crick agreed they would have to act fast, before Pauling realized his mistake. They must not throw away the opportunity. Jim was due to see Hayes in London two days hence, on Friday, 30 January 1953, and so he must show Linus' model to Maurice.

To the shock of many who read Watson's memoir of the race to discover the structure of DNA, *The Double Helix,* Jim and Francis toasted Linus's mistake. Such competitiveness should not be written about, many said. Reviewing *The Double Helix,* however, the Nobel Prize–winning biologist Peter Medawar scoffed, "In my opinion the idea that scientists ought to be indifferent to matters of priority is simply humbug."[143] The tut-tutters were ignoring history, the sociologist Robert Merton concurred: the battles over priority between Newton and Leibniz in the seventeenth century made those of today look tame. Pauling, unlike the British, was not observing any "no poaching" signs.[144] Watson himself did not look back sheepishly. A scientist "who retains the ambition of his youth" always hopes to "pull off something smashingly big."[145]

They were, Crick reflected later, "thinking about the problem at the right time. . . . We weren't the least afraid of being very candid with each other. . . . We pooled the way we looked at things. . . . We didn't leave it that Jim did the biology and I did the physics. We both did it together and switched roles and criticized each other, and this gave us the great advantage over the other people who were trying to solve it."[146]

"JUST A BEAUTIFUL HELIX"
Friday, 30 January 1953, Tea Time
With the Pauling manuscript burning a hole in his pocket, Jim arrived in the basement of King's late in the afternoon. Not finding Maurice, he barged in to see Rosalind. She was bending over a light box, looking at X-ray patterns from DNA. Unknown to Watson, she had already begun building models of the drier, more crystalline A form. Watson also did not know that she had been in touch with Pauling's collaborator in Pasadena, R. B. Corey, ever since Corey visited King's in May 1952.[147]

As Rosalind's future colleague Aaron Klug later remarked, she was far closer to solving the structure of DNA than her rivals realized. She had taken the clearest photographs yet of DNA. She had worked out the position of the phosphorus atoms of the DNA backbone. And she had found the two forms of DNA. For Klug, who won the Nobel Prize for Chemistry in 1982 and later became president of the Royal Society, Franklin's weakness was isolation. "She needed a collaborator and she didn't have somebody to break the pattern of her thinking, to show what was right in front of her, to push her up and over." He also said, "You know, in a way, Watson was her collaborator."[148] Watson was more pungent. Forty years later he said, "She would have been famous for having found DNA if she'd just talked to Francis for an hour."[149] She would have solved the structure "because she had the data to do it. She just needed someone to talk to."[150]

Rosalind didn't really want to look at Linus's manuscript, which Watson thrust at her. Linus said his three chains were in helices, and she had strong evidence that the A form was not a helix. Besides, Linus had worked from the crystallographer W. T. Astbury's prewar photograph of a mixture of A and B forms. Irritatingly, the paper cited a photograph of Wilkins's, even though Pauling hadn't seen it, but made no mention of her work. A bigger irritation was that Corey had not sent her the manuscript—and here was Watson, turning up with it like a bad penny. She told him not to be so hysterical: "Of course it's wrong. It's [DNA is] not a helix."[151]

Jim began to criticize Rosalind's data that showed that DNA was not a helix. Repeating Crick's arguments without really understanding them, he told her that her data were a fluke. From an ignoramus like Watson this was too much. "She got very angry at me," he recalled. He was afraid she would hit him.[152] Grabbing the manuscript, he raced for the door of her office—and ran into Maurice Wilkins.

Shaken, Watson went along the hall to Wilkins's room, where Wilkins felt safe but was trembling with anger. Wilkins mentioned that Franklin almost hit him once. He reached in a drawer and pulled out Franklin's Photograph 51 of the B form, taken the previous May (Jim later concluded that it had actually been taken by Franklin's postdoc, Raymond Gosling).[153] Maurice had the picture because Rosalind was turning over DNA material as she finished preparations to move to Bernal's lab at Birkbeck.[154]

It was a bombshell. Maurice could not have shown a picture to anyone with greater responsiveness to visual clues. The "very beautiful photograph," the "wonderful X-ray picture of the B form," with its prominent diamond pattern of spots, "unbelievably simpler" than any before, was unquestionably a helix, a perfect helix. According to later legends Watson and Crick had "stolen" DNA from Rosalind Franklin, but in later years Watson repeatedly denied this: "I didn't go into the drawer and steal it. It was shown to me and I was told the dimensions." There were the 10 bases separated by 3.4 angstroms over a distance of 34 angstroms. "I knew roughly what it meant." He admitted that seeing the picture was "the key event" psychologically in motivating him to drop everything and race to discover the real structure. It was "just a beautiful helix."[155]

At dinner in the Soho section of London, Jim told Maurice that in the emergency caused by Pauling's erroneous model, he must begin building models at once. Wearily, Wilkins said that he'd do it after Franklin left, within two months.[156]

STEEPLECHASE
Friday, 30 January 1953, on the Train to Cambridge
On the way back to Cambridge, Jim reflected on the DNA image in the photograph: "[I]t had to be a helix—and Pauling had published his dreadful structure."[157] In the margins of his newspaper he wrote down everything he could remember from his disastrous yet triumphant visit to King's. He asked himself how many chains DNA had. By the time he got back to Clare College, he had gone over the broad-brush technical arguments about how many bases there would be in a given distance along the DNA, and he had decided on two chains.[158]

That Weekend

Francis Crick had a slight hangover from a dinner party the night before, and he was irritated Saturday morning when "the former bird-watcher" told him he could crack the structure. But then Watson spelled out for him what he had written on the margins of his newspaper. Astbury's assumption before the war that the bases were separated by 3.4 angstroms was correct. Ever the devil's advocate, Crick agreed that two chains were more likely than three, but that the issue must not be prejudged. Watson's hunch might prove wrong.[159]

Watson lost no time telling both Max Perutz and Lawrence Bragg about Franklin's B-form picture and the erroneous Pauling model. Apparently Bragg had not studied the Pauling manuscript closely. Watson's report cheered Bragg a bit, "because, you know, Pauling was usually right." It was time to get going. The King's people had done nothing with models in over a year. The initiative must stay on this side of the Atlantic. Bragg gave Watson and Crick the green light to forge ahead.[160]

First Week of February

Ever conscious of the role of the DNA bases in encoding genetic information, Jim wanted them on the outside of the sugar-phosphate backbone, whence they could easily communicate their messages. So he started out building models on that basis. Otherwise, how could the bases be read? How could DNA link up to its surrounding packaging, consisting of proteins called histones? He recalled, "I would do it maybe three hours a day. . . . It was hard to get at it in the morning. . . . By the time you'd get in there was morning coffee, and then go for lunch. We'd have our walk. And then I'd come back and build models. . . . Francis was sort of working on his thesis." The thesis was "dreary," so Francis was easily diverted and we "talked all the time."[161]

"Not the slightest ray of light" appeared. With the backbone in the middle, the phosphates and sugars "packed" so that they were either too close or too far apart. At dinner on 4 February Crick and Watson argued extensively: Should the backbone be on the inside? Finally, when the issue came up for the umpteenth time at their office, Crick looked up from his thesis and asked impatiently, "Why not put the backbone on the outside?" Watson said that this would be "too easy," too free of constraints, allowing too many different models. But Crick said, "Then why don't you do it?"[162]

Sunday, 8 February 1953, Lunchtime at the Cricks'

Francis and Odile Crick invited Maurice Wilkins to their home in Cambridge for Sunday lunch. After dithering for several days, and after several pestering phone calls from Crick, Wilkins accepted. His excuse for his inability to make up his mind to accept the invitation was that at a seminar on 28 January, Rosalind "effectively refused to answer questions" about her talk. This had made him feel even more "lousy" and he feared he would be "too depressing."[163] From the moment of his arrival, Crick bombarded him with questions, but learned nothing more than what Watson had told him a week earlier. Crick had not seen Franklin's B-form picture, and Wilkins brought no X-ray images. Back in the DNA business himself, Wilkins was not opening up.

At lunch in the Cricks' home Wilkins talked "very slowly," Watson recalled. Crick and Watson and Linus's son Peter kept telling Wilkins that he had to start building models right away. "Our chief argument," Peter Pauling recalled, "was that if he did not, my father would certainly have another attempt and would get the correct structure." Wilkins repeated that he and the team he was assembling would start fashioning models only after Rosalind Franklin left.

So in search of a green light from Maurice, Jim and Francis changed their tactics: "Then came a most important question," according to Peter Pauling. "Maurice was asked, if he was not going to start building atomic models immediately, whether he minded if Jim and Francis tried to build atomic models. I do not remember the answer. Jim reports that Maurice said he did not mind, and we must accept that."[164] They sought Maurice's permission out of a sense of courtesy toward Maurice as a fellow scientist—their friend had studied DNA for years.

Crick recalled that although Wilkins gave his assent "slightly reluctantly, there was no doubt about it; I remember it quite vividly, and I think it was in our dining room at Portugal Place. I remember the scene because we were somewhat edgy about it."[165]

Wilkins did not know that Watson had already been building models for a week.

Second Week of February

On 10 February, having spotted an error that had caused Pauling to hypothesize three chains instead of two, Franklin at last resumed work on the B form. She was reviewing the evidence that it was helical.[166]

Soon after Wilkins's lunch with the Cricks there occurred one of the most controversial events of the drive to determine DNA's structure. Crick had learned, probably from Wilkins, of the report that Randall had handed out to the Biophysics Committee of the Medical Research Council two months earlier. Either he or Watson asked Max Perutz, a member of the committee, to see it. The document was not confidential, Perutz reasoned, so he "saw no reason for withholding it."[167]

The report did not contain pictures of the two different forms of DNA Franklin had been examining, the dry A form and the wetter B form. There was one confirmation of what Watson learned on 30 January: the

B form had a 34-angstrom "repeat." But Franklin also gave information that Jim could have jotted down in November 1951—but didn't. The "unit cell" of the A form was about 20 percent shorter than in the "more biologically important" B form. "While not necessary," Watson recalled, the information was "very, very helpful." Since Jim's knowledge of crystallography was "not much solider" than it had been in late 1951, he might easily have been mistaken again. But Perutz's committee membership gave them the data Jim had missed. "It was a fluke that we saw it," Watson wrote in 1969.[168]

Rosalind's careful measurements meant far more to Francis than they did to Jim. Francis had a flash of recognition similar to the one Watson had when he saw the Franklin photo. According to Horace Judson, it was Crick's "turn to find in Franklin's data what she had failed to see."[169] Franklin was not communicating with Crick, who was a friend of Wilkins's. Working with horse hemoglobin for his thesis, Crick had found a double-stranded space group of the type that Franklin had also found, but she did not appreciate the implications. It possessed a type of symmetry called "monoclinic C2," which specified that the two helical chains ran in opposite directions. To Crick, "This was the crucial fact." He now was certain that there were two chains. "One chain must run up and the other down," that is, the sequence of bases on one strand of DNA must be complementary to the other. In ten turns, then, the rung-like pairs of bases would be repeated, implying a rotation of 36 degrees from one base pair to the next.[170] Crick told the historian Robert Olby:

> I don't think I would have thought of running them in the other direction but for the clue of the C2 symmetry. . . . It was absolutely crucial to have this idea. . . . In my view it is an obvious enough idea for a trained crystallographer, but for someone like myself, who was a beginner, it needed a clue to get to that point, and the clue was Rosalind Franklin's data.
>
> It was very difficult for Jim to grasp the fact that the chains were running in the opposite directions. For example, when we got the chains the other way [antiparallel] he could not build the second chain the other way up. It was just too difficult! Somehow his mind didn't work that way.[171]

Crick didn't think faster than Watson. It was just that "Watson did not know the meaning of monoclinic C2." Jim did not "understand space groups" and thought, "Well, I don't want to be so constricted yet."[172] He was still reluctant to use the base-ratio information from Chargaff, who was "a bad man." Still, the question of how the bases would bind to each other had to be answered. Jim and Francis knew they must turn their attention to the chemistry of the bases.

20 February 1953

To Watson, having chains with identical, not complementary, sequences still looked attractive. He kept trying to make "like-with-like" pairs of bases. Working late on the evening of 19 February, he thought he had a "pretty" scheme. The next morning, after taking breakfast at an establishment called The Whim, he did not go straight to the office but instead returned to his room to reply to a letter from Max Delbrück at Caltech. His primary purpose was to send along a manuscript by himself and Hayes about bacterial mating, and ask that Max submit it to *Proceedings of the National Academy of Sciences*.

But he included remarks about DNA. "I am extremely busy, largely working on DNA structure. I believe we are close to the solution." It was a sentence bound to ring bells in Pasadena. Pauling had made "several very bad mistakes," Watson wrote, and had chosen the wrong type of model. But at least he was trying models, which the King's people weren't. No longer was Watson avoiding DNA work because "the King's group did not like competition or cooperation." Now, with Pauling on the case, "I believe the field is open to anybody." He added, "Today I am very optimistic since I believe I have a very pretty model, which is so pretty I am surprised no one has thought of it before." The model accounted for Franklin's X-ray data.[173]

Fourth Week of February 1953

Jim arrived at the office and tried out his "pretty" like-with-like model, but his scheme at once came crashing down and left him teetering on a high wire. He had told the people in Pasadena about it, and it didn't work. He needed a good chemist's advice and luckily, he got some soon. Jerry Donohue, a scientist from Pauling's lab with no deep interest in DNA, was visiting on a sabbatical. He was in a position to correct a crucial mistake of Watson's. Relying on a widely used textbook, Watson had been using a particular model, a so-called "tautomeric form," of how oxygen atoms bound to the carbon atoms at what chemists denominate position 6 of two of the bases. These were guanine, a purine; and thymine, a pyrimidine. Jim assumed that the form was "enol," in which an extra hydrogen atom is attached to a particular oxygen atom. But Jerry said that the textbook was off base. The actual form was "keto," in which oxygen does not have the extra hydrogen. There was another problem, this one involving adenine, a purine, and cytosine, a pyrimidine. For binding the chains, these needed to be in the so-called "amino" form, with NH_2 groups at the right places, not NH as in the "imino" form.[174] So Jim's model was "nonsense." The particular tautomeric form governed which hydrogen bonds could form between bases. With enol, it wouldn't work. With keto it would. Donohue's intervention was vital. "It was just good luck," Watson recalled. "If he had not been with us in Cambridge, I might still have been pumping for a like-with-like structure."[175]

Tuesday, 24 February 1953

Preparing to leave for Birkbeck College, and what would prove to be distinguished work on viruses, Rosalind Franklin closed her latest DNA notebook—"only two steps away from the solution," in Crick's words.[176] One of these steps was chains running in opposite directions and the other was base pairing. The day before, she had started a new page with the heading "Structure B, Photograph 51."[177]

Friday, 27 February 1953

After Jim, Francis, and Jerry Donohue had argued for possibly a week over the enol versus the keto form, Watson finally accepted Donohue's advice, somewhat morosely. For Francis Crick, it was his "most vivid memory" of the discovery. Crick was standing at his desk, Donohue and Watson were standing by the blackboard. If what Donohue had been saying was correct, "then it should be possible for the bases to go together." Watson spent the afternoon cutting representations of the four bases in stiff cardboard. But he didn't try fitting them together just then. He had agreed to go the theater, and he went.[178]

BASE PAIRING

Saturday, 28 February 1953

Arriving at their office before Francis, Jim cleared his desk to try forming base pairs with hydrogen bonds, using the cardboard cutouts. First, he tried more tinkering with adenine linked to adenine and so on. But it didn't work. Then he linked an adenine to a thymine, and a guanine to a cytosine. The two pairs, linked by hydrogen bonds, were "identical in shape." With hydrogens in fixed locations, there was a stable complementary relationship between the two chains. DNA had the capacity to self-replicate.

Donohue entered, and Watson asked him if he had any objections to the linkups he had made. Donohue said no. Jim's morale "skyrocketed."[179]

DNA's "central role in cellular existence" had become "unambiguously clear."[180] Neither Jim nor Francis had ever "anticipated that the answer would come so suddenly in one swoop and with such finality."[181] It was less than a month since Jim had seen Rosalind's picture 51. In one hour, he had gone from nothing to understanding the structure of DNA.[182]

About 40 minutes later, as Francis came through the door Jim told him his astonishing news: he had a credible model structure of DNA. "Not by logic but serendipity," as Crick put it, Watson had found base pairing that worked.[183] He was using neither Chargaff's rules, which,

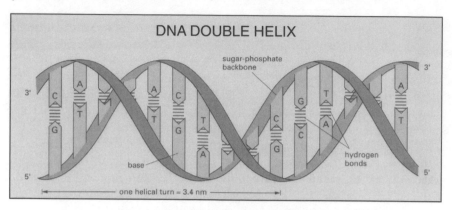

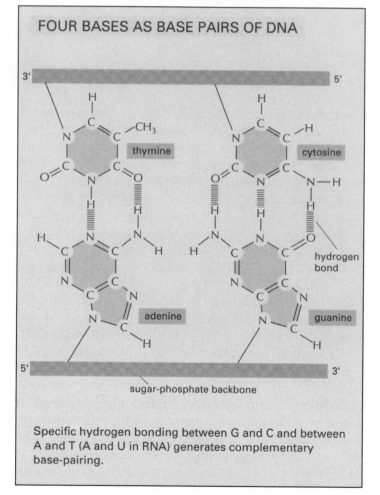

Schematic of DNA. CREDIT: Copyright 1994. From *Molecular Biology of the Cell* by Bruce Alberts et al. Reproduced by permission of Routledge Inc., part of The Taylor & Francis group.

Donohue noted, he was still "cheerfully disregarding," nor the crystallographic evidence that had moved Crick two weeks before.[184] Yet not only were Chargaff's ratios explained, but, even more exciting, the model also made possible a far better scheme for duplicating DNA than Jim's like-with-like models. They had stumbled on what one scientist later called "a scientist's dream—simple, elegant, and universal for all organisms."[185] Francis and Jim "became enormously excited." The structure was informative about a basic mechanism of life. One could begin "to think of the gene in terms of chemical structure." But Watson and Crick couldn't claim to be wonderful scientists. "It wasn't very difficult science," Watson reflected, "just a wonderful answer."[186] "We essentially guessed the structure," but it would prove to be "as correct as evolution."[187]

Sure it was luck, as happens in most discoveries, Crick wrote, "but the more important point is that Jim was looking for something significant and immediately recognized the significance of the correct pairs when he hit upon them by chance."[188] By "trial and error," Jim and Francis had pulled together an array of insights about DNA that had been accumulated over many years by Levene, Todd, Caspersson, Avery, Hershey, Chase, Chargaff, Wyatt, Pauling, Corey, Perutz, Kendrew, Cochran, Crick, Vand, Stokes, Astbury, Bell, Wilkins, Gosling, Furberg, Franklin, Gulland—and Donohue. But, as two historians later commented, "It was far from being a trivial matter to bring all these factors together and satisfy them uniquely."[189]

In 1984, Watson recalled his reaction as "near ecstasy." In 2000 he reported thinking at the time, "Boy, it was pretty. If it was right, it was going to be very important."[190]

4

A BEAUTIFUL
MOLECULE: BEING BELIEVED

A genetic material must in some way fulfill two functions. It must duplicate itself and it must exert a highly specific influence on the cell. Our model suggests a simple mechanism for the first, but at the moment we cannot see how it carries out the second one.
JAMES D. WATSON, COLD SPRING
HARBOR SYMPOSIUM, JUNE 1953

CONVINCING THEMSELVES

Urgently, even ruthlessly, the brash collaborators had pulled together and orchestrated the fragments of knowledge about DNA. The double-helix structure fit the data, but was it the real one? Was it right?

Jim awoke on Sunday morning, 1 March 1953, feeling "marvelously alive." Little did he imagine that in less than four years he would evolve from a scientist doing his own experiments to a scientific impresario who spotted key problems, selected people to work on them, and then pushed to get them accomplished with an intensity and an influence few scientists have matched. It was a much subtler and far more sustained effort than the extraordinary tour de force he and Francis had just pulled off.

As he walked to the Cavendish, he took delight in the beauty of King's College Chapel and the newly cleaned Georgian-style Gibbs Building. The sight of the buildings and their green surroundings made him reflect "that much of our success was due to the long uneventful periods when we walked among the colleges or unobtrusively read the new books that came into Heffer's bookstore."[1]

Francis felt less relaxed. Once he "saw the base pairs and their symmetry," Jim recalled, "he gave up working on his thesis" and turned to the vital work of making sure the double-helix hypothesis held up. The first metal model was "the key thing, because you could have the right base pairs and still not find a satisfactory structure." It took them about a week to be sure of the structure. A six-foot-tall version of their model was not ready until the beginning of April, when they sent their first paper to *Nature*.[2]

When Jim arrived at their office that Sunday morning, an anxious Francis was already there, testing again, with the help of a compass and ruler, whether Jim's cardboard base pairs—each with one purine and one pyrimidine—fit within the sugar-phosphate backbone of the double helix. They did.

John Kendrew and Max Perutz came in to see if Francis and Jim still thought they "had it," and Francis gave each a characteristically buoyant and booming lecture. Crick later recalled, "We did have a constant stream of visitors, so much so that Jim got sick of me because I was in a state of—somewhat of euphoria explaining all this, and Jim would have to go out of the room. He couldn't bear to hear it all over again." Jim rejoined, "It was so obvious. You didn't have to talk about it."[3] As he would do often in the coming days during such lectures, Jim slipped out. He went to see if the Cavendish Laboratory's workshop could hurry and finish the metal representations of the bases for the next, more precise, but still tentative stage of model building. It was finished in two hours. An hour later Watson fitted the metal bases together, so that "they satisfied the X-ray data and the laws of stereochemistry," forming a right-handed helix with one chain running opposite to the other.[4]

Then followed a nerve-racking quarter of an hour. Now it was Crick's turn to inspect the freshly fashioned model. Every time Francis frowned, Jim's stomach felt queasy. But Francis couldn't find anything wrong. The next morning was easier. While Jim sat on top of his desk dreaming of the letters he would write, Francis kept busy tightening the metal model on its support stands. By Tuesday evening, the refinements were finished and Jim and Francis were sure that the sugar-phosphate backbone would permit the base pairing they proposed. Jim and Francis still hadn't seen the latest X-ray evidence from King's College, London, but they told each other over lunch that "a structure this pretty just had to exist."[5] They could begin drafting their "note" for *Nature*.

Such modest confidence still left Jim and Francis with many worries. It might be the biggest scientific event since Darwin's *Origin of Species,* but had it been a fluke? "We had done something very good. Now what do you do?" After such "fantastic good luck,"[6] Watson "worried about whether I would be part of the next step."[7] They wondered how the double helix "could be definitively proved correct and then how we might determine the way in which the genetic information within it was used to order the amino acids within polypeptides."[8]

Invaluable then and later was Erwin Chargaff's discovery of the 1:1 ratio of A to T and G to C, the finding he could not or would not interpret. When Chargaff died at the age of 96 in 2002, Watson commented, "The base composition was an essential clue for finding the structure of DNA; there's no doubt about that. We could have come up with the answer, but no one would have believed it."[9]

THE HIDDEN FUTURE

Having found the elegant solution to the puzzle of how DNA duplicated itself, they knew they were now entering a different world. Instead of one question, they now confronted many—and for years afterward they would struggle to find the right way to use molecular biology to ask and answer those questions. Instead of the elegant simplicity of the double-helix structure they now confronted the bewildering complexity of how the DNA is duplicated and how it directs cells to make proteins. How did the coded information, the instructions stored in the library "stacks" of the cell's nucleus, get copied and reach the relatively distant manufacturing "suburbs" of the cell? DNA had two tasks, one strategic and the other tactical. The strategic job was that of faithfully making new DNA copies once in a cell's life cycle. The tactical task went on constantly. This was the making and unmaking of the many thousands of different kinds of molecules the cell needed to carry out such business as digesting food and building the many structures inside itself.

To go forward and answer such questions, Jim and Francis realized that they would have to enlist skeptical biochemists, even though Chargaff kept sneering for decades that it was "a lot of nonsense." Crick observed long afterward, "We had to get involved with the biochemists, but it wasn't clear to the biochemists that they had to get involved with us."[10] And the topics of cell duplication and function were myriad. On his sixtieth birthday, in 1988, Watson reflected that the period after finding the double helix was "the most difficult" in his life.[11]

What Jim and Francis knew at the time has been summarized by the South African biologist Sydney Brenner, later a colleague of Crick's for many years:

> It was known at the time that genes specified the structure of proteins as previously enunciated by George W. Beadle as the 'one gene–one enzyme' hypothesis. Frederick Sanger, in his own pioneering work [at Cambridge], had proved that proteins had a defined chemical structure in the form of their amino-acid sequence, and, as it was known that proteins had fairly complex three-dimensional structures necessary for their functions, the problem could be simplified to what Francis Crick called the Sequence Hypothesis: the one-dimensional sequence of bases in DNA specifies the one-dimensional sequence of amino acids in the polypeptide chain, and the amino-acid sequence determines how the chain will fold up.[12]

Detailed understanding of how proteins are made and function lay in the unknown future. Much choppy water had to be navigated over the next 13 years, a period of many key discoveries. Scientists from Paris, Cambridge, and Harvard competed to find the short-lived "messenger" RNA molecules that are copied from the DNA sequence of bases in each gene and that specify the makeup of proteins. A second key discovery was that of "transfer" forms of RNA, postulated by Crick, which escort each amino-acid component of a protein to the point where it hooks up to the messenger at the point of assembly. A third class of RNAs helps to form the ribosomes, the tiny, globular genetic tape copiers that preoccupied Watson for years. Also discovered, with important contributions by Crick and Brenner, were the actual three-base codes, or "codons," for each amino acid. The field of genetic control opened up, and Harvard colleagues of Watson found the special proteins called "repressors," whose existence had been postulated by François Jacob and Jacques Monod in Paris, which shut down genes when the products they specify are not wanted. This is specially important in the highly specialized cells of complex organisms.

Writing in 1968, the virologist Gunther Stent called this 1953–66 period the "dogmatic" phase of molecular biology and wondered if the field, with hundreds of scientists working in it by then instead of dozens, was running out of gas.[13] It wasn't. Still more discoveries unimaginable in 1953 followed. In the 1970s and beyond, biologists mastered the recombinant techniques for transferring genes freely into rapidly dividing cells, for cloning genes, and for sequencing or "reading" the genes. They stumbled on the startling fact that the genes of complex organisms, including humans, occurred in fragments of active sequences that were dotted between long repetitive stretches of apparently inactive DNA, and they identified genes involved in specific types of cancer.

CONVINCING COLLEAGUES

Early in March, Wilkins—too late—wrote to say he was ready to start building models. When he came to Cambridge on 12 March and saw the Watson-Crick model, he was convinced that this was the correct approach. To Crick's delight, Wilkins did not ask for joint authorship of a paper about the model and went back to London to tell "everybody" and review the evidence that the A form was not a helix. He found that it was a helix.[14] On 13 March, Wilkins wrote Crick, "I think you're a couple of old rogues but you may well have something." As he was finishing the letter, Franklin's associate Gosling came in to say that he and Franklin had already written a paper on DNA structure.[15] Franklin, who had just moved to Birkbeck College, also came to see the model. She, too, accepted it at once, while continuing to be confident in her "solid data," not just a hypothesis, that would be published in detail soon.[16] The old antagonism on her part evaporated. Many years later Watson said, with little charity, that Franklin knew that "she'd been sort of destroyed by her prejudices."[17] To each, Crick and Watson sent the draft of their first paper, which was completed before they saw the papers from the Franklin and Wilkins labs.

Though the merits of the Watson-Crick model were evident, there still was anguish at King's. Randall, a friend of a top editor at *Nature*, insisted that papers from the Wilkins and the Franklin lab appear at the same time as Watson and Crick's. The issue of acknowledgments was tricky. Watson and Crick did not feel they could admit that they had seen Franklin's evidence, including the famous Photograph 51, a fact that she was never to learn, nor could they, without revealing Perutz as the source, say that they had seen the Medical Research Council report. Indeed, in their paper they said they "were not aware of the details of the results" presented in "the following communications" from Wilkins and Franklin. Franklin's biographer, Brenda Maddox, says that Watson and Crick were "being economical with the truth." And Franklin "did not know the part her work had played." Her boss, Randall, wrote her soon after she left for Birkbeck that she must stop working on DNA, although in fact she continued for months afterward to collaborate with Gosling.[18]

The community Watson and Crick had to reach first was not large. "There weren't that many people around who were that interested," Watson commented in 2002; some "immediately said it was important, but many more said, 'Let's wait and see.' "[19] Watson later estimated the audience for DNA at about 50 scientists: 15 in England, 5 or 6 in France, and the others scattered. Even of these 50, few were doing research that could have solved the structure. Most had not faced the fact that the standard genetics of the past wouldn't arrive at the gene, whereas "molecular structure" would.[20] Did the whole world stand up and cheer? Crick asked later. "Not a bit of it, not a bit of it!"[21] Neither Watson nor Crick was ever asked to give a talk about DNA at Cambridge, Jim recalled. This was not just English reserve, but a kind of irritation. At this time, "[M]ost biochemists worked on proteins. . . . Someone coming along and saying, 'DNA is more important,' was unpleasant." Besides, "There was the feeling that we didn't deserve it—because we hadn't done any experiments and it was other people's data."[22]

Among the first colleagues Jim and Francis had to convince that DNA was a double helix was the head of the Cavendish, Bragg, who had told them to go ahead with their model building just a month before. Bragg had feared another Pauling victory and was correspondingly exasperated with the failure of both Wilkins and Franklin to do any model building for more than a year. Coming in for a look after a bout with flu, Bragg saw the point of the complementary structure of sugar-phosphate backbones linked by purine-pyrimidine pairs. But he hadn't heard of Chargaff's ratios. Jim reviewed the evidence for a 1:1 ratio for adenine and thymine, and for guanine and cytosine. Bragg became increasingly excited as he saw the "potential implications for gene replication." He agreed that the moment to call the people at King's had not arrived, but he urged that Alexander Todd be invited over as soon as they had worked out the atomic coordinates.

A Rockefeller Foundation representative who visited the Cavendish just then felt the enthusiasm: "If the structure were the correct one, it would present a very good skeleton on which to affix some of the modern theories of chromosome duplication."[23]

A few days later, Todd saw the model and declared it "biochemically plausible."[24] He said that Francis and Jim "had correctly followed" his advice about the chemistry of the backbone of sugar and phosphate."[25]

Meanwhile, early in March, Pauling gave a seminar at Caltech about his proposed triple helix, and the next day he wrote Watson not only to invite him to a big protein conference he was staging in Pasadena in September but also to ask—somewhat nervously, Watson thought—for details of the now-demolished "pretty" like-with-like scheme of 20 February.[26] But now the double helix had been found and Pauling soon learned of it, from an excited letter Jim wrote to Max Delbrück on 12 March.[27] When he got Jim's letter, Max immediately gave Jim's news to Pauling. Jim had asked Delbrück not to tell Pauling right away. But Delbrück, the enemy of secrets, couldn't wait to discuss the news with his colleagues, and knew that gossip would reach Pauling almost at once. Besides, he'd promised Pauling that he would let him know any news the moment it arrived. He took Jim's letter to Linus, who "was effectively convinced within five minutes."[28]

Linus wanted to see the new model—and get a look at the evidence from King's College. Early in April, on his way to a conference in Brussels, Pauling stopped off in Cambridge to see his son Peter and look at the model on the ground floor of the Austin wing of the Cavendish Laboratory. He told Watson and Crick that they had the answer.[29]

Both Pauling and Bragg were going to the Brussels conference on proteins, one of the esteemed Solvay conferences (named for the Belgian chemist Ernest Solvay), which had been held for decades. Bragg described both Perutz's work on the structure of hemoglobin and the Watson-Crick discovery. Pauling told the audience that the model he and his colleague Corey had worked out the previous December was probably wrong and that the Watson-Crick model was likely to be "essentially correct." He said, "I think the formulation of their structure by Watson and Crick may turn out to be the greatest development in the field of molecular genetics in recent years."[30]

Watson's letter of 12 March made a major impression on Delbrück. Today, the historic letter with Watson's simple drawings of base pairs in the B, or "paracrystalline," form is preserved in the Caltech archives between transparent sheet protectors. Many people have seen it, for Watson reproduced it in his scandalously successful 1968 memoir, *The Double Helix.* "While my diagram is crude," Watson wrote his mentor, "in fact these pairs form two very nice hydrogen bonds in which all the angles are exactly right." The only X-ray information used was the distance of 3.4 angstrom units between base pairs, found years earlier by W. T. Astbury. Still, Jim and Francis needed more detailed measurements from King's. In a day or two they would send their note to *Nature* "emphasizing [the model's] provisional nature and the lack of proof in its favor."[31]

On 22 March, Watson sent Delbrück a draft of the *Nature* paper. It already contained the famous observation, "It has not escaped our notice that the specific pairing we have postulated immediately suggests a possible copying mechanism for the genetic material." Jim told Max that Wilkins and Franklin would do "all comparisons of the experimental data with [the Watson-Crick] structure."[32] Crick and Watson had already begun a longer article, on genetic implications, which was published in *Nature* on 30 May.

But Jim was feeling restless: "I have a rather strange feeling about our DNA structure," Jim told Max. "If it is correct, we should obviously follow it up at a rapid rate. On the other hand it will at the same time be difficult to avoid the desire to forget completely about nucleic acid and to concentrate on other aspects of life."[33] In part because the enthusiasm about the double helix had reached an almost unbearable pitch, Watson had rushed over to Paris for a week right after his 12 March letter to do an experiment with the American microbiologist Harriet Taylor, wife of Boris Ephrussi. As good luck would have it he ran into a scientist, G. R. Wyatt, who had found a special form of cytosine that also conformed to the Chargaff ratio.[34] Jim told Max soon after that, "Paris . . . as expected is by far the most interesting city I shall ever know. The weather was pleasantly warm and everything seemed incredibly lovely."[35] At a moment of triumph, romantic longings grew sharper, but his confidence with women did not. As he remarked in *The Double Helix,* he was "too old to be unusual."[36]

Delbrück's fascinated reaction to the DNA structure was, "It was obviously right," although the details of how the structure would code for genetic specificity was "still very, very obscure."[37] In the middle of April Delbrück wrote Watson: "The more I think of it the more I become enamored of it myself." Although he was concerned about how the DNA unwound and opened up to do its strategic and tactical work, he thought that if the model was correct, "all hell will break loose, and theoretical biology will enter into a most tumultuous phase." Many dead ends of classical genetics might open up again.[38] The same day, he wrote his old mentor Niels Bohr in Copenhagen, "Very remarkable things are happening in biology. I think that Jim Watson has made a discovery which may rival that of Rutherford in 1911," the year Rutherford formulated the nuclear theory of the atom.[39]

ANNOUNCING THE DISCOVERY

About three weeks before Watson and Crick's first paper about the double helix was published, colleagues car-pooled over from Oxford to see their model against a white-painted brick wall in the basement of the Austin wing. One of them was Sydney Brenner. "The moment I saw it . . . I just knew this was the answer to things. . . . Bang, there it was." Looking back nearly half a century later, Brenner said the fashioning of the double-helix model was "when molecular biology actually started."[40]

The opening sentence of Watson and Crick's paper in *Nature* on 25 April 1953 was very modest: "We wish to suggest a structure for the salt of deoxyribose nucleic acid. This structure has novel features that are of considerable biological interest."[41] Forty-seven years later, in the White House celebration that took place when the sequence of 3 billion base pairs of human DNA was nearly complete, President Bill Clinton declared that the second sentence, which he recited, was "one of the greatest understatements of all time."[42] There was one important omission from the article, a sentence acknowledging Franklin's crucial contribution, perhaps because of the intense focus on the model-building approach. Half a century later Watson said, "We probably should have had one sentence [on Franklin]."[43] There was more to it than that.

The famous sentence near the end, about DNA replication, was the result of a tussle between Watson and Crick. That sentence read: "It has not escaped our notice that the specific pairing we have postulated immediately suggests a possible copying mechanism for the genetic material." Watson was all for British understatement and brevity. His view was, "The less said the better."[44] He reflected later, "I was trying to be English."[45] Crick was less hesitant and more aware that if they didn't say at least something, they were sure to be "scooped" on an implication that had hit them the moment the bases fit together. Forty years later, Crick recalled, "Jim was very nervous about saying anything, and I said, 'Well, we've got to say *something*!' Otherwise people will think these two unknown chaps are so dumb that they don't even realize the implications of their own work!"[46] Crick wrote the sentence.[47]

Drafting of the note to *Nature* had begun the first week of March, as soon as Watson and Crick thought they "had it," and was complete in the third week, when Wilkins received a copy and Rosalind Franklin and her associate Raymond Gosling completed the draft of their note to *Nature*.

Wilkins agreed that all three papers, one from his group, one from Franklin's, and the one from the Crick and Watson be published together, and Bragg, who was close to *Nature*'s editor, promptly approved the idea. Later, Watson, sending Delbrück the draft of the more detailed 30 May paper, wrote that when they saw the two 25 April articles from King's, they found them even more supportive of their ideas than they expected.[48] Crick and Watson wrote Pauling on 21 March, enclosing their draft.[49] A week later, on 28 March, Betty Watson typed the final draft announcement.[50] The sketch of a double helix was drawn for it by Odile Crick, whose usual subject was nudes. Crick told her later that the illustration was her best-known work.[51] The paper was sent off on 2 April and was published on a very fast track, just over three weeks later.

Looking back, biologists regarded the appearance of that paper as epochal. For many years, the X-ray crystallographer W. T. Astbury had used a term he had coined, "molecular biology." Delbrück's circle of geneticists realized that what they "had been doing all along was molecular biology."[52]

The 30 May paper on the implications of the DNA double helix for genetics was, Crick thought, "much better" than the first.[53] For Peter Medawar, "The great thing about their discovery was its completeness, its air of finality. . . . The discovery . . . was logically necessary for the further advance of molecular genetics."[54] Next, Crick and Watson turned to writing the report Watson would deliver at the annual Symposium on Quantitative Biology at Cold Spring Harbor in June,[55] and a fourth and more detailed paper for the *Proceedings of the Royal Society*, which appeared early in 1954.[56]

THE FIRST PUBLICITY

Over the last half century, the discovery of the double helix has been compared to Einstein's theories of relativity in terms of their earth-shaking significance. This perception often goes along with the hyperbolic idea that the first half of the twentieth century belonged to the atom, and the second half to the gene. But the first public notice of DNA in 1953 was far quieter than the reaction to the news, in 1919, that a prediction of Einstein's had been confirmed. News of the double helix was reported in an article in the Cambridge newspaper *Varsity*. This was followed over the next few months by a story in the *News Chronicle* of London, and two in the *New York Times*. For the pictures, apparently taken for a *Time* article that never appeared, Watson wore new clothes that Odile Crick had insisted he buy.

By contrast, the public unveiling in November 1919 of observations confirming Einstein's ideas of general relativity was a bombshell. Headlines blared in London and New York and a great physicist hailed Einstein's work as "one of the greatest—perhaps the greatest—of achievements in the history of human thought." Einstein's "distinctly shocking" news made him "suddenly famous."[57]

Much more modest press notice for the DNA double helix occurred on 15 May 1953. Sir Lawrence Bragg addressed a meeting at Guy's Hospital in London and spoke excitedly of the discovery as perhaps explaining how "the printing-off process" of heredity worked. The speech was covered by the *News Chronicle* and the *New York Times*. The headline in the *Times* was "Form of 'Life Unit' in Cell Is Scanned; Sir Lawrence Bragg Reports Gain in Study of Structure of Nucleic Acid Molecules." Mentioning Watson and Crick, Bragg said, "The investigation of these biochemical substances is in a thrilling stage at the present time. Such structures give us a hint not of the answer to the mechanism of living bodies, but of the direction from which an answer may come."[58] A month later, a *Times* feature drew a comparison that resounded for decades: Crick and Watson had unraveled "the structural pattern of a substance as important to biologists as uranium is to nuclear physicists." DNA was "the vital constituent of cells, the carrier of inherited characters, and the fluid that links life with inorganic matter." In the story, Linus Pauling admitted that the Watson-Crick structure looked better than his. Although both his and the Watson-Crick theories were highly speculative, and understanding of "molecular genetics" still lay ahead, the double-helix model, Pauling said, "looks very good."[59] Ironically, the story made no mention that Watson had made his first public talk on the matter less than 40 miles from Times Square, at Cold Spring Harbor, a few days earlier.

The attention made Watson nervous. More confirmation of the double helix was needed. He wrote Max Delbrück a few days after Bragg's London talk: "It is all rather embarrassing to me since the Professor is frightfully keen about it and insists on talking about it everywhere," even though his first reaction to Crick's and Watson's findings had been, "It's all Greek to me." Watson felt that the cheerful Bragg had "got out of control and I spend much of my time de-emphasizing it since I have not infrequent spells of seriously worrying whether it is correct or whether it will turn out to be Watson's folly."[60]

Besides, there was continuing sensitivity about whether putting a theory together from the data of other people—who didn't necessarily understand the implications—constituted piracy. In October 1953, when the upper-crust Third Programme of the British Broadcasting Corporation approached Crick about doing a talk, Watson wrote sharply to Crick that such "self-publicity" could lead him to break off collaboration: It was "in bad taste. There are still those who think we pirated data." He added that "a few enemies are worse than a few admirers," an odd comment in the light of later decades Watson spent in cultivating his image and reveling in having enemies. Crick went ahead. He said later, "It was a perfectly straightforward account of the structure."[61]

The following year, Crick alone was asked to write on the double helix for *Scientific American*. In those days such an article constituted the popular canonization of a discovery.[62] This annoyed Watson. It was tit for tat. But there were compensations. In June 1954, probably with the help of George Beadle of the California Institute of Technology, Watson was featured in *Fortune* magazine as one of "Ten Top Young Scientists in U.S. Universities," along with the physicist Richard Feynman and the cosmologist Allan Sandage. Watson was photographed holding a model of the DNA molecule. *Fortune* commented about its list: "None has yet won a Nobel Prize, but their elders will be surprised if none ever does."[63] (Both Watson and Feynman received Nobels.) Beadle referred to the article in his recommendation that month for an increase in the salary Watson was drawing from Caltech. Probably even more to Watson's taste was *Vogue* magazine's inclusion of him among 18 talented young people in its annual back-to-school issue. The caption for the photograph was "James Dewey Watson (left) at twenty-six is the co-propounder of an important new theory in genetics, analyzing nucleic acids, the basic determiners of heredity. A scientist, with the bemused look of a British poet, he both teaches and carries on his investigations at the California Institute of Technology."[64] Francis was annoyed.[65] Jim was charmed that "my, not his, picture was in *Vogue*."[66]

CHANGE AT COLD SPRING HARBOR

When Jim arrived in Cold Spring Harbor to tell the double-helix story at the Symposium, the grounds looked a little different from three years before. As would happen many times in the future when Jim was director, the twin laboratories on the site had not only won environmental victories with the help of their neighbors but also had put up two urgently needed new buildings. Under the leadership of the labs' director, a geneticist from Yugoslavia named Milislav Demerec, money was raised to build a 16,000-square-foot building containing stark, concrete-walled new laboratories (ultimately named for him) through the sale of the old Eugenics Record Office and surrounding land. Now the main building could be used exclusively as a library. A school bought part of the property; Demerec bought one house lot; and a neighbor, Amyas Ames, bought several lots. Ames became head of the Long Island Biological Association (LIBA), a group that had supported the Biological Laboratory at Cold Spring Harbor since the mid-1920s and owned much of the land adjoining the Genetics Department of the Carnegie Institution of Washington.

The second new structure was the 250-seat Scandinavian-style auditorium with no parallel surfaces, no vertical walls, a baffled rear wall, and a "wave-shaped ceiling," which was completed just weeks before the Symposium. The Carnegie Institution of Washington granted $100,000 for this auditorium, which was named for the institution's president, Vannevar Bush. Up to then, the largest place to meet had been the Fireplace Room in the Blackford dining hall and dormitory, which could hold only 100, and for the 1951 Symposium, 300 scientists had shown up. Ironically, the number registering for the 1953 conference exceeded the new hall's capacity; the money to build a yet larger successor was not raised until 30 years later. To make room for the new buildings, cabins that had been built on the harborside lawn near Blackford were taken down and moved to the hill behind the laboratory called James. Now, Cold Spring Harbor was ready for the next phase of genetics, which was "to understand the gene at the molecular level," Watson later pronounced. The lab could "participate in the revolution that took off with the discovery of the double helix."

The laboratories at Cold Spring Harbor were not only selling land for cash but also buying land to protect their environment, and, as Demerec put it, "insure the privacy that is essential for efficient operation of the laboratories." From Rosalie Jones, a neighbor Watson later called "irascible" and "a very formidable opponent," 23 acres were bought to prevent her breaking them up into half-acre house lots. To pay for this purchase, eight acres belonging to the lab were divided into four lots and sold, one of them to Alfred Hershey. All of LIBA's land was transferred into the jurisdiction of the surrounding Village of Laurel Hollow, whose zoning forbade lots smaller than two acres. Demerec already had thwarted a plan to turn the laboratory's Sandspit into a town beach, and some years later he defeated a plan to dredge the southwest corner of the harbor and turn it into a boat anchorage.[67]

"FOR A MOMENT, THE
ROOM REMAINED SILENT"

The double helix, Crick later became convinced, was just waiting to be discovered—surely in two years or so. But he acknowledged that with other discoverers, "it might not have been pushed as Jim and I pushed it."[68] Others were also pushing. Max Delbrück worked hard to serve up the Watson-Crick report to the eighteenth Cold Spring Harbor Symposium, whose subject was viruses. He obtained a grant so that Watson could fly from London and back and scheduled him to speak on the DNA structure, not on his work on bacterial genetics with William Hayes.[69] To be sure that everyone paid attention, Delbrück distributed copies of the 25 April papers to each participant. In his introductory remarks, Delbrück gave "special mention" to the proposed DNA structure, which he judged to be "of such relevance to many questions to be discussed at this meeting." He scheduled Watson's talk about halfway through the meeting, when attendance would be near its peak and the listeners not yet totally exhausted. With so much buildup, Watson's talk "could be very brief."[70]

But Jim could not avoid dealing head-on with the problem of DNA duplication. The issue gnawed at Delbrück, who was a legendary skeptic and "trenchant" front-row questioner, so Watson had to be ready for him. In mid-May Max had written Jim that he bought the basic structure, but did not think it was coiled the way Francis and Jim asserted. Max could not see how DNA could be unwound for duplication without getting hopelessly tangled. His view was diametrically opposite to Crick's, who, with the confidence born of detailed work on the exact structure, thought the problem "not insuperable." Jim devoted a quarter of his draft for the Symposium presentation to a detailed analysis of Max's objections.[71]

When he flew over the Atlantic to New York at the end of May, Jim carried with him a transparent cylinder that contained a foot-high wire model of DNA, which his friend Tony Broad had fashioned at Cambridge. "Even in its plastic case it was beautiful!" he recalled decades later.[72] He probably brought it to buttress his arguments about exactly how the DNA was coiled.[73] The flight, Jim recalled, happened just as Edmund Hillary and Tenzing Norgay reached the summit of Mount Everest and Queen Elizabeth II was crowned.[74] He had been away from the United States for three years. Although, in the informal style of Cold Spring Harbor, his shirt hung over his shorts and his tennis shoes lacked laces, his accent was partly Anglicized and his talk was dotted with English slang. One participant in the conference, Seymour Benzer, the Brooklyn-born physicist-turned-geneticist, thought Jim had crafted his appearance for the shock value.[75]

Among the pilgrims to the "Mecca" of Cold Spring Harbor that June was François Jacob, who had been so struck with Jim's eccentric appearance and style at Oxford the year before. He regarded Delbrück's invitation as "a membership card to the club." The atmosphere in American scientific meetings, he observed, was "warm, easygoing. . . . No constraints, no ceremonies. Everyone sat where he felt like sitting, next to whom he pleased. Without barriers or hierarchies, but with what was then unimaginable in Europe, young students who did not hesitate to challenge the official stars. All of these boys and girls were ravenous and elbowing their way along. A sort of horde unleashed on science like a pack of greyhounds after a cardboard rabbit."[76]

The April note in *Nature*, Jacob recalled, "had not electrified me or anyone else" at the Pasteur Institute. "The crystallographic argument was over my head." Only at Cold Spring Harbor did Jacob appreciate "the virtues of the double helix." He looked back on Watson's talk as "the star turn" of the Symposium. He described it:

> His manner more dazed than ever, his shirttails flying in the wind, his legs bare, his nose in the air, his eyes wide, underscoring the importance of his words, Watson gave a detailed explanation of the structure of the DNA molecule; breaking into his talk with short exclamations [about] the construction of atomic models to which he had devoted himself at Cambridge with Francis Crick; the arguments based on X-ray crystallography and biochemical analysis; the double helix itself, with its physical and chemical characteristics; finally, the consequences for biology, the mechanisms that underlay the recognized properties of genetic material; the ability to replicate itself, to mutate, to determine the characters of the individual.
>
> For a moment, the room remained silent. There were a few questions. How, for example, during the replication of the double helix, could the two chains entwined around one another separate without breaking? But no criticism. No objections. This structure was of such simplicity, such perfection, such harmony, such beauty even, and biological advantages flowed from it with such rigor and clarity, that one could not believe it to be untrue.[77]

Not everyone present was completely convinced. Hershey remained skeptical about dogmatic interpretations of his Waring blender experiment of the year before. He doubted that "DNA will prove to be a unique determinant of genetic specificity," and did not mention the Watson-Crick model in his own paper at the Symposium.[78]

Watson was moving beyond the phage workers' "black box" approach to viruses as packages of genes to the "primarily biochemical and molecular approach" of the future.[79] As he stood on the narrow stage running across the front of Vannevar Bush hall, before him in the front row were Delbrück and Szilard and others accustomed to "interrupt speakers whose thoughts had gotten out of hand."[80] He used a long pointer to reach the images blown up on the tilted screen above. These included X-ray pictures of the A form by Wilkins and his colleague H. R. Wilson, and of the B form by Franklin and Gosling. To the assembled virologists, Watson began, the importance of DNA structure for virus replication was obvious. Hence, "we shall not only assume that DNA is important, but in addition that it is the carrier of genetic specificity of the virus."

Jim contrasted the rigid, amazingly regular chains of sugar and phosphate and the "completely irregular" sequence of A, T, G, and C. Yet, together they formed the crystals of the A form and the paracrystalline B form. It was by assuming that each chain was a helix, as the X-ray evidence indicated, that he and Francis could arrange the adenine-thymine and guanine-cytosine base pairs inside the DNA fiber. Pairs of two purines would extend too far, and two of the smaller pyrimidines would be too far apart to form hydrogen bonds. The pairs of bases, one purine with one pyrimidine, in any sequence, were "stacked roughly one above another like a pile of pennies." He noted that work suggesting hydrogen bonds between the bases went back six years. Chargaff's ratios, he said, were growing ever closer to 1:1 as biochemists continued their observations. Given the irregular sequence of A, T, G, and C, Watson considered the ratios "a striking result" pointing to hydrogen bonding between complementary bases. "We believe," he said, "the analytical data offer the most important evidence so far available in support of our model, since they specifically support the biologically interesting feature, the presence of complementary chains."

The Watson-Crick model did not depend on some complementary tie to proteins. It did give a clear answer about DNA's strategic role, duplication of genetic information, but not, Jim acknowledged, about its tactical function. "A genetic material must in some way fulfill two functions," he told the virologists. "It must duplicate itself and it must exert a highly specific influence on the cell. Our model suggests a simple mechanism for the first, but at the moment we cannot see how it carries out the second one."[81]

Of the difficulties with the duplication scheme that Delbrück had raised in his letters, Watson considered untwisting the two chains "fundamental," and "formidable." How long a stretch of DNA must unwind, and what mechanism prevented tangling when the two chains unwound? He hoped that a "zipper-like" mechanism would be found. It was "a thoughtful defense." Delbrück did not object during or after the talk.[82]

One of many tantalizing unsolved problems was how to fit DNA into the head of a virus. It would have to "fold up into a compact bundle as it is formed." Another was how to explain spontaneous mutations. Jim and Francis's first guess was that the mutations could be caused by a change from the usual arrangements of hydrogen atoms in the keto form Jerry Donohue had insisted on.

At the end of his talk he said, "The proof or disproof of our structure will have to come from further crystallographic analysis." In fact, definite X-ray proof lay decades in the future. To be sure, the structure accounted for the existing "crystallographic regularity," but more work was needed to show that genetic specificity lay in the DNA alone, and to trace how the structure "could exert a specific influence on the cell."[83]

The conversations afterward ran the gamut. The biologist Barry Commoner, who also had been working on DNA, came up to Jim to question the double-helix idea. A lifelong mutual dislike began when Jim told Commoner that his recent DNA studies were "hopelessly wide of the mark." Maurice Wilkins had told Jim that Commoner was "a fool [who] didn't understand anything he was saying" and so "I told him that."[84] Long afterward, Jim recalled Commoner as "the only one who didn't like it."[85] Leo Szilard, who had fundamental patents in the nuclear-energy field, asked if the structure should be patented. Watson later said the model then had no practical use to justify patenting. As for copyrighting, "We couldn't read the messages."[86] Among those chatting with Jim was the young bacterial geneticist Norton Zinder; though he felt that his own Symposium paper, on how bacteriophages carry genes from one bacterium to another, had been overshadowed, Zinder was to become Jim's lifelong friend, rival, adviser, and troubleshooter.[87]

Another problem that seemed to lie long in the future was the way one DNA strand could be assembled onto a complementary strand. Later Watson could not recall "any serious discussion" at Cold Spring Harbor of the role of enzymes in catalyzing the duplication of DNA. For one thing, "the thought that anyone might soon make functional DNA in a test tube was too far out for serious minds." Only three years later, the biochemist Arthur Kornberg, the self-styled lover of enzymes, was hard at it. Kornberg thought that Watson and Crick were naïve to think that the complementary DNA chain would come together spontaneously on its template. In 1956, he had discovered an enzyme that could do the job.[88]

5

NOW WHAT?
THRASHING AROUND

Our characters were imperfect, but that's life.
JAMES D. WATSON, SANDERS THEATER,
HARVARD UNIVERSITY, 11 MARCH 2002

"DEAR ROSALIND"

The fun of telling people about the structure of DNA soon faded. "Even the most profound leaps forward become boring if lectured on too often," Watson reflected in 1981. "After giving some lectures on the double helix, I had gone stale and desperately needed some new thought to emerge so as to regain the feeling of being alive. But wanting to be clever and achieving it are not the same."[1] When people asked him how long the double helix had made him happy, he replied, "From the first of March to [the] first of September." He recalled, "After we found the double helix, nothing seemed to *happen*. . . . Three months would go by and you didn't read anything interesting. . . . Things were going slowly."[2]

In the years ahead, with a personality less developed and more romantic than Crick's, Watson struggled for relevance. While Crick and Sydney Brenner seized the leadership of solving the coding problem, François Jacob and Jacques Monod in Paris built on their previous genetic experiments to propose a strong hypothesis of how genes are controlled. For several years Jim thrashed around trying to deduce the three-dimensional structure of DNA's close relative, RNA, and hit a dead end. But as a scientist he showed himself more malleable than Crick, more ready to engage with the experimentalists, and become a different kind of biologist than Crick.

For the next dozen years, Francis and Jim were leading characters in the struggle to find a "central dogma" of the pathways for genetic information to travel from DNA through RNA to the making of proteins. And, despite many roadblocks and blind alleys, a new spirit sprang up. "For whereas the pre-1953 Phage group had been groping for the still unimaginable, test and elaboration of the clearly stated central dogma were now the principal research agenda."[3]

Preoccupied until he was nearly 40 with finding girlfriends and a wife, Watson experienced many agonies typical of young manhood. But between the ages of 25 and 28 he was turning from a player in a scientific orchestra into a conductor, from one who collaborates with at most a few others into an Olympian spotter of talent and problems who created a lab at Harvard, then a lab of labs at Cold Spring Harbor, and finally led the startup of an empire of labs, the worldwide program to sequence the entire human genome.

Why would a cocky, awkward, young scientist dedicated to ideas and basic discovery, brilliant in a specific problem, migrate into the deal making, power mongering, and fund-raising that an impresario must do. So many layers of guile are needed to pull people together and lead them, often indirectly and covertly. Perhaps Watson's conversion arose from the same fierce determination that ruled his studies and his focus on finding the DNA structure—a passion to solve the very most important problem—the nature of the gene—and to push his way right up to the ringside and never leave. If the problem shifted to how the gene directs the life process, he must move with it. If the former distaste for biochemistry had to be abandoned, it would be abandoned.

Also, Jim already had experience in the type of thinking that a scientific leader—or prophet—must do. During the dash to the double helix, the intellectual process was inherently executive—a pulling together of bits of theory and data from many corners of biology, chemistry, and physics, fearless of established habits and boundaries.

In this period, moreover, he faced the frustrations that afflict many discoverers and inventors. Between 1953 and 1956, he swung for the fences in trying for an X-ray diffraction structure for RNA, as a key to solving the next grand problems of how life worked—and struck out. After the amazing serendipity of finding the double helix, he remained ready to take risks to solve important problems. But this time, Jim wasn't lucky. The means to capture RNA's structure proved to be lacking. The problem was not solved for more than 40 years. It was natural to shift, almost without knowing it, into another role that guaranteed that he not only would never be bored but also would be at the galactic center.

This cool process went on underneath the romantic misadventures and scientific frustration that he chronicled in his gossipy 2002 memoir,

Genes, Girls, and Gamow. In that book, the rationalist and leader are masked as they were in *The Double Helix.* Watson continued his literary campaign to portray himself as an outrageous prankster instead of as a cold-eyed intellectual opportunist on an imperial scale. His development into a kind of elemental force in biology is hardly visible—except in an incident in 1955 involving Rosalind Franklin.

In July 1955, visiting Franklin's lab during a year back at Cambridge, Jim learned that her stingy, even hostile, principal sponsor, Britain's Agricultural Research Council, was hesitant to buy a new X-ray diffractometer that she needed to maintain speed in her studies of tobacco mosaic virus (TMV). Watson was in a position to help. The next week, with his usual delight in getting on in high circles, he had dinner at the home of Victor and Tess Rothschild, a former monastery called Merton Hall, across the Cam river from St. John's College. It happened that Victor was chairman of the Agricultural Research Council and knew of Rosalind's difficulty. During dessert, with his accustomed brashness, Watson managed with no apparent difficulty to turn the conversation from the bratty behavior of Linus Pauling's son, Peter, and daughter, Linda, to Franklin's need for the diffractometer. He learned that the view at the ARC was that Franklin's proposal would do better if she allied herself with a senior British plant virologist.

Jim hastened to write Rosalind a short, friendly, but remarkably high-toned letter that shows intense preoccupation with and grasp of the intricacies of securing needed backing. It is evidence that Watson already was thinking like an entrepreneurial research manager finding money for hungry-bird researchers and is worth quoting in full:

July 22, 1955
Dear Rosalind

I have had a long talk with Victor Rothschild about your ARC grant. His reaction was very sympathetic and he indicated that he would write [Sir William] Slater [secretary of the ARC] immediately. Apparently he knew your situation very well and has been talking with Slater about how much additional funding you could usefully absorb. The idea of having one or two people attached to K. Smith's group appealed to him, and I think that [J. D.] Bernal [to whose laboratory Franklin had shifted when she left King's two years before] should submit a detailed proposal along these lines. However, with August holidays coming up, nothing is likely to happen until Sept. Perhaps it would be a good idea if we could talk again before anything positive is done, so that the mistake of applying for too little could be avoided?

Unfortunately I plan to leave for the continent on Monday [25 July 1955] and so it's not likely that I shall see you before Sept. 1.

I shall most likely stop in Tübingen on my return from Switzerland.

Jim.[4]

This was more than just a supportive letter from one virus researcher to another. Before the double helix, Rosalind had brushed him and Francis off as amateurs. Yet here he is advising her on how to shape her proposal, with particular emphasis on how much money she could "usefully absorb" and "the mistake of applying for too little," a constant theme in Jim's later career. It is not known whether the prickly Franklin bridled at this offer of help, or whether she ever sent him a draft of her proposal.

ROMANCE, A SUMMER IN CAMBRIDGE, AND THE GENETIC CODE

After the June 1953 Symposium on viruses, Jim intended to stay at Cold Spring Harbor for a week and then return to England. He postponed his departure for a further week when he learned that the famous taxonomist Ernst Mayr, of the American Museum of Natural History, who had summered with his family at Cold Spring Harbor for a decade, would soon arrive from New York City. After supper one evening, Jim and the Mayrs went to the movies in Huntington and afterward came upon a square dance off Route 25A. Jim danced for the first time with the Mayrs' daughter Christa, who would be starting at Swarthmore College in the fall. He had last seen her when she was 14, but "she had changed."[5] The night before his flight to London there was a square dance on the lawn at the lab. No longer feeling awkward together, after the dance the two walked along Bungtown Road to the Sandspit beach, where they talked until sunrise.

Jim neither kissed Christa nor even took her hand. "I wanted to touch her but fearing a negative reaction, avoided even holding her hand as we walked back to the lab's center. By then, the sun was just rising and Blackford Hall was starkly beautiful. . . . With butterflies rumbling through my stomach and not wanting to awaken her parents, we silently bid each other adieu on the Mayrs' doorstep." It was the beginning of a mostly long-distance relationship that lasted for two and a half years, in which Jim exhibited a tendency to emotional overinvestment in the early stages of friendships with women whom he thought he might marry. This contrasted with his ability to undertake lighthearted friendships with other women, such as Linus Pauling's daughter, Linda. Jim's account of the romance indicates that Christa found the friendship rather overwhelming.[6]

The fevered emotional state of this young man was still on display in 2002, when he told an interviewer, "I divide men into those who think of women ninety percent of the time and those who think about them ninety-nine percent of the time. I was a ninety-percenter." He and Francis were certain they were making "this great revolution," and were becoming famous, but he felt bored that no one was inviting him to parties— "because most people didn't know what we were doing was important."[7]

After returning to Cambridge for the summer, Jim wrote Christa affectionately that the "sleepless night" at Cold Spring Harbor had helped him sleep on the long transatlantic flight. He was working on the long, technical paper on the double helix for the Royal Society while Francis got on with his thesis. His chatty letter mentioned a radio program with music by Bach, Bartok, Fauré, and Schumann, and then he said he longed to use a water pistol. A few days later he went to Sydney Brenner's house in Oxford for the birthday party of Brenner's stepson—and ate all the sweets.[8]

In the wake of their stunning discovery, Jim and Francis worried constantly about what to work on next. They had speculated in their second DNA paper, which appeared on 30 May, that "the precise sequence of bases is the code which carries the genetical information." They meant that the bases strung out along the DNA chain would directly specify the amino acids strung out along protein chains. Their words elicited a swift and galvanizing reaction that July from an unexpected quarter—the noted, migratory, and highly playful physicist George Gamow. He had been shown the paper when he was visiting Berkeley. When Crick and Watson received Gamow's "bizarre" letter, in which he offered ideas about how chains of amino acids could be assembled into proteins right on the DNA, they were startled into thinking seriously about the code. How did the sequence of bases in DNA code for the sequence of amino-acid subunits in proteins?

Gamow was born in Odessa in 1904 and was later a colorful member of Niels Bohr's lab in Copenhagen; he was a key figure in developing what the astrophysicist Fred Hoyle derided as the "Big Bang" theory of the origin of the universe.[9] Jim remembered that Gamow had first become famous in physics in the 1920s for figuring out "how alpha particles get out of the nucleus." At Copenhagen, the 24-year-old Gamow sat in the front row at the physics seminars at Bohr's institute. Later in his career he wrote the "Mr. Tompkins" series of popular books about science. Gamow was "sort of an entertainer," full of card tricks and limericks, "just sort of trying to be one up on you." In science, too, Gamow "was clearly trying to have fun."[10]

Wading excitedly into biological theory, Gamow, who signed his letters "Geo," suggested, with the help of drawings, how the amino acids could fit directly into diamond-shaped holes along the DNA and thus line up to form proteins. He was ignoring a great many things. His biology was patchy, with no reference to the location of protein synthesis far from the DNA in the nucleus of the cell or to the possible role of RNA, much of which lay in the cell's outskirts. He said nothing specific about chemistry, nor did he consider the energy needed to drive the cell's processes. Although Geo hit on the idea of code units, or "codons," three bases long, and saw the possibility that amino acids might have more than one "codon," he proposed an overlapping code that prohibited many particular pairs of amino acids in the sequence. This did not fit with the latest evidence on the sequence of amino acids in insulin, the protein that Fred Sanger was deciphering. And biologists already knew that proteins were not assembled right on the DNA in the nucleus but relatively far away, in the cytoplasm.

Geo offered a list of 20 amino acids as potential subunits of proteins that Francis found "a little screwy." So Jim and Francis, looking over Gamow's letter at the Eagle pub, started drafting their own list. No author had done this in the textbooks of the day, but Crick and Watson looked at the problem "with the eye of the outsider." They dropped as improbable some amino acids Gamow had mentioned and added others. Their list also had 20 amino acids. It was guesswork, but it turned out to be right.[11]

With the Watson-Crick list of the 20 "proper" amino acids that are plausible in nature, Brenner wrote decades later, "the problem of the genetic code could be specified as how to transform sequences of the four elements in DNA into sequences of twenty elements in protein." At first, the task looked as if it might be easy. It "had the flavor of a problem that could be solved by thought alone, like breaking a cipher; and it would have been a great triumph if it had been done that way."[12] Instead, a lot of hard work in labs lay ahead.

Back in March, Crick had written his son Michael that "DNA *is* a code. That is, the order of the bases (the letters) makes one gene different from another gene just as one page of print is different from another."[13] In July, despite the many holes in Gamow's theory of amino acids fitting into grooves along the DNA, Crick thought his letter was provocative,[14] but Watson was at first dismissive.[15] He recalled filing the letter away unanswered because neither he nor Francis would be in Cambridge in September, when Geo planned to be there.[16] But in the fall of 1953, Crick met Gamow in New York and insisted that Watson meet him, which he did in Washington shortly thereafter.[17] Meeting Jim and Francis at intervals over the next two years, Geo kept thinking and talking. Crick, later aided by Brenner (who joined him in Cambridge in 1956), was seduced into spearheading the international effort to crack the code, which was worked out by 1966. But at the start, Crick recalled, "Our ideas . . . were totally wrong. We thought that RNA had some structure with twenty cavities. . . . People have forgotten what . . . we didn't know at the time."[18]

That summer, the young man with an itch to use a water pistol in Cambridge, who would be taking up a postdoctoral fellowship in Pasadena in the fall, took part in a cruel practical joke on Crick—in part to pay him back for all those supercharged explanations of the DNA model the previous spring. On the stationery of an international conference in Stockholm that Linus Pauling and his son, Peter, were attending, came a purported invitation with Linus Pauling's signature, forged by Peter, for Crick to come to Pasadena as a visiting professor. Crick, who was scheduled to spend 1953–54 at Brooklyn Polytechnic, wrote at once accepting. A complication was that Pauling had already invited Lawrence Bragg. When Bragg heard of Crick's invitation he found it awkward. How would he and Francis divide up the DNA story? He discussed the problem with Max Perutz and Perutz confessed to the hoax. Bragg told Perutz to tell Crick, but Perutz didn't. So Bragg called Crick in and told him. Crick speedily sent the fake letter to Linus Pauling. This led Pauling to wonder whether he had indeed invited Crick, but then he spotted a split infinitive in the forged letter and knew it wasn't his work. He reduced his son's allowance slightly as punishment. Odile Crick and the wives of scientists in Cambridge were less forgiving and Watson felt a chill of disapproval.[19] The memory of the trick continued to delight Jim. Long afterward he observed, "Most practical jokes are aimed at your friends because of some excess." They would not destroy the victim but reform his character.[20]

Just before leaving England for two years in Pasadena, Watson attended a biology conference on Lake Como in Italy and there met the biologist Jeffries Wyman, the American scientific attaché in Paris. When Watson told him he was dreading two years in Pasadena, Wyman suggested he go to Harvard instead.[21] It was the germ of an idea that came to fruition over the next three years.

"MUCH TRAVAIL" IN PASADENA

Watson was beginning several years of traveling, around California, around the United States, and back and forth across the Atlantic. When asked later why he traveled so much then, he said it was partly "to have someone to talk to. At Caltech the number of people who were interested in RNA was at a given time no more than three." At Stanford there was nobody. In the world the total number may have been 25. "So you traveled."[22]

Arriving in New York by ship in September, Jim went on to his uncle and aunt's home in New Haven. Visiting the Mayrs in Cambridge, Massachusetts, he missed seeing the college-bound Christa, before going on to spend a week at his parents' "modest wooden home" less than a mile from the dunes in Indiana. There, the big topic was the impending marriage in Tokyo on 25 September of his sister to the intelligence official Robert Myers, like Betty a graduate of the University of Chicago.[23]

The plane from Chicago that brought him to Los Angeles descended through a "dirty yellow smog," a bad portent for his time at Caltech.[24]

On the way to the campus by taxi, he could not help being fascinated by the numerous "old wealth" mansions along the route. It was the beginning of what he thought of as exile. He was going to a university where all the problems people studied seemed to be Pauling's problems. Besides, he found Caltech dreary. Still lacking a girlfriend, he had not yet learned to drive and was "trapped" in the university's faculty club, the Atheneum, where he stayed until he could move into an apartment.[25]

A major problem was his changed relationship with Max Delbrück, his mentor in the phage group. Jim was suffering from "those post-DNA semi-blues";[26] once again he would not do what was expected of him. The double helix had carried Jim away from "the innocence of the phage group."[27] Max, whose interests had shifted away from the tiny viral packages of genes to a light-seeking organism called *Phycomyces,* hoped that Jim would take over the supervision of his graduate students working on bacteriophages. Camping trips with the Delbrücks, such as one for four days over Thanksgiving, continued.[28] But now that he had the double helix, Watson was bored by phages, just as he had become bored with the bacterial genetics work with Hayes. He turned back to RNA structure as the next key step.[29]

The big protein conference to which Linus Pauling had invited Jim the previous March was being held in September; it was to be attended by Bragg, John Kendrew, and Max Perutz from Cambridge; John Randall and Maurice Wilkins from London; Hugh Huxley, who was doing a stint at MIT; Crick, spending the year in Brooklyn; and Pauling, Robert Corey, Delbrück, George Beadle, and Watson from Caltech. Beadle had become head of Caltech's Biology Division in 1946, and he was determined to raise it to a new level. In 1958 he shared (with E.L. Tatum) a Nobel Prize for the discovery of the "one gene, one enzyme" phenomenon.

At almost the same time, the new science of DNA took a fresh and significant turn. Seymour Benzer, the former physicist who had listened to Jim in June, was back at Purdue University. In preparing a demonstration for his class he had found a way to link the Watson-Crick insight with the classical mapping of the genes in viruses. He found defective mutants of a gene that normally allowed viruses to take over bacteria rapidly. The mutants would allow him to prove that the genes were discrete entities that could be deciphered. Over the next years Benzer would spell out many hundreds of bits of the gene called rII, each a site of damage, in greater and greater detail. The picture built up on an ever-longer chart that unrolled like a Torah scroll at conferences where Benzer dazzled those looking on. It was a distant precursor of the Human Genome Project.[30]

At Pauling's protein conference, both Watson and Alexander Rich, in Pauling's lab, found the facts and concepts being reviewed to be too familiar. They were learning nothing new. Their thoughts turned to X-ray studies of RNA, of the sort Watson had tried the year before. Somehow RNA mediated the transfer of genetic information from DNA to the site of protein assembly. The hypothesis was that RNA copied from DNA would assume a three-dimensional shape that would link up the unique side-groups of the amino acids.[31] Jim , who had found temporary space in the Pauling lab, prepared RNA fibers and Alex zapped them with X-rays.

Jim was determined to get the structure of RNA, which he thought essential to understanding how DNA directed protein synthesis. The work on coding that Geo Gamow had spurred was less likely to solve the problem. Watson was frank about his concern for his career. He had written to Max Delbrück on 25 March, "Right now I am not thinking about my future—instead I should like to get RNA once, then worry about where I should be."[32] He wrote again on 1 June 1954 that the RNA work was "at a standstill. We need a cute idea or a much better X-ray photograph and neither possibility seems in the air." The base ratios were not complementary, even though X-ray pictures hinted at a DNA-like structure. "I still feel that RNA is the most important problem for us to crack, but it will probably come from inspiration and not from solid concentration." Perhaps a month in the summer with Francis would help, "if for no other reason than we're probably the only people with the solid conviction that until we solve it, we shall know nothing about proteins."[33]

Despite "much travail," Watson recalled ruefully later, he and Alex never observed "ordered diffraction patterns like those of DNA." There seemed to be no way to decide whether RNA had one or two chains. "Clearly there was some ordered structure in RNA, but we saw no way to get to it."[34] At an April 1954 symposium organized by Pauling on the structure and function of nucleic acids, Jim and Alex had to confess, "About the functions of RNA, we possess little definite information."[35] Although they occasionally thought they had a "cute" or "pretty" result, their repeated attempts over two years never bore fruit. The reason they didn't get very far, one scientist observed, was that there were several different types of RNA, and none of them had a highly stable structure.[36]

As 1953 gave way to 1954, Delbrück grew ever more exasperated with Jim, who was now going his own way. Decades later Jim recalled, "Max disapproved of me so much." George Beadle, however, was determined to keep Watson at Caltech as part of his plans for the Biology Division. Sensing that Jim should be "cut free of any further dependency on Max,"[37] "Beets" Beadle was prepared to take over the relationship with Jim, whom he assigned an office near his own. He offered Jim a three-year senior research fellowship, and Delbrück, despite his dissatisfaction with Watson's progress over the past nine months, endorsed the plan. Jim's salary would increase from $5,000 to $6,000 a year, starting a few weeks hence. The trustees made it official. For this, Watson turned down a job at the National Institutes of Health in Bethesda, Maryland, which he had sought as an alternative to service in the U.S. Army. Delbrück, on leave in Germany, wrote Beadle from Berlin:

> Dear Beets:
>
> My frank opinion is that I would like to wash my hands of the responsibility for Watson. You know that I have been and still am one of the chief apostles of [the many who consider] Watson the Einstein of Biology.
>
> *But*, it is also true that I am tremendously disappointed in his personality during this last year. He expected everybody to be concerned with his problems and could never bring himself to show the slightest interest in anybody else's problems. Both as to scientific as to personal problems. This went so far but he completely repressed and forgot any commitments he had made. A *large part* of this was due to the fright he had of being drafted but it shows that he breaks down under stress. . . . *However*, there is another large part which is constitutive, namely that of being ruthlessly egocentric in scientific matters. That goes to such ugly details as publishing prematurely, putting his name as senior author, accepting too many public lecture invitations, etc.
>
> All in all, the question whether coddling or roughing Jim is in the long run wiser, for the good of Jim, for the good of CIT, and for the good of science, is a question I would prefer to leave to wiser men to answer.[38]

Beadle was running a long campaign of letters and appeals to keep Jim deferred from the draft. Again and again over two years, Local Board 75 in Chicago would classify Jim 1-A and Beadle would appeal, so that Jim's classification would return to 2-A for a few months. He told Jim's draft board that the double helix was "one of the significant achievements in this field in recent years," and that Watson's work could be "of the greatest importance in furthering our understanding of disease-producing bacteria and viruses." He enlisted Caltech's president, Lee DuBridge, in the effort, telling DuBridge that the double helix "is very likely the most important advance in biology and basic medical science of the last decade." On 14 June 1954 Beadle wrote yet again to Jim's Chicago draft board—and to Delbrück:

> Dear Max:
>
> Your desire to leave the solution of the Jim Watson problem to "wiser men" than you puts us on the spot. You certainly make the problem clear.
>
> On the basis that the enthusiasm for keeping Jim here for a stretch was practically unanimous in the [Biology] Division, we have

agreed to recommend him for a 3-year appointment as a Senior Res[earch] Fellow. Before doing this I had long talks with Jean [Weigle], Renato [Dulbecco] and others who know him well. Jean, especially, says he has really taken a new lease on life. I therefore hope we did the right thing.[39]

The same day he wrote Jim, who was by then teaching a summer course in physiology at the Marine Biological Laboratory (MBL) in Woods Hole, Massachusetts, a characteristically upbeat letter. Beets thought the appointment was going through but he needed to be sure Watson knew of Delbrück's feelings, and he wrote Watson: "I had a letter from Max and he feels like you and I thought he would, i.e., he's a strong booster for Jim Watson in *principle* but still expresses disappointment about the details of the last year. I've replied that the faith of the [Biology] Division is strong and that we hope and believe it is justified. I'm *sure* it is."[40]

On 28 June, as part of his campaign to continue Jim's deferment from the draft, Beadle wrote to a scientific manpower official in Washington, "I guess Jim will have to produce now, won't he?"[41] It was a hopeful thought, but Watson was not happy at Caltech—he acknowledged that the first of his six lectures on bacterial genetics was "dreadful"[42]—and soon the siren song of Harvard was heard again.

A "WISKIE TWISTY" PARTY IN WOODS HOLE
Watson spent much of the summer of 1954 at Woods Hole, where he was substituting for another scientist as instructor of the physiology course at MBL. His plan was to get the physiology students to repeat the Hershey-Chase experiment of 1952. His helpers were Delbrück's brother-in-law Victor Bruce and Matthew Meselson, a University of Chicago graduate and friend of Peter Pauling's who was just finishing his thesis with Pauling at Caltech and had just learned about the double helix from Delbrück. Meselson recalled that Max had tossed him a reprint of a Watson-Crick double-helix paper and briskly told him to read up on "the most important development in biology in the last ten years." Meeting Watson, Meselson told him about his already-developing ideas about exploiting heavy water and light water in bacterial experiments. Pauling had dismissed it as "nonsense," but Watson liked it at once.[43]

En route to Woods Hole, Jim drove across the United States with Leslie Orgel, an English theoretical chemist in Pauling's lab, who had been newly recruited to the coding problem, whom he left off in Chicago, and Alex Rich's wife, Jane, whom he dropped off in New York. Arriving at Woods Hole and settling into a brick dormitory overlooking Eel Pond, he quickly detected "an absence of Cold Spring Harbor–like intellectual intensity." For the shock value, he continued wearing the shoelaces of his tennis shoes untied during the morning lecture and sported a floppy hat night and day. His water pistol "was judged inappropriate."[44] Still, as Jim wrote Sydney Brenner, who was at Cold Spring Harbor on a visit, "The Physiology course is frightfully intense and I have never worked so hard in my life."[45]

One visitor at Woods Hole that summer was Rosalind Franklin, on an exhausting and overscheduled first tour of the United States; the money had been patched together from many sources. Alex Rich drove her down from Cambridge. Both Watson and Franklin now dwelt in the RNA world of viruses and ribosomes. Jim was pleased that her recent X-ray pictures of rod-shaped TMV supported the conclusion he had arrived at in 1952, that the protein subunits that formed the virus's coat were arranged in a helix. He offered her a ride to the West Coast, but she had too many speeches to make about her work with coal. They met again in Pasadena, and went together to dinner at the Paulings' house.[46]

One day, as Matt Meselson and Jim were standing at a window of a brick classroom building, Jim pointed over to a figure seated on the grass across the street and offering gin-and-tonics to passers-by. "See that guy down there? He's a real smart-ass. He thinks he can do anything, and we're going to test him to see whether he can do the entire Hershey-Chase experiment in a single lab session." With this raffish introduction, Meselson walked over to meet Franklin Stahl, who, it turned out, would appreciate some help with some equations for his genetics experiments with phages. Meselson knew enough math to write out the correct equations.[47]

Like any scientists, they fell to talking about the eternal question: "What next?" High on Meselson's list of postdoctoral projects was an idea for proving that DNA replicated as Watson and Crick predicted—the two strands separating, and a fresh complementary strand forming on each original strand. They thought they could get at the truth of this by an experiment in a centrifuge with DNA of normal weight and one with heavier isotopes of some of its atoms, and then look for a hybrid of intermediate weight. The hybrid would consist of a parent DNA strand and a new complement and thus confirm the Watson-Crick scheme. Watson regarded the idea as "finding a way to distinguish newly-made DNA strands from their parental template."[48] Stahl was coming to Caltech for a postdoc that fall, and said he'd join Meselson in the experiment as soon as Meselson finished his thesis. It was the germ of the famous Meselson-Stahl experiment that proved that DNA replication is "semi-conservative." In their first euphoria over the double helix, Crick and Watson had thought the problem "was over." Years later, in 1988, Watson told a Harvard audience that included Meselson, "It was just lucky that Meselson and Stahl did their clever experiment. Thank God they did it!"[49]

Both Francis Crick and Geo Gamow came to Woods Hole that August to do more work on the genetic code. As members of a joshing new organization called the "RNA Tie Club," they batted coding schemes around,[50] but Watson wrote Beadle that little resulted but "buffoonery."[51] The club had been born in Berkeley a few months earlier, chiefly to distract Gamow from his never-ending card tricks.[52] There would be 20 members, one for each amino acid, and each would buy a specially designed club tie from a Los Angeles haberdasher, along with a special tie pin to represent each member's amino acid. Gamow called himself the Synthesizer and took alanine as his amino acid. Watson, whose amino acid was proline, was designated the Optimist. Crick, tyrosine, held the title of Pessimist. Absurdly, this bit of foolishness turned out to be an excellent way for a few scientists to exchange mimeographed notes about interesting challenges. One, from Crick in 1955, about the escort forms of RNA, or "adaptors," as he called them, was, he thought, his "most influential unpublished paper."[53] Brenner later reflected, "The comic antics of George Gamow and the RNA Tie Club were only a façade behind which we grappled with the deeply important problems of fundamental biology."

Brenner recalled the situation at the end of 1954: Crick, struggling with coding, noted that nucleic acids did not have the chemical properties to act in the specific recognition of different amino acids. His solution was the "adaptor" hypothesis. Each amino acid would be coupled to a short nucleic acid by a specific enzyme; these adaptors would then recognize the coding sequences by standard base pairing.

"When he circulated it in 1955," Brenner recalled, "it was just poohpoohed by everybody."[54] But to Brenner, "It was the only one that made chemical sense."[55] Soon, transfer RNAs and enzymes for attaching each were discovered, and some years later, the 64 three-letter "codons" were deciphered. When he first heard the adaptor idea, Jim didn't like it, because it ruled out the RNA helices he and Leslie Orgel had sweated over during Jim's second and last year at Caltech.[56] But a few years later, in his Nobel lecture in Stockholm, Watson celebrated the radical concept that led to the discovery of transfer RNA.[57]

An element in the buffoonery was the invitation, purportedly from Gamow, to about 100 people, inviting them to a "wiskie twisty" RNA party on 12 August to celebrate Geo's arrival at Woods Hole. It was to be held at a cottage belonging to the Nobel Prize winner Albert Szent-Györgyi, where Gamow and his wife were staying. The hoaxers were Jim Watson and Andrew Szent-Györgyi (Csuli), the Nobel winner's nephew. The whole thing was news to Geo, as he made clear to Francis when the latter thanked him for the invitation. Letters of acceptance poured in to the main Szent-Györgyi house nearby and were duly brought to Geo.

Crick wrote later: "It did not take me long to discover that Jim was one of the perpetrators of the hoax. He did not usually play practical

jokes, but his mentor, Max Delbrück, was notorious for them. . . . I negotiated a treaty. Jim and Csuli . . . would provide the beer and [Geo]
would provide the whiskey. The party turned out to be a great success,
with almost everyone invited turning up for it."[58]

It was not just idle fooling around. Watson may have become a little
tired of Gamow. Crick's account hints at his delight in catching Watson
in another hoax—and the satisfaction of rescuing him.

JIM "WAS VERY INTROVERTED THEN"

In the summer of 1954 Sydney Brenner visited Cold Spring Harbor as
part of a trip that Demerec organized for him to see several key laboratories.[59] After Watson wrote him from Woods Hole, Sydney liked the idea
of accompanying Jim on his return drive across the country to Caltech.[60]
He would join Jim at Woods Hole, and both would travel to Cold Spring
Harbor for the annual bacteriophage course.[61]

Sydney had been working on Geo Gamow's ideas on an overlapping
code and was developing a definitive disproof that he circulated to the
RNA Tie Club.[62] He wrote Jim that Gamow's theory implied that any
amino acid in the protein sequence could be followed by only one of four
different amino acids. But Frederick Sanger's data from the protein insulin showed that one amino acid, glutamic acid, was followed by six.[63]

Arriving at Cold Spring Harbor, Brenner heard about the precision
gene-mapping ideas of Seymour Benzer, who was preparing to spell them
out in a talk there that Crick, Hershey, and Watson would hear at the
end of August. It was a way to match up DNA with the chemistry of proteins. Later, Sydney recalled his reaction: "Well, there it is. That's the way
to analyze the DNA and we can do the protein chemistry. So that was
the origin of how to solve the genetic code."[64] He told Jim that Benzer
was looking at the problem in terms of the Watson-Crick model and
Sydney was authorized to discuss the manuscript with Jim and Francis.
In the event, because Benzer suddenly had to be absent, Brenner presented the paper.

Going up to Woods Hole for a week, Brenner heard of the idea of the
Meselson-Stahl experiment, and debated with Crick and Gamow about
the genetic code. Crick suggested that Brenner join him at Cambridge,
England, and, after some two years back in South Africa, he did so in
1956.

That summer Jim told Sydney of his ambition to lead a laboratory.
Brenner later recalled, "He said there are only two jobs he would like.
One is to be head of Cold Spring Harbor; the other is to be head of the
virus lab at Berkeley. . . . He was lucky to realize the right one."[65]

Jim's car trip back to California with Sydney included visits to his aunt
and uncle in New Haven, Connecticut, to Doty's laboratory at Harvard

during Hurricane Carol (which Rosalind Franklin also experienced as she was visiting in Woods Hole),[66] to the Mayrs' farm in New Hampshire, to Woods Hole, to the Watson family home near the Indiana Dunes, and to Luria in Champaign, Illinois; and driving along the Million Dollar Highway in Colorado and through Monument Valley on the border of Arizona and Utah, made famous by the many movies that were shot there. Brenner considered Watson "very introverted then," but a lot of traveling had taught him that "you just leave people alone."[67] The day before Watson and Brenner arrived at the Watson family home, Jim's grandmother Nana had died at the age of 93. From savings from his fellowship, he was able to pay burial costs of $400.[68]

Back in Pasadena, Jim could no longer bootleg space in Pauling's lab because Alex Rich had gone back east to work at the NIH in Bethesda, Maryland. In his office across from Beadle's, Watson could get down to serious building of models of RNA.[69] He and Leslie Orgel spent a frustrating fall trying to make RNA chains that would fold so as to form cavities appropriate for amino-acid side groups. Linus, just as he learned that he would finally get his Nobel Prize for his theory of the chemical bond, warned Watson not to expect another big breakthrough soon. When Watson told him of a proposed structure for RNA, he advised waiting for stronger evidence before publishing.[70]

Jim's amusements included an occasional date at such local restaurants as the Stuffed Shirt and a trip to the Forest Lawn cemetery, which had just been mercilessly described in Evelyn Waugh's *The Loved One*. To escape the smog, he climbed 14,000-foot Mount Whitney from Lone Pine, hiked near Mineral King with Renato Dulbecco and his son, and climbed to 7,000 feet in the San Gabriel Mountains with Dulbecco and Manny Delbrück. He had dinner with the visiting Rosalind Franklin at the Paulings'. Also at the Paulings', André Lwoff suggested that Watson come to the Pasteur Institute for a year. Jim was sure he must leave Caltech, where he was "getting nowhere."[71]

GETTING JIM TO HARVARD

As soon as the DNA structure was announced, Paul Doty, a biophysicist at Harvard, concluded it was "obvious" that Jim Watson had to come to Harvard. There, a radical turn was beginning toward what would be called molecular biology. Doty, who had arrived at Harvard in 1948 and was given tenure only two years later, had shifted from his earlier interest in polymers to biological macromolecules, such as nucleic acids and proteins. He was interested in pulling them apart and putting them back together. On a visit to Cambridge, England, in June 1952, he had met both Watson and Crick and lunched with them at the Eagle pub in Cambridge several times when they were discussing their hopes to crack

DNA before Pauling and Chargaff. Doty was on friendly terms with McGeorge Bundy, a political scientist who that year had become dean of the Faculty of Arts and Sciences at Harvard and a very powerful number two to the new president, Nathan Pusey. Bundy played a controlling role in new appointments and programs. As Harvard recruited two leading biochemists from the University of Chicago, Konrad Bloch and Frank Westheimer, and arranged John Edsall's transfer from Harvard Medical School in Boston, a new graduate program in modern biological science was started. Knowing that there would be no interest in the Chemistry Department in bringing Watson on board, and despite "considerable resistance" to the idea in the Biology Department, Doty began talking with Bundy about finding a position for Watson there, and it came to pass.[72]

In the summer of 1954, which Doty recalled as "the critical year," the deal came together, albeit informally. On the beach at Woods Hole, Doty often talked science with such colleagues as Konrad Bloch, a later winner of the Nobel Prize. Bloch remembered that Jim was being recruited for Harvard that summer.[73] Jim went up to Cambridge to see Christa at her summer job in the Harvard Biological Laboratories, and then went over to see Doty in the nearby red-brick Gibbs Building on Divinity Avenue. The short walk to Doty's lab, Jim recalled, carried him "from the boredom of old-fashioned biology into the sparkle of chemistry well done." He was pleased at the idea that Harvard had too many "stars" in chemistry to be dominated by a single "pope," like Pauling at Caltech.[74] From Doty, Jim learned that Bundy was "unlikely to let biology continue its dreary path to nowhere."

In October 1954, Jim was invited to give a talk at Harvard—clearly a prelude to an offer of employment—and in January 1955 he gave it. It was hard to hear him. For Doty, it was his first exposure to Jim's "low volume." The classical biologists fixed on this as a reason not to hire Watson. The head of the search committee, a botanist, wondered whether Watson would ever succeed in lecturing to large groups.[75] He wrote Beadle, Delbrück, and the notoriously soft-spoken Fritz Lipmann of the Rockefeller Institute in New York, yet another future Nobel Prize winner, for reassurance. Beadle, whose plans to keep Jim at Caltech were failing, apparently did not reply. Delbrück praised Jim's "absolute determination to make the most important discoveries in biology," and said that Jim had been "incomparably successful." In a somewhat weary tone he added, "No telling what the next problem might be."[76] Lipmann scribbled a reply on the letter of inquiry: "Since he has plenty to say, people will get close."[77] Doty later recalled wryly that "the rapport between Watson and other biologists was not total. . . .[T]he good offices of McGeorge Bundy were taxed to the limit to prove that the appointment was unbiased."[78]

Harvard swallowed its doubts and, with strong backing from Ernst Mayr and George Wald, offered Watson a five-year assistant professorship, to start on 1 July, which he accepted. He checked out the available space in the Biological Laboratories. But then he came up once again with an unusual deal. He would take a sabbatical at the start of his five years, working with Francis in Cambridge, England.

REJECTION

During his sabbatical in Cambridge, Jim hoped to finish a project he and Francis had talked about for years: the general structure of viruses, which seemed to fall into two categories, rods and spheres. He and Francis were sharing the same ground-floor office in the Austin wing that they had occupied two years before. Jim delighted in the "conversational subtleties" he had missed at Caltech. But Crick was "not necessarily father-like" as Max Delbrück once had been,[79] and was intensely absorbed in work with Alex Rich on the three-chain structure of collagen. As with DNA, the key issue was how hydrogen bonds held collagen's three chains together.[80] So in the summer of 1955, Watson turned again to the rodlike tobacco mosaic virus (TMV).[81] This virus was now the key interest of Rosalind Franklin, who was on her way to working out a 40-angstrom diameter for the RNA in TMV. She considered Watson an "insignificant" player in the TMV field.[82]

At Cambridge, meanwhile, Alex Rich pursued his goal of learning more about RNA, and so he and Watson mounted their final X-ray assault on RNA. But Jim's models still "didn't fit with the X-ray facts." Rich and he saw RNA basically as a single-stranded helix, but could not see a role in carrying information from DNA to the protein factories. To get his mind off a looming "boring" result, he played tennis on the courts of Clare College.[83]

As planned Rich left to return to NIH, and now Watson and Crick did get down to virus structure; they took their ideas to a select conference on viruses at the Ciba Foundation in London in March 1956. Among the virologists present was Rosalind Franklin, who talked about X-ray studies of TMV structure and shape. In a discussion, both Crick and Watson supported Franklin's views.[84] Jim and Francis emphasized that the amount of nucleic acid severely limited the number of amino acids that a virus could code for.[85] The virologists were not at all comfortable with the dogma of genetic information flowing in one direction only, from DNA to RNA to protein.[86] But Watson and Crick were determined to press their search for carriers of genetic information.

During the year in Cambridge, Jim pursued his friendship with Christa Mayr, now spending her junior year abroad as a student in Munich. He visited her there, and the seven-year difference in their ages and the intensity of Jim's interest appear to have magnified constraint. Jim arranged for Christa to be invited to the Mitchisons' Scottish home, Carradale, for New Year's of 1955–56. No sooner had they arrived than

Christa made it clear that she did not love Jim nor could she ever. Jim felt he could not argue.

Rejection scenes are not usual in memoirs. They are too painful and embarrassing. But Watson, with typical literary precipitation, lays it out plainly:

> After the teapots were back in the kitchen and the main room now unoccupied, with family and guests off in other places, she no longer avoided my face and blurted out what she had to say—that she was not at all in love with me and knew her mind and needs well enough to know that she would always feel this way. As hard as she tried over the past year, she found it impossible to convert her liking for me into the deep love needed to share her life completely with another person. Forcefully said, these were not off-the-cuff remarks but came as if repeated over and over in her mind ever since leaving Munich.[87]

Some months later, Watson learned that Christa intended to marry a nuclear engineer whose child she was already carrying.[88]

6

HARVARD: "FEW DARED CALL HIM TO ACCOUNT"

I never could work under others. You only get somewhere with people if they feel they are working for themselves.

JAMES D. WATSON, 1988

"A LABORATORY OF SPECIAL INFLUENCE"

Back in the United States from England, Watson visited his sister, Betty, and her husband, Robert Myers, in Washington. Since their return from Tokyo the year before, they had been living in Falls Church, Virginia, with their young son, Timothy.[1] Their daughter, Holly, had just been born. Now Bob Myers was being posted to Jakarta, and could not take his black MG convertible, the very symbol of English sportiness, with him. Jim brought it for $1,500 and drove it to Cold Spring Harbor for the summer.[2]

At Cold Spring Harbor, Jim met his first graduate students and began preparing for many years of organizing others' research. Arriving in Cambridge in September, Jim had an awkward but polite final conversation with Christa at the Mayrs' house. He then drove his MG over to the Biological Laboratories, where he found he could not bear to be alone in his new office on the third floor. In evening light, he strolled to the elm-shaded Harvard Yard to the sound of katydids, but did not go on to the bright lights of Harvard Square.[3]

He was arriving at Harvard in a shrieky and petulant mood. Fiercely competitive, he loved facts as well as gossip. Determined to push forward to the next big thing, he had an unusual instinct for identifying what that was and who could get there. The old-style biology practiced by many at Harvard would not do. It was time to sweep beyond mere description of animals and plants and move into a new biology based on physics and chemistry. In the large red-brick Bio Labs, which Jim later described as "factory-like,"[4] the specific challenge was the broad, complex field of protein synthesis—which followed the direction of genes in DNA. How did the pieces of the cell's machinery work together? The early target was the structure of ribosomes (the sites of protein assembly). Then Jim's lab entered the rivalry to find messenger RNA (the carrier of the gene's message to the assembly point). The next concerns were the identification of factors essential to starting or stopping the copying of DNA into messengers, and the isolation of repressors, which shut down genes when they aren't needed. In the process, Watson was moving away from the sort of biological theory building that fascinated Francis Crick and jumping into biochemistry.

To go into such matters, Jim fired up what one colleague called "a laboratory of special influence" in modern biology.[5] "He had such strong ideas," another said, "and they really shaped the way . . . of thinking about science, who we choose to be the next generation of scientists, who we think is going to be good."[6] In a few rooms on the third floor he pushed for an atmosphere that inspired ambitious experiments. "It's important," he said later, "that you establish conditions where people will really become important early in life, because it gives them the self-confidence to think big."[7] The main relaxation for Jim's growing group was an Anglophile tea—the secretaries and technicians at one table and the researchers at the other. The tea usually was prepared by one of the numerous women undergraduates who worked in the Watson lab.

In the still small world of molecular biology in 1956 and subsequent years, Paul Doty observed, "Jim's personal style of engagement left a telling mark. . . . [He brought] an extraordinary vigor and innovation" to finding able students, building a molecular biology department, and teaching." To Doty, Jim "was a constant source of freely shared ideas and stimulating criticism . . . [shaping] the research of many others [and functioning as] the unseen co-author of many, many papers."[8]

Doty was one of a small group of scientists who were determined to move biology at Harvard beyond nineteenth-century-style description and classification, and, through reductionism, to focus on specific pieces of the puzzle of life that were amenable to experiment. Among these were John Edsall, who had just come over to Cambridge from Harvard Medical School, in Boston, and Konrad Bloch, who shared the 1964 Nobel Prize in Physiology or Medicine, just two years after Jim. Aware that Harvard was "a remarkable desert" in molecular biology,[9] they had started a program of higher studies in biochemistry only a year before Watson's arrival, and looked to Watson to attract the needed new talent. Doty recalled that "the principal characters" were beginning to move and assemble in Cambridge. These efforts led to formation of a separate Department of Molecular Biology and Biochemistry in 1967.[10]

MAKING ENEMIES

Watson's attitude as he began his stormy years at Harvard was confident but also nervous. He was working in a building whose exterior loudly proclaimed allegiance to the classical biology with its frieze of beasts marching just under the roof, elaborate portals decorated with metal images of insects, and, to the right and left of the doors, giant metal sculptures of rhinoceroses. Through the 1960s, members of Watson's laboratory would push each other—and sometimes Jim—up on top of the right-hand rhino's back. Together with others arrayed around the rhino's feet, they posed joshingly for group pictures that were lovingly preserved.[11]

"I don't think I ever seriously worried whether we'd win, and change the direction at Harvard," Watson recalled, "but there were days which were pretty bleak—which often led to an outburst on my part. But I think you often get somewhere by slight exaggeration." This included using obscenities, but he claimed he didn't use them for minor matters. At Harvard, he recalled in 1988, he had to raise a crisis about every two years, and Harvard became convinced he was "ungracious." Six or seven times in his life, Watson recalled, he "had to create a situation where if I lost I'd be out of a job." Unless you do this, "Someone will run over you."[12] Later, he admitted that sometimes he "went beyond what I should have done."[13]

Jim excoriated the tradition-bound Harvard Biology Department. McGeorge Bundy, ever in the thick of faculty disputes, wrote wryly: "Watson's outspoken disapproval of the practices of our department of biology has not been made more palatable by the fact that he is usually right."[14] Watson regarded Bundy as his "Harvard protector."[15]

When Biology Department members gathered, Edward Wilson recalled, the tone was an "edgy formality like Bedouin chieftains around a disputed water well." Watson, along with George Wald, John Edsall, Matthew Meselson, and Paul Levine were certain, in Wald's words, that "there is one biology, and it is molecular biology." If Watson passed Wilson in the hall, he usually did not speak. He denounced Wilson's suggestion of hiring an ecologist. Watson's behavior was appalling to Ed Wilson, even if Watson felt "that he was working for the good of science and a blunt tool was needed." Wilson admitted that the double helix "towered over all that the rest of us had achieved and could ever hope to achieve. It came like a lightning flash, like knowledge from the gods. . . . [I]t injected into all of biology a new faith in reductionism." Life "might be simpler than we thought." Without this achievement, however, Watson "would have been treated at Harvard as just one more gifted eccentric, and much of his honesty would have been dismissed as poor judgment."[16] In his memoirs Wilson gave this description of Watson, "the most unpleasant human being" he had ever met, at work:

> He arrived with a conviction that biology must be transformed into a science directed at molecules and cells and rewritten in the language of physics and chemistry. What had gone before, "traditional" biology—my biology—was infested by stamp collectors who lacked the wit to transform their subject into a modern science. He treated most of the other twenty-four members of the Department of Biology with a revolutionary's fervent disrespect.
>
> At department meetings Watson radiated contempt in all directions. He shunned ordinary courtesy and polite conversation, evidently in the belief that they would only encourage the traditionalists to stay around. His bad manners were tolerated because of the greatness of the discovery he had made, and because of its gathering aftermath. In the 1950s and 1960s the molecular revolution had begun to run through biology like a flash flood. Watson, having risen to historic fame at an early age, became the Caligula of biology. He was given license to say anything that came to his mind and expect to be taken seriously. And unfortunately, he did so, with casual and brutal offhandedness. Few dared call him openly to account.[17]

One day in 1958 Watson learned that Wilson was being awarded tenure before him—Wilson asserts that it was "an accident of timing," Harvard's move to keep him from taking an unsolicited offer from Stanford. According to one of Watson's students, it was "a big, big day in our corridor." Watson could be heard coming up the stairwell to the third floor, shouting "Fuck, fuck, fuck." Another student recalled that the decision to give tenure first to Wilson became known on the day when Daniel Mazia of Berkeley was in town to give a seminar. At dinner that evening at Paul and Helga Doty's, Mazia urged Watson to move to Berkeley, where he would be appreciated. "But Jim was too angry to bother to listen." Colleagues stormed to McGeorge Bundy to extract another tenure position for Jim.[18] Carroll Williams, the chairman of the Biology Department, began to wonder whether Jim should be at a university at all.[19] In 1960, the aggrieved nonmolecular biologists formed their own committee on macrobiology. This was another step toward the fission of the Harvard Biology Department.[20]

Some years later, Jim was shouting again when Harvard refused to give him a salary increase after his Nobel Prize. When he left Harvard in 1976, he calculated that the university had underpaid him a total of $50,000.[21] Nonetheless, Harvard paid him a salary for years while he was getting his Cold Spring Harbor enterprise off the ground, shuttling back and forth between the campus and Long Island—and no doubt boosted substantially the royalties of *The Double Helix* in 1968 and later by forcing its publication elsewhere.

Changing biology at Harvard meant fighting over every new appointment to the faculty, and Jim often was exasperated. The resistance he generated was understandable because he was violating the university doctrine that no research topic is any more important than another.[22] One reason Watson fought for "great colleagues" was because the students, who "really should work on something important," deserved the best professors. The young people would, of course, bring much tension. They had to be reexamining or overturning the ideas of seniors who had started up the same way. Watson's stern doctrine was that good institutions must "reward intellectual success even when it leads to the effective academic redundancy of many of its older members."[23]

The needs of his students and postdocs led to many feuds with others in the Biology Department. One concerned access to the Bio Labs, which up until then had been locked at night.[24]

Many Biology Department faculty meetings left him fiercely disgruntled. In wrangles over new appointments Watson declared, "We should not waste any quota positions on embryologists." As a compromise, when Harvard hired several new molecular biologists, including Matt Meselson, embryologists did receive Harvard offers. "Fortunately," Jim recalled, "the animal person turned us down. The plant person came and was just as bad." Embryology and developmental biology had, in Watson's view, "gone through a wonderful descriptive phase" but at that stage the study of development of living creatures was impossible to understand. To be sure it was important, but "the time had not come to do it." For about 10 years Watson gave an annual lecture "on why developmental biology should be avoided until the basic facts about genes and their expression were better known." The true flowering of developmental biology did not occur until the 1980s.[25]

Watson recalled that except for when he had to attend faculty meetings, "when I was in that building, I was always very happy."[26] He didn't worry much about where the next research grant would come from. "My main anxieties always related to whether our science was going well."[27]

Others were worried also. One graduate student, who was charmed into collaborating with Watson on his early experiments on ribosomes, said that in 1957, "it was widely thought that Watson would be a flash in the pan, the smart stuff had come from Francis Crick, and Watson would be rather less like lightning and more like a lightning bug."[28]

A "CONFIDENTIAL" STYLE OF LECTURING

Lecturing was a problem. Watson wrote "beautiful lectures" but delivered them badly.[29] Doty observed that Jim's early lectures "were so soft-spoken that people referred to his style as 'confidential.' "[30] A problem for Watson was that in lectures, he "had to teach something I didn't know." This was "the only occasion where I had to think." It gave him nightmares. He dreamed of forgetting everything halfway through, while knowing he would also have to give a second lecture.[31] Still, one teaching assistant found his graduate lectures on viruses "incisive and insightful."[32]

Not all students were swept off their feet. An undergraduate whose main fascination then was what could be done with bacteria recalled, "He mumbled. And he went on a bit, and he would wander off into stories about people. It certainly wasn't a wonderful course in terms of the science, per se. . . . It was just an event."[33] In 1968 a graduate student told an interviewer from the *New York Times*, "Watson is a phenomenon sui generis. In class he has no exceptional wit, or showmanship, and none of the charisma of the old-time prof, but he's clear and intuitive."[34]

Another student of the early 1960s was warmer. Watson was the most visible advocate of the molecular approach to biology on the faculty, and he had most to do with pulling would-be scientists into a crusade. Jim took undergraduates seriously; he wanted to teach them, as many professors do not. He was the opposite of the smooth, spellbinding star lecturer. One student observed, "The best that can be said of his lectures was that they were unpredictable." Instead of marching smoothly along well-determined paths, the talks amounted to "directed random walks." He often spoke to the blackboard. Easily distracted, and distracting his students with "legendary gestures," Watson dropped in bits of scientific gossip that could entertain but added little to his ideas and arguments.

Describing Jim's teaching in 1973, the science historian Horace Judson observed that although "the mumblings and false starts never ceased, what Watson was talking about was exciting, and his way of seeing it, dubious and depreciative, was exciting. He spoke ad lib and seemed caught up in the subject, thinking it through afresh, so that each lecture grew spontaneously from live and immediate findings, questions, and speculations. . . He was funny and scandalous about the shortcomings of scientific ideas and of scientists."[35]

For many undergraduates, Jim was a "big hero." Well before his Nobel Prize and his best-selling books, students attended his lectures and others he advised them to hear, and were eager to work in the new molecular biology labs. They were captured by both matter and manner. The message was simple and powerful: Studying the biological macromolecules, DNA, RNA, and protein, was "the only productive scientific path forward in biology." Watson highlighted the hope of a rigor in biology like that of the physical sciences. But he also conveyed his message in an "iconoclastic and aggressive way," not the worst tone to take with young people who had a world to make.[36]

One undergraduate whom Watson attracted into science was Nancy Hopkins, a graduate of the Spence School, a girls' school in New York—

what was in those days referred to as a "finishing school." Later, Hopkins became a full professor of biology at the Massachusetts Institute of Technology, moving from viruses to a leading role in developmental biology with her work on zebrafish. Her first interests at Harvard were mathematics and architecture, but by 1963 she was aiming at medical school. The only way to graduate on time with the required credits was to major in biology. She took Biology 2, which was taught by several professors, including Wilson and Watson.

The professor for the first two weeks was not inspiring. He seemed to think biology was only physiology. The third week, Hopkins took her usual seat at the far right of the second row, close to the lecturer. Watson came in to start his part of the course.[37] She had scarcely heard of him or his Nobel Prize.[38] She was mesmerized, even if Jim dropped his voice and turned toward the blackboard for the punch line. Many years later she recalled that hearing Watson lecture determined her career in biology. "That was it!" She learned to lean forward to hear Jim. She raced to class and sat on the edge of her seat.

The students knew that a revolution was going on and Jim brought it into the classroom. He "made you feel you were in the place of greatest excitement." Amid intense competition, the three-letter "codons" of the genetic code were being worked out. Molecular biology was only 10 years old, but Hopkins was enthralled by the idea that it could explain cancer, memory, aging, everything in biology.[39] "I went into this classroom, and fifteen minutes later my life had changed totally. I saw unfolding in front of me the answer to life and all of the questions I'd ever had about it."[40]

Hopkins recalled, "He was so excited." His urgent enthusiasm for molecular biology made students look on biologists like Sydney Brenner as "rock stars." And Jim was the friend of these stars, had lunch with them, and introduced students to them at parties at his house. Watson seemed to feel "a kind of equality" with students to whom he would tell that morning's gossip from the frontier of biology. Sometimes the gossip was not about science. Harvard's president, Nathan Pusey, had ruled that Watson was spending too much time away from the university and ordered him back. Watson denounced this as "Boy Scout stuff." Hopkins recalled, "It was amazing to be told that the president of the university was a jerk." Jim actually seemed to believe that "the smartest person in the room might be the guy washing the dishes." Nancy wanted to work in the lab, and she did. Watson told her that she was "just like me—a one-track mind."[41]

GOADING YOUNG RESEARCHERS

Watson gave a great deal of scope to his graduate students, seeking to imitate Salvador Luria and Max Delbrück. The students, he thought, were very intelligent or they would not have gotten anywhere near, and could "run at their own pace." Typically, they had heard enough of his lectures "to know what I thought were the big problems to be solved over the next several years. They also knew what fields I thought had no effective chance of soon moving forward." The idea was to get the students going on "problems bound to yield sufficient data for a thesis but that nonetheless had potential big payoffs." In their sink-or-swim years in Jim's lab, they also had the benefit of alternate mentors right there, such as the Swiss biochemist Alfred Tissières and Walter Gilbert, a theoretical physicist who turned biologist. They also had the chance to visit and share techniques and equipment with labs such as Doty's in Cambridge and others at Harvard Medical School across the Charles River. They knew that they were working "primarily to advance their own careers." Though terrified of Watson, they weren't working for anyone but themselves.[42] "I never worked for anybody," Watson recalled on his sixtieth birthday. "I never could work under others. You only get somewhere with people if they feel they are working for themselves."[43]

It was intense. Mario Capecchi, a graduate student of Watson's in the 1960s, recalled, "It was a blast. He always introduced the right mixture of paranoia, so that we worked our asses off." To Capecchi, Jim "just kept on going and going, sort of reminiscent of that pink bunny" who marches beating a drum in the Energizer battery advertisements. Capecchi, who was later the leader of "gene-targeting" research at the University of Utah, and received a National Medal of Science in 2002, recalled what he thought was Jim's key technique of mentoring: "You have to have an enemy. It doesn't matter whether that enemy is real or not," but it was vital to "getting your graduate students, [and] your postdocs, to work as hard as they can. . . . It's not that you're being competitive. You simply want to get incentive." Of course there were battles "as to what he thought was fair play and what I thought was fair play, because we came from very different worlds." But Capecchi "always realized that he was right in the end, and we always went by what he said."[44]

Another graduate student, William Haseltine, commented about Watson, "He is not your idea of an isolated, remote scientist. . . . He is driven by anger at people. He is motivated by anger. He is vindictive. He isn't just passive in his anger. He's active in his anger. He demonizes competitors. He ridicules them." The attitude in the Watson lab was that "only a few Englishmen in Cambridge" were in the same league. All the others were "fools of one kind or another."[45]

From the beginning, Watson's approach to young researchers was intuitive. Robert Risebrough, a Canadian-born graduate of Cornell, was working at Cold Spring Harbor in the summer of 1956 and joined Watson that fall. He wanted to move on from ornithology and Jim advised him to start by learning how to work with the exceedingly tiny virus Phi X 174. Also at Cold Spring Harbor, for two weeks that summer, was Julian Fleischman, from Philadelphia, who had been doing course work at Harvard in the newly launched biochemistry program. When Watson briefly visited Harvard in 1955–56, Fleischman had asked about working in his lab. Watson said yes, and also advised him to spend some time at Cold Spring Harbor.[46]

His method extended to undergraduates whom he accepted as researchers. In 1961, Alfred Goldberg of Providence had finished his sophomore year at Harvard and was working at Cold Spring Harbor on a dull project involving the fruit fly *Drosophila,* the classic tool of genetics. Even Fred Goldberg, who had studied much less biology than his fellow undergraduate researchers, knew that *Drosophila* was no longer on the cutting edge of biology. But the summer was "truly exhilarating" anyway. Goldberg attended the phage and bacterial genetics courses, which were animated by the new insights in protein synthesis. One of the lecturers at Cold Spring Harbor that summer was Jim Watson. On a walk to the beach, Jim said Goldberg should look him up in the fall.

In a single year new findings had turned basic thinking in biology upside down. Messenger RNA had been discovered. Contrary to "the dogma that one ribosome encoded one protein," it was found that ribosomes were not specific to genes but operated like tape recorders.

Such a "global change" wasn't happening so fast in any other field. In one of the Cold Spring Harbor courses that summer of 1961 was a postdoc from Marshall Nirenberg's lab at NIH, Heinrich Matthaei, who played Ping-Pong with some of the undergraduates. A prevalent topic was the question of when, someday, the genetic code would be solved. Matthaei pulled a couple of his classmates aside and told them, "Aha. You'll find out in a couple of months. We've solved it already." Nirenberg's announcement that he had found the codon for one amino acid was the sensation of the international congress of biochemistry in Moscow that August.

When Goldberg, now a junior, showed up at Jim's lab at Harvard, Jim gave him a desk and a problem, which was how magnesium associates with the newly found "messenger" RNA. This research, of the kind graduate students normally do, eventually turned into a paper in the *Journal of Molecular Biology* in 1966. Goldberg remembered being "surrounded by tremendously good people." There was much independence in Jim's lab and no hand-holding. Fred had to think for himself. He had started by checking in with Jim once a week with questions about how to obtain one thing and another, but Jim replied, "Fred, there are some people I like to see once a week; some I like to see once every two weeks. I think you're one of those." Jim's "uncompromising commitment to excellence" and his "cutting manner . . . when he encounters nonsense," was "eye opening." To survive there, Fred remembered, "you had to have an ego."[47]

Jim was always on the lookout for graduate students and postdoctoral fellows in labs that he visited. A visit to Salva Luria at the University of Illinois in the spring of 1958 was typical. Watson gave lectures, and argued about ribosomes with the ebullient Sol Spiegelman. On hand was Masayasu Nomura, a shy but self-assured and determined Japanese scientist with a nearly impenetrable accent. Nomura had come to the United States as a postdoctoral fellow to learn "as many approaches as possible" to such problems as the transfer of information in living systems. He was working in Spiegelman's lab and Watson asked him to go over to the campus Union for coffee. Of Spiegelman's research Watson said, "This is not the way to do science. You should come to work with me." He didn't say just for the summer. Many years later Nomura recalled this declaration as "the first shock. It's unthinkable in our polite society in Japan."

For decades thereafter, at the University of Wisconsin and the University of California at Irvine, Nomura was a notable student of how RNA and a host of proteins are put together in forming the ribosomes. In the summer of 1958 and again in 1959 he worked in Jim's lab at Harvard. Mistakenly expecting to learn physical chemistry, Nomura instead was introduced to ribosomes, which were Watson's early obsession at Harvard. He also absorbed Jim's opinions about "good" science and "bad" science in patient and frank conversation about "serious and important science philosophy."

The only air-conditioned room in the Watson lab belonged to Ernst Freese, another physicist turned biologist, who convinced Jim that his independent work on bacteriophage genetics absolutely required a constant temperature. Neither Nomura's workspace nor Jim's office next door, both facing west toward the afternoon sun, was air-conditioned. Every day Nomura repeatedly had to go up one flight of stairs to John Edsall's laboratory to make one-hour "runs" on the only nearby ultracentrifuge. The experiments usually kept him at the lab until midnight, and Nomura's wife complained. To help Nomura rest or do something else, Jim amazed him by offering to do a centrifuge run for him.

But Watson could not swerve Nomura from his own program of study, which would take Nomura next to Seymour Benzer's lab at Purdue. Nomura should come and stay at Harvard, Jim said again and again, not do his planned stint at Purdue. "You are going backward," Jim told Nomura in a phone call. "It's classical genetics and you should move forward" into biochemistry. "Jim, you had all [of] classical genetics," Nomura replied. "So for you it's going backward. But I never studied genetics. I don't have any experience. So even though it's backward for me, it's important." This exchange did not cool their relationship. In 1961, after Nomura had returned temporarily to Japan, Watson visited the Nomuras on his way home from the famous biochemistry congress in Moscow. Many years later he made Nomura a trustee of Cold Spring Harbor Laboratory. Nomura told Watson he had no aptitude for that kind of duty, but Watson rejoined, "Oh, just come and enjoy New York."[48]

Not all postdocs had such a good experience with Watson. In the genetics department that Max Delbrück ran in Cologne in the 1960s, Jim found Daniel Wulff, a Santa Barbara native and Caltech graduate who later was a professor at the University of California at Irvine and the State University of New York at Albany. Wulff didn't know why, but Watson arranged with Delbrück that Wulff would come to Harvard in 1963. At Harvard, Watson asked Wulff to study whether a substance killed cells by poking a hole in them. He found that it didn't and wrote up a paper for the *Journal of Bacteriology*. When Jim looked at the draft he said, "This is so bad I can't believe it." It was typical bluntness. "He intended me to take his criticism and write a paper that would be more to his liking. But I never did and he never pressed it. . . . That was not a productive time for me."[49]

The researchers Jim recruited, who had done and would do many more experiments than he did, found him acutely sensitive to snags in their way, even if he was the sort of father who just threw children into the pool to start them swimming. Fleischman, who played tennis and went skiing with Jim, found his interactions with him "very informal. Whenever something cropped up, we would sit down and talk about it. . . . I never felt that he was unavailable or wasn't giving sufficient attention to anything."[50]

One graduate student who clashed with Jim, Charles Kurland, admitted that there was "good communication and good discussion" in the Watson lab, and that Jim "was in some ways the absolute best experimentalist I ever came across—in his mind." Nonetheless, "He was a disaster in the laboratory. He would walk into the laboratories and things would start falling off of shelves."

In 1958, Chuck Kurland was going ahead with his marriage to a non-Jewish woman. His mother wrote Watson asking him to do something to prevent this. Chuck heard from Jim that she was very, very upset and had written to him, but Jim did not show him her letter, nor his soothing reply. Chuck was irritated, but Jim said, "No, no, you have to understand, she's frightened." All Chuck knew at the time was that Watson "had done something very positive and very kind." The marriage went ahead and Kurland eventually was reconciled to his mother. Only decades later, when his father died, his youngest brother sent him Watson's letter from his father's papers. In a "thoughtful. . . compassionate" letter, Jim urged Chuck's mother not to worry. He didn't know Chuck's wife-to-be very well but said that she was, as Kurland now recalls it, "a very fine woman, very attractive, and gives a nice appearance, and would certainly be a calming influence on me." The marriage endured.[51]

Not all of Jim's graduate students recalled working very closely with Jim at Harvard. David Zipser, who grew up in Manhattan and went to college at Cornell, concentrating on genetics and botany, before working at Caltech for a year, was recruited by two graduate students, David Schlessinger and Kurland, and he came, but not for any defined project. Zipser recalled, "I don't think they recruited me for anything. I mean, things weren't that specific. . . . Everybody was interested in everything. Molecular biology had a small number of facts and a very, very strong paradigm, so you didn't have to know any facts. Unlike botany, which had nothing but facts and no paradigm. And since I have a bad memory for facts and a very good understanding of paradigms, it was ideal for me. I found it very simple." Zipser soon plunged into such problems as the true size of the protein called beta-galactosidase, important for studies of the controls on gene expression. He showed that the protein was twice as big as the prevailing theories said was maximum for proteins. As he was finishing his thesis, Jim sent him to work with François Jacob and Jacques Monod in Paris—without telling the people at the Pasteur Institute of his plan—before he moved to Columbia. In 1970 he joined Cold Spring Harbor Laboratory.[52]

After his first few joint efforts on the structure of ribosomes in the late 1950s and a paper or two on messenger RNA, Jim followed the policy of not putting his name on the papers. One student wondered whether this was from generosity or conservatism.[53] Another, feeling rejected, understood only much later that this was part of a Watson policy of making young researchers into stars.[54] "I believe in the now," Jim later said. "To hell with being discovered when you're dead."[55] Having abandoned hands-on experiments, he knew that nothing that he would ever do in research would match the discovery of the double helix.[56] According to a young coworker, "Jim knew what he had accomplished with the DNA structure. He knew that he had done the most significant thing anyone could possibly ever do in twentieth-century biology."[57]

With "an uncanny instinct for the important problem, the thing that leads to big-time results," picking up tips from journals or even "out of thin air," he pushed his students toward "impossible-to-solve problems and away from safe topics."[58]

Evening group meetings evolved into a tradition. Younger researchers presented their work so far to the group, and faced a barrage of questions from Jim, Matt Meselson, Wally Gilbert, who developed into coleader of the Watson lab, and Mark Ptashne, who worked in both rivalry and cooperation with Gilbert on the gene-control proteins called repressors. Not only were the younger researchers studying many papers, as many as

250, "to understand what was wrong and where they had missed and where the new opportunities were," they also were seeing each other get the same type of scrutiny—live.

One graduate student recalled the array of "totally brilliant" scientists, backed up by a "wolf pack" of younger researchers who would jump in if they smelled blood. He couldn't imagine a more daunting front row. "Typically Wally would be looking like he was listening, staring up in the sky. Jim would be looking completely distracted. Mark Ptashne would be practicing his shadow violin. And Matt Meselson would sit there taking notes. Well, you can't imagine a more intimidating front row. . . . [It was] like having Babe Ruth and Ty Cobb and Ted Williams sitting there looking at you while you're doing batting practice." Although they appeared not to be, "they were paying attention. They would compete with each other to tear you to shreds if you made a mistake." It turned out to be "wonderful training."

Alongside the pressure was Watson's drive to "make you a star." Graduate students or postdocs should strive for "world-class recognition." When the younger scientist made a discovery, "Jim understood immediately the importance of the discovery. He understood immediately how to present it to the scientific community and to create a star out of these scientists. That's something you don't get anywhere else. I never saw it anywhere else."[59]

Of this relentless grilling Watson himself said it was "very important to criticize" if students didn't know how to speak. He had learned this many years earlier under the withering criticism of Leo Szilard and from the phage group, where "the bad guys just stopped coming around." They were torn apart. The criticism was given not out of nastiness but to teach students that "books are wrong; facts aren't facts; there are bad scientists." You couldn't let students give a bad seminar. "If someone can't take it, they can't be there anyway." Although he had to be careful whom to take, he was lucky that the majority of his Harvard students wrote "super theses."[60]

Others found this apparent mania for Nobel Prizes unappealing, "a self-serving conceit." A great result that called for cracking a bottle of champagne could, and sometimes did, turn out to be not so great. "Jim did many things which . . . the stronger minds simply ignored."[61]

Watson's behavior was the opposite of smooth and the environment was rocky. When a graduate student complained of too little direct attention, Watson told him brusquely that if he wanted to leave without a thesis, he could. He did this with such vehemence that the student was shaking, pale, and near tears when he emerged to talk to a colleague. Happily, the student went on to finish his degree.[62] Jim only paid attention when he was interested. If he was bored, he would ignore the speaker—a behavior that terrified the ambitious young scientist. He often asked rude questions at meetings. His conversations in the hall or the briefest of between-experiment visits to the scientist's workbench were short and always with a purpose. As a goad, Jim spoke of an interesting recent result or piece of gossip, or reminded people of where their work fit into a larger picture. He was an unusual coach of people who were expected to produce the best, the newest, the most important results in rivalry with other labs. He was not forging a team to work on pieces of his problem.

Watson was and remained in a hurry. Then and later he had the view that the fantastic complexity of the living world, and the urgency of fighting diseases that resisted standard medical care, called for impatience in defining central problems and then pushing on at once with the next step. Almost by definition, the people to do this are young, doubting, and rebellious. For them, a leader's renown cannot be enough. At any given stage of development of ideas and equipment, the problems loom as difficult, if not impossible. In stumbling toward some basic insight, the young researcher must be willing to put in years of obsessive effort, surprises, upsets, reversals, the relentless skepticism of colleagues, endless abandonments of cherished hypotheses, mistakes, and failures. Equipment must be adapted, literature searched for a telling detail, and people in many disciplines questioned or recruited into collaborations. Neither the fame of Watson's achievement nor obedience to authority seems sufficient to explain the scope of his influence. Fame often goes to figureheads who do not deserve it. Watson's authority with young researchers appears to have been based upon the depth and range of his intellect.

VIRUSES AND CANCER

Fame came to Watson and Crick, first among scientists and aspiring scientists and then among the general public. As Watson pushed his way up the ladder at Harvard, becoming tenured associate professor in 1958 and full professor in 1961, he was frequently invited to lecture, and awards began to accumulate. More and more people were tuning in to the revolution that the double helix discovery had started.

Among the questions that caught Watson's interest at this time was the relationship between viruses and cancer—a realm frighteningly more elaborate than that of bacteria and their viruses. Not long before, viruses with genes stored in DNA had been found to cause cancer in laboratory animals. Could RNA viruses do the same? Could viruses be implicated in human cancer? It was an old interest of Salva Luria's, and on his visit to Urbana in 1958 Watson became excited again. Van R. Potter of the University of Wisconsin lectured on the biochemistry of cancer. Unlike bacteria, "higher" cells "needed specific signals to divide," Potter observed. In fact, the majority of cells in our bodies are not dividing but are in an apparent resting state where DNA synthesis is not occurring. Cancer cells are dividing all the time, synthesizing DNA without letup. Meanwhile, leading enzymologists like Seymour Cohen and Arthur Kornberg had begun to show that the DNA of phages carried code for many of the enzymes needed to reproduce the viral DNA in an infected host cell.

Watson recalled, "This discovery opened my mind to the possibility that animal viruses might have similar genes, and soon I was telling everyone within reach that the cancer-causing (oncogenic) capacity of the DNA tumor viruses must arise through their possession of genes that turn on DNA synthesis."[63]

In 1959, preparing a lecture for Harvard students, Watson remembered Potter's lecture and wondered whether an invading cancer virus would carry the code for enzymes that directed incessant growth in higher cells. These extra enzymes would jolt a normally quiescent cell into active division. Perhaps the alien enzymes would constantly signal for making more DNA and set off incessant division. "If true, I thought we might have our first real clue as to how viruses cause cancer. . . . That viral carcinogenesis might have such a simple answer dominated my thoughts all that spring of 1959."[64]

Later that year, Massachusetts General Hospital named Crick and Watson winners of the hospital's triennial Warren Prize, and the two lectured at Boston's Museum of Science. Each spoke of his latest enthusiasm, Crick about small "adaptor" or "transfer" RNAs (tRNAs) that pulled amino acids to where proteins were made on a copy of the message in the DNA. Watson took this opportunity to speculate about viruses and cancer. The overflow audience considered Crick's talk exciting. Watson agreed. At that juncture "Francis's tRNA" was about the only thing "making us happy."[65] But Watson ran into the general conviction that cancer was a researcher's graveyard because it was just too complicated. Watson recalled ruefully that Crick's talk "had the virtue of being not only elegant but also right," while his "had to seem more hot air than future truth."

> I left the Museum of Science Lecture Hall depressed at the thought that I had appeared at least an order of magnitude less intellectually powerful than Francis. Clearly I might have given a more convincing talk if I had a plausible hypothesis as to why RNA viruses also sometimes induce cancer. For, as opposed to the situation with DNA, resting animal cells are constantly making RNA. Conceivably there were two very different mechanisms through which the DNA and RNA viruses caused cancer. The other possibility was that my idea, although pretty, was just wrong.[66]

Watson's idea "failed to excite anyone else, in large part because there was no way to test it." But the topic would again engross Watson less than

10 years later. For now, it was back to the RNA molecules that help drive protein synthesis. In that field, "there were hard facts to think about."[67]

INCREASING FAME

Meanwhile, the 1959 Nobel Prize for Physiology or Medicine was awarded jointly to two biochemists, Arthur Kornberg of Stanford University and Severo Ochoa of New York University, whose laboratories had found enzymes that could duplicate nucleic acids in the test tube. In retrospect, some have suspected that the Nobel Committee sought to honor more traditional biochemistry before honoring the buccaneers who had found the double helix. But, as Watson pointed out later, the double helix had only received its first important confirmation in 1958, with the Meselson-Stahl experiment. According to Watson, after this people said, "Our copying scheme is undoubtedly right. So that made the work more important."[68]

The key question in biology, the science journalist Jerry Bishop wrote in April 1960 in the *Wall Street Journal,* is how the body makes proteins.[69] In May of 1960, an article in a special supplement to *Medical News* celebrated "certain fundamental similarities" underlying the "enormous diversity of life," among them "the action of four-unit DNA molecules which produce twenty-unit proteins." In the past 5 or 10 years, "The study of how it works has become the most fruitful area of biological research." The excitement in biological laboratories appeared to match that of earlier days in nuclear physics.[70] Clearly the molecular biologists' claim to primacy had worked through to the press.

The analogy with physics cropped up again when Watson was included in a *Fortune* magazine list of "great American biologists." An accompanying article said that, whereas physicists were baffled by a blizzard of elementary particles and chemists worried if they had run out of frontiers, biologists were sure that, armed with the theory of DNA replication, they were "in the midst of an exciting period of tremendous advance." Listed with Watson were 10 others, including several who influenced Watson greatly: Sewall Wright, Herman J. Muller, Alfred H. Sturtevant, George Beadle, Max Delbrück, Alfred Hershey, Arthur Kornberg, Seymour Benzer, Joshua Lederberg, and Fritz Lipmann. The caption under Watson's picture read:

> James Watson, thirty-two, now at Harvard, won scientific acclaim seven years ago when he and Francis H. C. Crick of England jointly proposed an ingenious "model" for the structure of DNA. This model has been as fruitful for biology as the Bohr-Rutherford model of the atom was for physics forty-odd years ago. As a boy, Watson was a bird watcher, "a pleasant way to get some science when you're young but not as an adult." He obtained his Ph.D. at Indiana and taught at Caltech before going to Harvard.[71]

In the fall of 1960, Crick, Watson, and Wilkins received an Albert Lasker Award for medical research. The annual prizes had been established by Mary Lasker as part of her decades-long campaign to increase funding for cancer research. Even then, Lasker Prizes were regarded as predictors of a future Nobel Prize.[72]

In 1961, George Beadle, Watson's advocate at Caltech, was installed as president of the University of Chicago. Among those receiving honorary degrees on the occasion was a 33-year-old alumnus, James Dewey Watson.[73] It was the first honorary degree of many that Watson would receive, including, ultimately, one from Harvard.

On 14 September 1962, *Life* magazine named Watson one of its "Red-hot 100."[74]

Journalistic treatments of Crick and Watson's work grew more nuanced. The accounts frequently mentioned the idea of altering human heredity. They also covered progress in deciphering the genetic code. One article in October 1962 referred to the 1953 discovery as "one of the great leaps in the history of science," which "was far more a product of the imagination than of laboratory experiment." The double helix was "a bold guess as to how DNA's various subunits were arranged and how the molecule is put together." The article also emphasized the role of Rosalind Franklin's data. Franklin was described as "a gifted scientist whose career was cut short soon afterward by untimely death." Franklin had died in 1958 of ovarian cancer.[75]

In *Scientific American* the same month, Crick offered a general conclusion: The genetic code is "read" from a fixed point in nonoverlapping groups of three bases. Nirenberg and others had already found most of the triplets that code for one amino acid or another.[76] On 11 October, a newspaper reported from a meeting in Cleveland that "all living things use the same chemical molecules for the transmission of hereditary information," and quoted Severo Ochoa: "I think it's pretty safe to conclude that the code is universal."[77]

NOBEL PRIZE

At his home in Cambridge on 18 October 1962, Watson was awakened in the usual way for Nobel Prize winners. A Swedish reporter called with the news that he would share that year's Nobel Prize for Physiology or Medicine with Francis Crick and Maurice Wilkins (who was on sabbatical that year at the Memorial Sloan-Kettering Institute for Cancer Research in New York). Three is the maximum number of people who can share a Nobel Prize. Watson recalled wondering how the prize would have been divided if Franklin had lived.

Soon after, his students and lab colleagues got the word. Undergraduate Goldberg received an excited call at his dormitory room from his parents in Providence, after a news bulletin, and he rushed to the Bio

Labs.[78] In a classroom where Watson insisted on lecturing as usual in Biology 150, "The Biology of Viruses," he joked, "There is a great deal more levity than usual today."[79] The photographers could not wait for a press conference scheduled for 10:30 and invaded the classroom. A student, perhaps Goldberg, had scrawled on the blackboard, "Dr. Watson has just won the Nobel Prize!" The Associated Press picture of that moment ran in next day's *New York Times*.[80] Goldberg recalled that the press photographers demanded that Watson be photographed in the lab, with his white coat on, "which is something none of us had ever seen." [81]

Reporters at the press conference had a jolly time recording Watson's appearance. One wrote, "He was wearing white socks, sort of run-down shoes, and a mixture of clothing."[82] Another wrote, "While he wore a neat, white, button-down shirt and tie . . . his brown worsted suit was chalk-marked and his scuffed shoes [were] non-descript."[83]

The questions were predictable: Would DNA conquer cancer? Did humanity already have control of heredity? What will DNA do to religion? Asked about hobbies, Watson rejoined, "I don't collect stamps, if that's what you mean."[84] One reporter visiting Watson's office picked out three titles on the shelves, *Polynucleotides*, *Biosysthesis of Proteins*, and *Carcinogenesis*—the names neatly summarized Jim's major preoccupations of the moment.[85]

"We were lucky," Watson said. "I don't think it was any great intellectual insight that led us to this. I don't think it took any profound thinking."[86] He also said, "We thought [that] DNA was important and that we ought to know its structure. Crick and I thought we could guess the structure if we went about it in the right way—and I suppose we did. . . . The DNA structure was ready to be solved. I can't imagine two years going by without someone else making the discovery."[87]

Announcing the prize in Stockholm, Arne Engström of the Royal Caroline Medico-Surgical Institute, the Karolinska Institute, which selects the winners of the Nobel Prize for Physiology or Medicine, said that the "determining of the structure of the substance that is responsible for the forms that life takes [was] a discovery of tremendous importance." He added that it had "no immediate practical application."[88]

A profile in the *New York Times* described Watson as "a child prodigy and Quiz Kid, college student at 15 . . . extremely brilliant even by professorial standards, boyish and balding, but not overly modest."[89] Students mentioned his mumbling in class but praised the clarity of his exposition. His tendency "to be somewhat intolerant" of classical descriptive biology was noted. The student-produced *Confidential Guide* to Harvard courses for that year found his lectures in the introductory course Biology 2 "a joy," even though he was hard to hear. "The course avoids an orderly textbook approach. . . . Students appreciated the opportunity to hear of the latest research in animal and plant genetics."[90] Admitting that he was not surprised at receiving the award, he said his father, who had lived with him since 1957, "was more certain than I." Jim told the newspaper, "It is an important thing we have accomplished, but we have not done away with the common cold—which I now have."

Would his work lead to "improvements" in humans? "I'd say that if you want to have an intelligent child, you should have an intelligent wife."[91] He was searching for one insistently, but would not succeed until 1968. A reporter asked whether the fame of the Nobel Prize would make him more successful with women. Watson's plaintive reply was "I like women but they don't seem to like me."[92]

The traditional champagne was broken out by 11 A.M., and an hour later his excited colleagues swooped Jim off for a lunch that lasted three and a half hours.[93]

Paul Doty and his scientist wife Helga threw Jim a party in their big house on Kirkland Place. Edsall, describing the discovery for *Science*, wrote, "No fundamental revision of the picture has been required since the early formulation of Watson and Crick." The discovery had triggered "thousands of subsequent papers," including those of Meselson and Stahl, and Kornberg.[94] Max Delbrück wrote to Crick congratulating him, and Crick replied, "As Jim always thinks of you as his scientific father I always regard you as my scientific uncle."[95]

So Francis and Jim and Maurice got to go to one of the grandest parties anywhere, along with Max Perutz and John Kendrew, who shared the prize for chemistry. Jim brought his sister and widowed father. They had hoped for two days in Copenhagen, but bad weather diverted their flight directly to Stockholm.[96] Though photographs show him "tall, attentive, and haggard in the background" in Stockholm, Jim was exultant. In a voice so soft that few could hear him, Watson said that when the double helix discovery was made, "We knew that a new world had been opened and that an old world which seemed rather mystical was gone."[97]

The lectures and ceremonies went on for almost a week, but the key date was 10 December, the anniversary of Alfred Nobel's death in 1896, not long before the days are shortest, and moods are gloomiest in far-northern Stockholm. After a Nordic morning twilight of two hours, the sun rose. Four hours later it set again, followed by two more hours of twilight. As Watson and Crick, preparing for the award ceremony, put on their white tie and tails at a Stockholm shop that rented formal clothing, Watson was not gloomy. He asked Crick, "You know what a Nobel Prize is good for?" Crick asked him what. Watson: "Getting dates."[98]

The ceremony was held, as is traditional, in an already dark late afternoon in the great Concert House off Stockholm's Haymarket. The winners and their Swedish scientist sponsors, dressed like all the men at the ceremony in white tie and tails, sat on a flower-drenched stage. An orchestra played between the awarding of the separate prizes. When a winner had been praised in his own language, he stepped to the front of the stage, bowed to an audience gleaming with bright-colored taffeta gowns, medals, and the special hats of Swedish university professors, before descending a few steps to a vast blue carpet, decorated with the golden crowns of the Swedish flag. There, he met the old King Gustav VI, shook hands, and, to the popping of flashbulbs, received his medal and check.

After this the many guests, the king, and the new winners were driven over to Stockholm's majestic Town Hall, with its Venetian campanile, on the shore of one of Stockholm's many lakes. Passing through the vast Blue Hall (so called despite its lack of the blue tiles intended to cover the brick interior), they mounted a grand staircase to the floor above for the banquet for about 600 in the Golden Hall. This was covered, walls and ceiling, with golden Byzantine-style tiles. After all other guests were seated, a fanfare of trumpets announced the king and queen and the winners of the science and literature prizes as they moved into the room in glittering procession. The winners and the members of the royal family sat at an immensely long table that ran down the middle of the room.

As the banquet reached the dessert course, one could hear in the Blue Hall below a buzz of conversation and clattering coffee cups. It came from hundreds of invited university students. Toward the end of the dinner, a group of students entertained the company in the Golden Hall with student songs. Then all swept out to a large balcony and saw the students below, some of whom, wearing peasant costumes, were performing Swedish folk dances. The winners descended the grand staircase, to be greeted by a student who addressed them. Then the orchestra struck up the Blue Danube Waltz and the dancing went on until 2 A.M.

Francis did the twist with one of his daughters and Jim danced with a Swedish princess who wore a low-cut gown.[99]

7

MANIFESTO AND MARRIAGE

The real reason I came here was to get a wife, and I got one! I always wanted to marry a student here, and I married a student here. The conventional argument was, I was looking for good genes. But she was very pretty.
JAMES D. WATSON, SCIENCE CENTER,
HARVARD UNIVERSITY, 30 SEPTEMBER 1999

DEFINING THE FIELD:
MOLECULAR BIOLOGY OF THE GENE

In the early 1960s, Watson felt driven to define where a new kind of biology was headed and, in the process, to capture the young people who would take it there. The attempt led him to write a manifesto for molecular biology. He was convinced that "biology has as sound a basis as was provided [to] chemistry, about 1932, by the explosive development of the quantum theory of the atom." So it was time for new textbooks "to give the biologist of the future the rigor, the perspective, and the enthusiasm that will be needed to bridge the gap between the single cell and the complexities of higher organisms." Once that gap had been bridged, biologists could secure "hard facts" about such issues as cancer, development of complex organisms, and neurobiology. "Today's most challenging biological problems," he asserted, were "the structure of cell membranes, the nature of cancer, the fundamental mechanism(s) of differentiation, and how the ability to think arises from the organization of the central nervous system."[1]

For six years he had been struggling to teach the new subject to introductory biology students at Harvard. The first of his three aims was "to convey the excitement of the recent discoveries of molecular genetics."

The second was to relate the findings to "the basic problem of biology—the nature of cells and how they divide." The third was to show "how our ideas about molecular genetics have developed out of the work of classical geneticists and biochemists." The resulting textbook, *Molecular Biology of the Gene,* which came out in 1965, was arguably as influential as the memoir he published in 1968, *The Double Helix*— and it was much more universally praised among scientists. Decades later, Sydney Brenner remembered the book as "an absolute breaker . . . a real watershed thing . . . a great synthetic piece of work. . . . It was the book of the new order [and] broke with traditions in a big way. . . . It really taught a lot of people the subject. I've always said he should be more proud of that than. . . *The Double Helix.*"[2]

The textbook was published just as molecular biology was consolidating its first big successes. It sold so well that royalties in the first year equaled Jim's Harvard salary and helped him pay down his mortgage.[3] A notable feature of the book was what Brenner called the "Massachusetts declarative style." Many headings in the book were assertions in the form of sentences: "Some genes are neither dominant nor recessive"; "Specific cellular reactions require a specific enzyme"; "Chemical bonds can be explained in quantum mechanical terms"; "The gene is (almost always) DNA." The dogmatic tone was comforting. The field had arrived—and needed recruits.

There was work to do. Molecular biology mattered, Watson held, because of the insights it might provide for combating cancer or infectious disease. Basic science and the conquest of disease were entwined. Looking at the beginning of normal protein synthesis provided a model of how the specialization of cells—the process called differentiation—got going and shaped the development of animals and plants. Cancer was an example of development gone wrong, of uncontrolled cell division, arising from mutations in just a few of the cell's genes. These could be driven by internal factors or by invading chemicals, radiation, or infection. This was just one example of induced protein synthesis. Another was the immune system's mobilization against foreign material, which involved a controlled proliferation of cells to make antibodies. Looking at such biological systems in crisis fed back into picking apart the workings of normal cells. According to David Botstein of Stanford University, the book "was revolutionary in intent and revolutionary in practice. . . . He opened the door to this science . . . on a grand scale." Referring to Watson's aggressive personal style Botstein added, "It's sort of like Wagner, right? Do you have to love the guy to love the music?"[4]

This charming, almost naively enthusiastic textbook was not inscribed on tablets of bronze. Instead, it conveyed a tone of urgency and up-to-dateness, and introduced a host of puzzling unsolved problems. "May" and "might" appeared throughout the text. One had to think as well as experiment. "It is immensely simpler to imagine that DNA replication involves strand separation and formation of complementary molecules on each of the free single strands," as Meselson and Stahl had demonstrated in 1958.[5] Phrases pointed to gaps in knowledge: "Until now, no one has been able[6] . . . This point has not yet been proved[7] . . . We do not yet possess any evidence[8] . . . Nothing is known[9] . . ." But there was recent news, such as the near-final cracking of the genetic code. Only five years earlier, there had been "little optimism that we would know details of the genetic code."[10]

Jim took the reader on a brightly written tour of such topics as classical Mendelian genetics, the obedience of cells to the laws of chemistry, the chemist's view of the bacterial cell, and the importance of the weak atomic interactions that he and Crick found so crucial in the structure of DNA. He went on to discuss the use of energy in the cell and the formation of proteins on templates, how genes are arranged along chromosomes and how they work, how DNA is organized and duplicates, how RNA is formed and influences the making of proteins, and how viruses reproduce. Other chapters described the genetic code and genetic control. The final two chapters focused on immunology and cancer. In writing of cancer, he gave particular attention to the role of viruses. To be sure, this happened "in cells that we are only beginning to study at the molecular level." But there was a more important fact: "At last the biochemistry of cancer can be approached in a straightforward, rational manner."[11] So much for the "mystery of life" following unknown and perhaps unknowable laws.

He kept attacking the idea of special laws for biology. "The growth and division of cells are based upon the same laws of chemistry that control the behavior of molecules outside of cells. . . . There is no special chemistry of living cells."[12] The biologist's real law was to pick what was important and amenable to experiment. Reduction of problems to the simplest level was inevitable. When biologists in the 1930s and 1940s turned toward "chemically simple molds, bacteria, and viruses," Watson noted, they were seeking a way around a discouraging fact: the detailed structures of key proteins weren't known, so that it wasn't possible to go very deep into the relationship between genes and proteins. He referred to the tiny organisms as "biological items more suitable for chemical analysis" than proteins.[13] In the 1960s, this bet was paying off: "The chromosomes of the bacterium *E. coli* and the phage T4 are quickly becoming the best understood of all genetic material."[14]

The chemist's "secrets of life," the essentials for a cell to grow and divide, were three: (1) a solid membrane to hold molecules within the cell, (2) enzymes to tear food apart and build the components back into new molecules, and (3) sources of energy from the food, or in plants, from the sun. All of these "secrets" rested on the existence of gene-specified proteins.[15]

He plunged into what might seem arcane topics for a beginning biology student, devoting a whole chapter to such "easily broken" weak chemical bonds as hydrogen bonds. Citing sources such as Pauling's classic textbook on chemical bonds, Watson wrote, "Weak bonds are important not just in deciding which molecules lie next to each other, but also in giving shape to flexible molecules such as the polypeptides [proteins] and polynucleotides [DNA and RNA]."[16]

Although much remained to be discovered, Watson wrote, one could assert, "One fifth to one third of the chemical reactions in *E. coli* are known." Thus one could hope to work out all the reactions in 10 to 20 years.[17]

The tone of certainty came out constantly in such sentences as: "There is no reason to believe that any genetic information is carried in other than nucleic acid molecules," or, "Some DNA molecules are larger by several powers of 10 than any other biological molecule."[18]

Jim dedicated the book to Salva Luria, his former teacher and one of many who had commented on the manuscript. Luria was touched.[19] Two who worked on the manuscript with him, Nancy Hopkins and Joan Argetsinger Steitz, became coauthors of later editions, which kept appearing until 1988. For the editing, Watson enlisted young Radcliffe students such as Dolores Garter and Ellen Glass, whom he thanked "for continual advice, not always followed, on what is a grammatical sentence." A student in the lab observed that Garter "made his style flow more, become more natural and perky." [20] For Jim the writing was a struggle. Sometimes his coworkers would receive sections scribbled on scraps of paper.[21] But Hopkins thought that Jim was aiming at "the right level of student," and over the years became a "superb" writer. "He had such a strong intrinsic style. Any one of us could edit in a Watson accent." The words still sounded like Jim.[22]

The book delighted readers, including many in his own lab. Mario Capecchi, who worked in Jim's lab for several years in the mid–1960s, recalled, "This was not only a good text, but it was an innovation. It was an up-front, in-your-face, didactic approach."[23]

Watson himself was self-deprecating. It was "a very simple book," he observed in 1988. " [There was] very little to put in it."[24] The introduction to the massive fourth edition of 1988 noted that back in 1965, "There were few practicing molecular biologists and not too many facts to learn. So what we knew about DNA and RNA could be explained to beginning college students." But in 1988, "No molecular biologist knows all the important facts about the gene."[25]

DOES THE RIBOSOME CARRY THE MESSAGE?

In the years when Watson wrote *The Molecular Biology of the Gene,* he also got started on what he called "Honest Jim," the memoir that became

The Double Helix. All the while he continued to work in a maelstrom of competition over who would unlock the key steps in the making of proteins, a central question of biology. It was a brambly, anxious frontier, where the principal topic was the other nucleic acid, RNA, the middleman on the road from DNA to protein.

The Watson lab members' annual silly climb onto the back of the rhinoceros statue was a small break in the tension of the lab. It allowed them to make fun of "that preposterous building" and its "unforgettable" rhinoceros guards.[26] The group faced spirited competition with other labs, including those in Paris and the other Cambridge. They were all on the same speedy gossip network, but their mood was competitive. Each wanted to be the smartest and the first. What was the best way to think about protein synthesis? they wondered. Which experiments were feasible for picking apart the cell's workings? "I always had the feeling," one young assistant recalled, "that Jim likes to set up enemies to make it into a race. . . . He doesn't want it dull." The attitude was, "Everybody should be in there all the time doing this work. . . . You had to get it done, and you had to get it done fast."[27]

In those early days, Watson's lab on the third floor of the Bio Labs was laid out like "a railroad apartment," one student recalled. On one side of Jim's "large, Spartan office" lay experimental rooms which his graduate student Bob Risebrough and postdoctoral fellows Ernst Freese and Alfred Tissières began using in 1956, and on the other a large double lab with workbenches around three sides and an "island" in the middle, "dividing it into two awkward working areas." The big room was the place for tea and arguments with visiting luminaries.[28]

As they worked, they were certain, as Jim had written on his slip of paper years before in England, that RNA had to be the intermediary between DNA and proteins. Jim's fluorescent desk lamp at Harvard bore a scrap of lined paper with the same slogan, "DNA makes RNA makes protein."[29] As has been mentioned, most of the cell's RNA was found in the ribosomes of the outer region, the cytoplasm, where protein synthesis was concentrated. The Harvard lab's pursuit of the ribosome's structure arose from the hope that the RNA of a particular ribosome was a template, carrying the blueprint for a particular protein. Jim's slogan attracted a beginning graduate student, David Schlessinger, to join Jim's Swiss colleague Tissières "to investigate the ribosome's complexity."[30]

Schlessinger perceived the state of the new science of molecular biology as "perilous." Traditional biologists were mobilizing to hold the line for "real biology." The new *Journal of Molecular Biology* appeared "intermittently," Schlessinger recalled, "and Jim would come through the labs to see if any of us had finished anything" to help fill an issue.[31]

Surely ribosomes were important in protein synthesis, the constant task of cells that was directed by the sequence of bases along the backbone of DNA. Watson recalled in his Nobel Prize lecture: "In all our early ribosome experiments we . . . assumed that the ribosomal RNA was the template. Abundant evidence suggested that proteins were synthesized on ribosomes, and since the template must be RNA, it was natural to assume that it was ribosomal RNA."[32]

Chasing what Masayasu Nomura called "the Holy Grail of biology at the time, namely the mechanism of information transfer from the gene to protein,"[33] Jim and his coworkers needed a little hilarity like their rhino climb. In the middle and late 1950s, Watson recalled, "There were periods when, you know, the science wasn't going well and at the same time I didn't have a girlfriend. Those were tricky times."[34] Still plagued by a sense of "doubt and anticlimax," Jim feared that the DNA discovery was a fluke that happened because he had been in the same room at Cambridge with Francis Crick. Somehow, he must prove that he "could survive without Francis."[35] Watson was heartened by Meselson and Stahl's proof in 1958 that DNA divides the way the double helix model predicted—he was sure that this was a key event on the road to the Nobel Prize. But with little else "making us happy," he was worried about not being in on "the next step."[36]

For that next step, many bits of knowledge from different labs were already scattered on the table. As we have seen, Avery's work on bacteria and Hershey's on viruses proved that the genes resided in DNA, whose structure had been determined by Jim and Francis, using both Franklin's and Wilkins's X-ray pictures and Chargaff's base ratios. The Watson-Crick model of 1953 showed dramatically how DNA could be duplicated and also specified what each generation of cells did before they divided. DNA evidently carried the design for the thousands of different kinds of proteins needed to catalyze the cell's constant cooking—at a temperature far lower than an oven's. The amino acids were strung together in a sequence that folded up—making structures like the ones Perutz and Kendrew were probing with X-rays—so that the protein could do its job of cutting or pasting. Fred Sanger had deciphered the exact linear amino-acid sequence of insulin.

It seemed logical that the strings of DNA bases matched up with the strings of amino acids. But where did the events of protein synthesis occur? Apparently not in the nucleus, where the DNA was, but somewhere out in the cytoplasm. For that and other reasons, it didn't seem plausible that proteins formed directly on the DNA. There had to be an RNA intermediary. For a long time, the small but growing club of molecular biologists clung to the idea that the RNA of the ribosomes held messages specific to the various proteins. The belief persisted in the face of increasing trouble. For one thing, one would expect the number of ribosomes, made of RNA and proteins, to rise sharply when the cell started making a new protein. But the increase was not sudden. Worse, the ribosomes were very stable and very homogeneous in size. Their RNA seemed to vary little between species and to match very poorly with the nuclear DNA. How could the ribosomes be specific to proteins?[37]

Meanwhile, Crick and Brenner were having fun practicing "theoretical biology" without a license. They stuck to genetics—"doing very well," in Watson's view[38]—by serving as sparkplugs in the search for the genetic code. In addition, Crick's idea of a special "adaptor" type of RNA to escort individual amino acids to the assembly line was soon confirmed in the laboratories of Paul Zamecnik and Mahlon Hoagland at Massachusetts General Hospital. There were "transfer RNAs" for each of the protein's amino acids, and their job was to line up on the ribosome so that the "activated" amino acids could be stitched together.[39]

The theory that there was one ribosome for each protein was called "one gene–one ribosome–one protein."[40] But Crick was bothered by the fact that the ribosome's RNA was buried within its globular structure. His "adaptor" would have a tough job lining up with the ribosomal RNA. He expressed this worry during a visit to the Harvard lab. "I'll never forget sitting at tea," said Charles Kurland, then a graduate student. "Francis was saying that . . . he couldn't understand how you could make a protein on a ribosome from ribosomal RNA. How do you get into this compact thing and read out the code? He understood pretty much that you had to have something other than ribosomal RNA to do the job."[41]

For some time, however, both Francis and Jim continued to believe that the RNA of particular ribosomes specified particular proteins. Crick made it a feature of his "central dogma," which created "excitement and urgency" during Masayasu Nomura's first summer visit to the Watson lab in 1958. At one of the numerous unpublished Gordon conferences held in New Hampshire preparatory schools each summer, Nomura heard Crick speculate that ribosomes resembled the spherical viruses he and Jim had worked on a few years before, with coats of many identical protein subunits.[42]

Taking a different path from Crick's, with his emphasis on theory, Watson plowed into directing and encouraging experiments, including many that involved the once-dreaded biochemistry. Having failed to find a three-dimensional structure for RNA, it was natural to turn to the structure of the snowman-shaped, RNA-laden ribosomes.[43] In Jim's view, "genetics and biochemistry began coming together."[44]

The plunge into ribosomes occurred while Jim was becoming more confident that the double helix discovery would amount to something. In June 1956, shortly before Jim came to Harvard, he and Francis spoke at a conference in Baltimore on "the chemical basis of heredity," which

Watson remembered as "the scientific world's first double-helix-based overview of genetics." Housed in a top-floor hotel suite, he and Francis "had never before been so well treated." Despite sour taunting from Erwin Chargaff, they felt intellectual currents running their way. Opening the conference, George "Beets" Beadle, the coauthor of the "one gene–one enzyme" theory and Watson's old sponsor, focused on what genes are, how they replicate, and how they direct protein synthesis. Genes, he said, were "segments along DNA molecules that most likely specified RNA templates for ordering amino acids in proteins." DNA makes RNA makes protein.

Knowledge about DNA and protein synthesis was opening up in challenging ways. Excitedly, the scientists at the Baltimore conference heard from Arthur Kornberg of Washington University in St. Louis about an enzyme that could synthesize DNA from components in the test tube. At about the same time, however, the puzzle about RNA as the carrier of genetic messages grew more tangled. Elliot Volkin and Lazarus Astrachan of Oak Ridge National Laboratory in Tennessee, following up an experiment of Alfred Hershey's years earlier,[45] found an unstable form of RNA in bacteria infected by a virus called T2. Existing for only a few minutes, this RNA had much the same ratio of bases as the T2's DNA. Soon, these short-lived molecules were to undercut the idea of highly stable ribosomal RNA as the template for specific proteins.[46]

Watson and Tissières kept trying. In 1958 they found evidence that ribosomes had two components, one heavier than the other.[47] This hunch was confirmed by Kurland, who was beginning a long career in what he called "ribosomology."[48] Nonetheless, the ribosome was fading as the principal means of specifying the makeup of proteins. Too many experiments pointed away from this notion. Kurland was shocked when the noisy Sol Spiegelman of the University of Illinois, in a lecture Jim sent him to hear, called the ribosome "a garbage pail."[49] It was becoming more difficult to maintain that ribosomes were specific to particular proteins. Over the next two years, ribosomes were recast into general-purpose structures analogous to tape players.

THE "PERFECT COUPLE"

If ribosomal RNA didn't specify proteins, and the little RNA adaptors were just escorts, what *was* the intermediate carrier of the genetic information? After many ideas came together, this question was attacked in detail in 1960 at both Caltech and Harvard, and the existence of messenger RNA as the carrier of specifications was demonstrated. For the Harvard experiments, Jim pulled in the physicist Walter Gilbert. Getting Wally Gilbert into biology, Watson later said, was his biggest achievement in science.

They had first met in Cambridge, England, among the green lawns and ancient riverside colleges. In the fall of 1955, Gilbert, a graduate of Harvard, was at Cambridge working for his Ph.D. in theoretical physics with Abdus Salam, like Wally a later winner of the Nobel Prize. Gilbert's wife, Celia, was a budding poet who enjoyed Jim's favorite novelist, Henry James. Wally and Celia attended a party probably organized by Marietta Robertson in honor of her astrophysicist father. Also at the party was Watson, "a lanky postdoc with a piercing gaze and blue eyes," who was back in Cambridge for a year. Wally had heard Jim lecture on DNA at the physicists' scientific club. "We must have talked of many things at that party," Gilbert recalled, "because we became friends then."[50]

A few months later, Watson met the Gilberts at another party, and Wally asked Jim what he would do to top the double helix. The half-serious answer was, find a rich wife. Jim began what became a long-standing practice of introducing the latest candidates for a wife to Celia, who sought out at least one rich young woman for him.[51] He also started showing up on the Gilberts' doorstep, obviously in need of food and companionship.[52] Learning that the Gilberts would return to Harvard just as he showed up there, and noting that Wally's pre-doctoral fellowship was only $2,700 a year, Jim offered Celia a job as his first technician.[53] From one of Jim's women friends, Belinda Bullard, who lived across the street, Celia borrowed a chemistry textbook and plunged into "covalent bonds, ionic bonds, anions, and cations." She wondered if the last was a dirty word.[54]

Spoofing her days in Jim's lab, Celia joked in 1988 about such tasks as posing for a brochure, peering nearsightedly at tobacco leaves to find lesions from tobacco mosaic virus (TMV), a principal focus for Rosalind Franklin in London before her death. Jim maintained some interest in TMV.

Continuing her satire, Celia quoted her "tedious and repetitive" lab notebook: "pour plates, autoclave, wash pipettes, order supplies." It was nerve-racking. When she seemed bored, Jim invited her to his lectures, but they were too difficult. "Jim drops his voice at the end of every sentence, then hisses and smiles as if we heard him." The expression "ten to the ninth" gave her a migraine. She made "a terrible mistake," measuring out agar—warm, wet agar whose smell nauseated her—in tenths of grams instead of grams so that "the plates wouldn't gel." Although Jim's first graduate student, Bob Risebrough, figured out the problem, Celia said she wrote, "I wonder if Jim will fire me." But Jim said he was "sure that the only thing I have to do is see that there is a good tea for him and Bob. Wants a full tea." He asked her to get a print of two ducks, mergansers, framed. She fancied, "They seem to have something to do with his *idée fixe* that he is going to get the Nobel Prize." She didn't know what.

Cecilia's "notebook" records that Jim, who remained "a graduate student for a long time in spirit," regarded the Gilberts as "the perfect couple." A refrain in the notebook was "Can he come to dinner tonight?" Another was "a new girl." One girl stopped returning Jim's calls. Another was "not interested but he doesn't seem to notice." Another thought that Henry James was a bandleader. One afternoon, feeling worn out with fighting Harvard, Jim asked to come to dinner when the Gilberts were invited out. Cecilia's notebook entry, "I feel terrible. . . . He says not to worry. He'll just go to Paul and Helga's [the Doty's]."[55]

MESSENGER RNA

The success in 1960 of finding a short-lived messenger RNA copy of one or more genes stored in the DNA had many fathers. Jim contributed to the solution of this mystery, but the exact amount of credit he should receive remains contested more than 40 years later. Clearly, he did not come up with the concept, but he shared in proving the existence of messenger RNA.

Such a molecule was first postulated in the crowded laboratories at the Pasteur Institute in Paris by the biochemist Jacques Monod and the geneticist François Jacob, struggling to explain an experiment they had done in 1958–59 with a visiting American, Arthur Pardee of Berkeley.[56] The Pasteur Institute group, including André Lwoff, worked on a top floor that they called "our attic."[57] Their central interest was genetic control, how genes were turned on and off.

The ability to switch on and off quickly is vital to all living cells, including the simplest. The cells cannot afford vast warehouses of thousands of different kinds of proteins, stored against the moment of need. Instead, the cells require an extreme version of what in the corporate lingo of the 1980s was called "just-in-time" inventory management. A particular protein is stitched together just when there is a job, often lasting only moments, for that protein to do. When no longer needed, it is broken up into its amino acids, which become available for the manufacture of the next protein. The genes must get a signal to start making a particular protein, and another signal to stop. In *Molecular Biology of the Gene,* Watson noted that cells make a "rapid adaptation to a changing environment," thus avoiding "wasteful synthesis." In bacteria, the rate of synthesis of many enzymes "depends on the availability of external food molecules."[58]

The Pasteur studies focused chiefly on two gene sets needed for survival. One, called the lactose operon, made it possible for bacteria to live on lactose (milk sugar) by cutting lactose in half. This set of genes included the blueprint for an enzyme, called beta-galactosidase, which was made only when the bacteria detected a supply of the sugar. Another gene in this set encoded the makeup of an enzyme called permease to make the bacteria's membrane more porous. The other gene set lay in a bacteriophage called lambda. A set of genes allowed the bacteriophage to lurk silently in the host genome, most of its genes switched off, for many bacterial generations and thus not kill the cell immediately, which most viruses do right after they invade a cell.

Pardee, Jacob, and Monod worked with the lactose system in what became known as the "Pajama" experiment. They exploited the fact that some bacteria have a "male" factor, z+, that turns it into a donor of genes to other bacteria that lack this factor. The z+ "males" formed a tube to transfer genetic material. Genes begin moving over in a string, one after the other. This process of donation to z– "female" microbes could be stopped at any time by putting the bacteria into a blender. So one could stop just before or after the genes for digesting lactose had been transferred. The results were puzzling. The donor did not send over its ribosomes to the recipient.[59] As soon as the lactose genes were injected into the female, the lactose-digesting enzyme was produced at the maximum rate in less than three minutes. Hence, the gene was "expressed" at once. But as soon as the injected male gene was broken up by its radioactive phosphorus "label," the synthesis of the male protein stopped. If the intermediate RNA had been stable, synthesis would have continued for some minutes. But this RNA was unstable. So it was an unstable RNA like Volkin and Astrachan's that carried the information, not the stable ribosomal RNA.[60]

Rumors of this result quickly reached Jim's lab and Nomura and Watson discussed it.[61] Soon afterward, Nomura went off to Sol Spiegelman's lab at the University of Illinois in Urbana. There, Benjamin D. Hall and Nomura showed that the Volkin-Astrachan RNA could adhere to the ribosomes but also separate from them, so that it could be distinguished from the larger molecules of ribosomal RNA.[62] This finding was just a step or two away from proving that the Volkin-Astrachan RNA was the messenger. Pardee's experiments also brought him close to the solution.[63]

The Paris scientists preferred the idea of the messenger to the discredited surmise that proteins were formed right on the DNA. They had already determined that the genes they studied were in sequences called "operons," under the control of an "operator" at the start. This was the site of initiating the transcription of genetic messages, or halting it. The operator was the site of regulation. Jacob and Monod began defining a messenger. It had to have a short lifetime in bacteria, and be variable in size to match the differing sizes of proteins. Its base ratio must match that of the DNA, and it must be able to attach to the ribosomes. The ribosomes needed to be nonspecific, able to be programmed. It all went against the consensus of the moment.[64]

Jacob and Monod carried their puzzlement to a small conference in Copenhagen in September 1959, arranged by Ole Maaloe, Jim's old

Copenhagen colleague from the "long winter of rain and darkness." Among the 30 attendees were Jim Watson, Francis Crick, Sydney Brenner, Seymour Benzer—and Niels Bohr, whose lecture on "Light and Life" had propelled Max Delbrück into biology 20 years earlier. The message from Paris, Jim recalled, was, "The synthesis of an induced enzyme took off maximally." To do this with ribosomes would take several generations of cells. He, like the Pasteur group, thought that synthesis of proteins directly on DNA was "a horrid possibility," because it contradicted so many facts.[65] During most of the presentations at the conference, Jim ostentatiously read a newspaper, perhaps hiding disappointment at the barrenness of all that work on ribosome structure. "When Jim's time to talk arrived," Jacob recalled, the whole audience pulled out newspapers and began reading. As for the Paris results, the response was disappointing. "No one batted an eyelash. No one asked a question. Jim continued to read his newspaper."[66]

Another influence on Monod's mind was the discovery in Severo Ochoa's lab of an enzyme that could assemble RNA molecules with a random order of bases, adenine (A), guanine (G), cytosine (C), and uracil (U, the RNA analogue of thymine). For Monod, who had hitherto refused to regard RNA as important, RNA was "an intruder in the organized world of his thoughts." But there was at least one in Monod's group who dealt with RNA, François Gros, back from training in Sol Spiegelman's lab in Urbana. He began looking at what would happen if he substituted various analogues, slightly altered forms, of U, A, G, and C in RNA. He found that in the first minute after he added a variant of uracil, called 5-fluorouracil (a tool of cancer chemotherapy), the bacteria began making an abnormal enzyme that had lost the ability to digest its usual sugar. This sudden effect also pointed toward the "short-lived intermediate," the messenger.[67]

It was time to prove that ribosomal RNA did not carry the design for particular proteins, and to demonstrate the messenger's existence. The crucial experiments were done in Matt Meselson's lab at Caltech and Jim Watson's at Harvard.

They were plunging deeper into the RNA world that still preoccupies biologists decades later. This world is dominated by three types of RNA: ribosomal, transfer, and messenger. The biologists knew about the first two, but were just stumbling onto the third.

The idea for the Caltech experiment took shape in Cambridge during a party at a conference on Good Friday, 1960. Jacob brought the puzzling findings from Paris and talked them over with Sydney Brenner. At the evening party, they began an intense conversation, with lively interjections from Francis Crick. They recalled Volkin and Astrachan's short-lived RNA and excitedly thought up a way to test whether such messengers could reprogram ribosomes. Crick later referred to "that fateful Good Friday."[68]

Brenner and Jacob each had the month of June open for trying it, and Matt Meselson had the best equipment and methods for the task. The idea was to see whether bacteria, making their own proteins before infection by a virus, could be set to making viral proteins after infection. To contrast the situations before and after, the bacteria were first grown for several generations in a medium with only the heavy isotopes of nitrogen and carbon so that their ribosomes would be heavy (and could be isolated in one of Matt's ultracentrifuges). The bacteria were then transferred to a medium with light isotopes. When radioactive phosphorus was added to label newly formed nucleic acids, the label bonded only to the otherwise unaltered heavy ribosomes. At the very end of a tense month, Brenner and Jacob had shown that the ribosomes stayed constant. In their infection system, the nonribosomal instruction-bearing RNA (the short-lived molecules that were soon dubbed "messenger RNA") must be at work.[69]

But what about uninfected cells? In that same spring and summer of 1960, Harvard was tackling uninfected cells by a "technologically less sophisticated" approach.[70] François Gros went over to Jim's lab for the spring and summer for what he later considered "more painstaking" work than the elegant Caltech experiment. At Harvard, minds had continued to change about the ribosome and had begun to focus on the short-lived RNA. Experimenting with the virus called T2, graduate student Bob Risebrough got results that pointed to "a new RNA class unknown until that moment, both because it comprises such a small part of the total RNA and because it is heterogeneous in length." With Charles Kurland, another of Watson's graduate students, Risebrough had started out studying ribosomal RNA by labeling it with radioactive phosphorus, sometimes for as long as an hour. But if they ran the labeling step for just two or three minutes, the RNA's makeup didn't match the ribosomal RNA nor the transfer RNA escort molecules. All at once, ribosomes had become "stable assemblage sites," or "molecular factories."[71]

"Jim was disappointed," Gros recalled. "To cheer us up and forget experimental results, which should have filled us with joy but, for the time being, left us down-hearted, we used to take walks on the Harvard campus and talk about RNA and women." Gros remembered the bachelor Watson, then 32, as "extremely romantic."

They checked their results over and over. Perhaps the short-lived RNA was worth studying. The plans Gros had arrived with were laid aside. In the "suffocating" heat of the "severe" Harvard Bio Labs, Kurland and Gros set to work. Realizing that they were making mistakes by working together until three in the morning, Kurland and Gros organized themselves into a day shift and a night shift.[72] Kurland, although he had more experience with some details of the experiment, was "very skeptical . . . because of my own misconceptions." Gros, on the other hand, was "the determined one. . . . There was a messenger there that we had to find."[73]

One day, near the end of May, Watson introduced Gros to an intelligent-looking but almost mute Walter Gilbert, who began following Gros "like my shadow."[74] Jim had told Wally about something "very exciting," a search for "an evanescent molecule" that he should look at. Wally noted the sharp contrast with the previous "dull, brute-force ribosome physical chemistry,"[75] and, "fascinated by it," got involved. He recalled, "I came over to the lab and looked at [the experiments] and joined in" with Gros and Kurland and Watson. Gros would hold a four-liter flask with the bacteria, and Gilbert would pour in the radioactive phosphorus used for labeling new nucleic acids.[76]

Jim operated the stopwatch. The process was stopped after 20 seconds by pouring the whole mixture into an ice bucket. Then, using a so-called sucrose-gradient method, the molecules were separated by size. The rapidly labeled RNA molecules were quite variable in size, as would be expected for messages for proteins of different makeup. They were bound to the ribosomes.[77] Thus, the messenger itself could be found in uninfected cells. The short-lived messenger, which took up 5-fluorouracil immediately, constituted about 2 percent of all the RNA in the cell (compared to 15 percent for all the transfer RNA). The RNA would link up with the unchanging ribosomal RNA and then fall off.

According to Kurland, Gilbert was important for both checking and continuing the experiments, after Gros had gone back to Paris and Kurland had quarreled with Watson and was told to concentrate on finishing his thesis. Kurland was upset that he didn't have enough help in the lab—"It was a matter of more hands!" To Jim's annoyance, Chuck Kurland urged closer collaboration by the Harvard team with the Jacob-Brenner-Meselson team. He thought the Caltech experiments "had an elegance and directness that we never even approached at Harvard."[78] Wally later looked back at this time of working with Jim as "probably the formative experience of my life."[79]

Watson's sense of competition was intense. During a Gordon conference in New Hampshire, Meselson, who would soon move from Caltech to Harvard, told Jim and Wally what Sydney and François were up to at Caltech. The Harvard experiments were not finished. Jim persuaded the Pasadena group to delay publication until the following spring, when the Harvard and Pasadena papers would appear together in *Nature*.[80]

GENETIC CODE

Finding messenger RNA soon bore fruit. It was possible to put together synthetic messengers and use them as crowbars to begin cracking the genetic code. In the spring of 1961, just as the papers about messenger

RNA (mRNA for short) and genetic control were appearing, and Cold Spring Harbor was dedicating its annual Symposium to mRNA, another set of walls started tumbling down. Marshall W. Nirenberg of the National Institutes of Health found that an artificial messenger containing only uracil (U) could induce production of proteins consisting only of the amino acid phenylalanine. This meant that the three-letter "codon" for the amino acid was UUU. It was this achievement in May that Heinrich Matthaei had mentioned at Cold Spring Harbor that summer. News of it created a sensation at the world biochemistry congress in Moscow in August 1961. The congress was attended not only by Crick and Watson but also by Matt Meselson, Wally Gilbert, and Paul Doty, who secured Jim a choice seat near Red Square to watch a parade welcoming the Soviet cosmonaut Gherman Titov back to earth.

Most Western biologists flocked to Moscow to show solidarity with Russian colleagues recently liberated from the tyranny of the geneticist Trofim Lysenko, a protégé of the late dictator Stalin. The visiting scientists didn't expect, as Jim wrote later, "anything but a chance to see communism in action." But Watson soon heard rumors of Nirenberg's "bombshell." Gilbert and Tissières attended the small session where Nirenberg presented his results. Meselson recalled the session. "I had never heard of him and he spoke at a very small room in Moscow. . . . Nobody in that room, so far as I know, understood the significance of what he had done, except me. I ran up after he was through, and I hugged him. I basically grabbed him and squeezed him, and I said, 'You've got to talk to a larger audience.' I went and found Francis." Over breakfast the next day, Jim passed the news along to a disbelieving François Jacob, but it was not a practical joke. The work was solid. Crick "hastily arranged a big lecture on the last congress day that let Nirenberg convince as well as stun most in the audience."[81]

Over the next few years, labs such as Nirenberg's in Bethesda and Severo Ochoa's in New York raced each other to work out the remaining 63 three-letter codons. By 1965, Watson could write in *Molecular Biology of the Gene,* "It seems likely that most of the essential features of the genetic code will be available to us by the end of 1965."[82] Many years later, he proudly recalled that the code "was solidly established by 1966, only 13 years after Francis Crick and I discovered the double helix."[83] It was a "big step" because biologists "knew how the cells read the DNA." But the scientists wondered when *they* would be able to "read" the DNA. They thought, "We can't isolate individual DNA molecules. We can't amplify them, and we don't know how to sequence them."[84]

But this took little of the edge off Francis Crick's triumphant lecture at the Royal Society in London in May 1966, where he displayed what became the canonical table of the codons. He happily reported that the

Cambridge group, having found "full stop" sentence-ending codons such as UAA and UAG, now had found the "capital letter" for starting the genetic message, a codon with the sequence AUG. This codon signaled to the enzyme for copying DNA into messengers, "Start here." New details of synthesis on the ribosomes had emerged. At one site on the ribosome, transfer RNAs latched their "anti-codons" to the codons on the messenger, bringing the amino acids into line for joining to the adjacent acid. Then the amino acid would be shifted to the second ribosomal site, where the transfer RNA would be dropped off. To be sure, Crick said, "We're in for a difficult time. The table is interesting chiefly as confirming all our previous ideas, our general scheme."[85]

A month later, Crick was the star of the 1966 Cold Spring Harbor Symposium, focused with the usual trendiness on the genetic code. Watson, fresh back from a sabbatical in the Geneva lab where Tissières now worked, enjoyed the show: "Happily we weren't dominated by too many facts, and we could even listen to the occasional nonsense and have fun."[86] Crick opened the conference with a speech entitled "Yesterday, Today, and Tomorrow." Watson wrote, "The Symposium was very much for Francis to dominate—rather in the way I held forth at the 1953 gathering and Jacques Monod and Sydney Brenner did in 1961."[87] The atmosphere crackled. In the back of Bush Auditorium, scientists were phoning their laboratories to get the latest data.[88] As usual, Crick sat in the front row, listened intently to every talk, interrogated, dominated, and ended the week "in a state of exhaustion. It appears that for him there is nothing between rest and overdrive."[89]

Crick's fiftieth birthday—8 June 1966—coincided with the Symposium cocktail party on the lawn below the Blackford dining hall. Before the lobster banquet, Watson was determined to celebrate the birthday in style.

> Knowing it was a more than noteworthy occasion, I had driven earlier with Paul Doty's student Bob Thach to Entertainments Unlimited in Levittown. There, from a book of pictures, we chose "Fifi" as appropriate for a coming-out-of-the-cake act [on the patio beneath] the Blackford porch. Luigi Gorini, in the know, had his camera on hand to record Francis' laughing [and enthusiastic] reaction to his birthday present.[90]

A couple of months later, Jim and Francis were together on the pine-dotted Greek island of Spetsai, the scene of John Fowles's novel *The Magus*, to teach in a course on the latest techniques of molecular biology. The recent triumphs of messenger RNA and the code added spice to the occasion. With some 60 advanced students from many countries, the course was organized by the French biochemist Marianne Grunberg-Manago and sponsored by the North Atlantic Treaty Organization. The faculty was housed in a small hotel near the boys' school, empty for the summer, where classes were given each morning and late afternoon. Crick told the students that ribosomes were like a rock on whose surface, in effect, information travels. In the evenings, another faculty member, Maxine Singer, recalled, students and faculty congregated at a disco down the road. She spotted Jim standing aloof on a little balcony, watching. One evening Francis and Jim were whisked off to the nearby villa of the famous Greek actress Melina Mercouri for a glitz break.[91]

But the event that stuck in participants' minds was the arrival by sea of Crick, his wife, Odile, and their two daughters. They had motored over from Italy in the cruiser with which they had just replaced a sailboat they previously shared with a wealthy Milanese molecular biologist. Jacques Monod, an avid sailor, had once watched Francis at the helm of a sailboat and cracked that this was the only occasion where he saw Francis Crick in a modest mood. Watson recalled that Crick had "several anxious moments docking in the harbor."[92] Singer remembered it more vividly. Class had finished for the day. All were at the harbor awaiting the Cricks' arrival. With the boat moored, Francis put his dinghy in the water to come ashore. But he dropped the line and the dinghy began drifting off. He had to plunge into the harbor to get hold of it again. A large onshore audience laughed.[93]

FORWARD GEAR

The messenger that had been so promptly exploited in cracking the code was an exciting discovery that Jim said "really changed my life"[94] and set the course of the Watson lab's activities for the rest of the 1960s. The challenge was "to understand how messenger RNA was made and then functioned to order the amino acids on ribosomes during protein synthesis."[95]

Such problems attracted Wally Gilbert back from the Harvard Physics Department in 1962 for another stint in Jim's lab (and a seat atop the rhino). The problem this time was how the cell could be so efficient, making so much of a particular protein when so little of the cell's RNA was messenger. To Jim and the others, this was "paradoxical."

Having helped isolate mRNA, Gilbert now turned to the interaction of the messenger with the ribosome, now called "a non-specific workbench." He started by studying the binding of Nirenberg's artificial UUU messenger to the ribosome. This led to an important discovery: The ribosomes lined up along the mRNA and moved down it like inchworms. From the larger, heavier globule of each ribosome, chains of a particular protein grew, one amino acid at a time. At the tip of each chain was a tightly bound kind of transfer RNA, then often called S-RNA, that looked important to Jim. Suddenly, Wally and his competitors saw "the whole thing as a moving process," with ribosomes picking up one end of the messenger and running along it. Gilbert had found "an answer for an old question: why is there so little messenger?" The assembly line of ribosomes came to be called polyribosomes and then polysomes for short. Publishing his results, Gilbert thanked Watson "for many conversations" and "for having introduced me to molecular biology."[96]

Given the scarcity of the mRNA that had been found in 1960, the finding the same year of a bacteriophage with RNA, not DNA, as its genetic material seemed like a miracle. Tim Loeb, a biologist in Norton Zinder's lab at Rockefeller University who was working with New York City sewage, came across the RNA phage that was called f2. It infected male *E. coli* bacteria. The RNA of f2 carried instructions for a few proteins that are essential for reproducing the virus, including the proteins that make up the virus coat, and an enzyme for duplicating the viral RNA. Daniel Nathans, then at Rockefeller, recalled "the excitement . . . when one of their isolates was found to have an RNA genome." Up to then, to study how RNA viruses reproduced, scientists were limited to viruses, such as polio virus, that infected complex, slow-growing animal cells. Now scientists could use an RNA virus that multiplied with the rapidity of bacterial cells.

Still better, Nirenberg had conducted his artificial-messenger code-cracking experiments with extracts from *E. coli*, not the intact cell. It began to look as if the RNA of f2 would be a plentiful "natural messenger" to use in "cell-free" experiments. Many questions that were slow and complicated to study would open up to simpler and faster research.[97] The field exploded. Competitors multiplied. Other RNA phages were found, which received names such as MS2 and M12, and the R17 phage (from the sewers of Philadelphia) on which the Watson laboratory concentrated for years.

The RNA phages drew particularly passionate interest because they could be used to explore details of how the synthesis of chains of mRNA got started, and how synthesis stopped. This was particularly important before 1966, when not all the 64 codons of three letters, including start and stop signals, had been worked out yet. With fervent encouragement from Watson, such Harvard researchers as Mario Capecchi, Gary Gussin, Jerry Adams, and Ray Gesteland (who later moved to Cold Spring Harbor) plunged in. Although Jim was, as usual, conversing intensely with his coworkers, his name did not appear on the flood of papers, causing confusion among non-Harvard scientists who asked Zinder who these unknown people were. Capecchi later recalled that going into RNA phages "took a lot of balls. Zinder had it all sewed up." Zinder was not making f2 available.[98]

It was an uncomfortable period for Zinder. Watson had muscled in. He recalled, "Our two sets of students became friendly enemies—and blamed Jim and me for giving away our secrets." A former Rockefeller colleague of Zinder's who had migrated to Harvard told him "that Jim was annoyed by our labs coming up with the same results at the same time. I wasn't all that happy either."

The two labs briskly competed over the start signal. Ever the gossip, Jim had heard from his British connections of a particular form of the amino acid methionine, N-formyl methionine, that seemed intimately involved in getting synthesis started. He thought he had acquired an unbeatable lead as the two labs raced to isolate this special methionine in the RNA phage coat protein. But when Jim visited Norton's lab and found out that Norton also knew about the special form of methionine, his face fell. It was a dead heat. The two labs began to exchange data and publish papers "back to back." Scientists in each lab were downcast that they hadn't arrived first.[99]

Watson was pleased with Capecchi, saying that he had accomplished as much while a graduate student as most people do in an entire career. Like many biologists, Capecchi had started out liking the elegant simplicity of physics, but, after graduating from Antioch College in 1961, decided to switch to molecular biology. He recalled, "Here, you had a new field being born. The predominant feeling was that anything was possible. You could ask any question. That's fairly unique. It doesn't occur that often in history."[100]

After a few terms in Alex Rich's lab at MIT learning the experimental ropes, he looked around for a place to do graduate study. "I was trying to decide between going to MIT or Caltech or Harvard, and so I came to Jim to get his advice. I told him my plight, that I had a difficult decision to make." Jim's answer, that he should come to Harvard, was expressed bluntly, "You'd be fucking crazy to go anywhere else." Capecchi recalled this as "fairly persuasive" and went to Harvard.[101]

In Jim's lab, Mario was "provided with what appeared to be limitless resources. I could not be kept out of the lab. We were cracking the genetic code, determining how proteins were synthesized, and isolating the enzymatic machinery required for transcription." Gilbert, who joined the lab around 1963, and Watson "complemented each other brilliantly because they had such different approaches to science. Jim was intuitive, Wally quantitative. As students, we received the benefits of both."[102]

Mario came away with a lasting impression of the incentive-creating role of a mentor. He reflected, "Watson is a bold person who says anything that comes into his mind. His bravado encouraged self-confidence in those around him. He taught me not to bother with small questions, for such pursuits are likely to produce small answers."[103]

Big questions continued to come up as Capecchi moved on to Harvard Medical School and then to the University of Utah. He was a leader in the 1980s in developing the "gene targeting" that allowed researchers to "knock out" genes one at a time in mice to explore their functions in detail. This achievement brought him the Kyoto prize in 1996, a share in an Albert Lasker Award for basic medical research in 2001, and his National Medal of Science in 2002. Watson was in the audience when he accepted the Lasker award, thanking "my mentor, Jim Watson, who taught me not so much how to do science, but the essence of science."[104]

Details of how protein synthesis started were still scarce in the early 1960s. How were mRNAs stitched together? The enzyme that raced along the DNA to make messenger molecules was called "DNA-dependent RNA polymerase." To start its work, this multicomponent enzyme has to find a particular sequence of DNA letters called the "promoter," just upstream of the operator, and wrap itself around the double-stranded promoter to begin pulling apart the two strands so that one can be copied into mRNA. But how does the enzyme know exactly where to go? In 1968, researchers both in Watson's Harvard lab and at Rutgers University found that one component of the enzyme was split off the five others right after the mRNA synthesis started. This component, which was called sigma factor, stimulated the making of far more mRNA than the five-component polymerase could make on its own. Apparently, the sigma factor allowed the enzyme to "read the correct promoters."[105] Once the sigma protein subunit had helped get synthesis started, it dropped off the "core enzyme," ready to lend a hand in the next copying action.[106]

Suddenly, in Jim's judgment, the field of genetic control moved beyond mere confirmation of the ideas of Jacob and Monod and escaped the predictable and boring. Two years later he wrote of "an excitement equal to that which accompanied the discovery of messenger RNA." He put sigma factor into the second edition of *Molecular Biology of the Gene*.[107]

The researchers involved in the sigma discovery were Richard R. Burgess and Andrew A. Travers at Harvard, and John J. Dunn and Ekkehard K. F. Bautz at Rutgers. Ray Gesteland told Travers and Burgess about Dunn and Bautz. Although they had reached the same conclusion independently, the Harvard and Rutgers teams worked together for a month and then sent in a joint paper to *Nature*.

Burgess came to Jim's lab from the Caltech lab of Robert Sinsheimer. He wanted to continue working on a virus called M13, but after a few months Jim redirected him to RNA polymerase. Although Jim thought of this as a "straightforward six-month project," Burgess found himself struggling for years to purify the enzyme. He purified and purified, and lent samples to Jeffrey Roberts, a lab colleague working on the virus called lambda. Roberts told him the samples didn't work so well. "What do you mean?" Burgess replied. "It's the best I ever made." Back to purification and voilà, sigma factor.

Soon after, Burgess went on to a postdoctoral fellowship in Tissière's lab in Geneva and then the University of Wisconsin, where he continued Jim's custom of popping bottles of champagne when students finished their theses or a key finding was made. He kept working on the problems of transcribing the information in DNA into messengers. Twenty years later, Burgess summed up Jim's lessons for achieving scientific breakthroughs: Make a combustion engine of research by attracting good people and surrounding them with more good people, put them to work on nonmundane problems, encourage intense interactions, don't bother them with daily monitoring, load them with credit, and make sure results are communicated clearly, freely, and quickly. The only time Jim got mad at Dick Burgess was when he failed to tell people at a conference in Germany about some results of Jeff Roberts's.[108]

A VIEW FROM THE PERIPHERY

An outsider, that is, a protein chemist, who joined the Harvard group in the 1960s took whimsical pleasure in what he observed on the third and fourth floor of the Bio Labs. At first Guido Guidotti, whose focus was hemoglobin, was reluctant to come. When he gave the usual trial seminar, only the chemists John Edsall and Konrad Bloch attended. After he refused an offer, Harvard invited Guidotti back and this time he met Jim, who had just come back from the Nobel ceremonies in Stockholm. When Guidotti entered Watson's office, Jim said, "I actually have some chores to do. Would you mind coming along?" The chores were to return some pictures that had been hanging in Jim's house to Harvard's Fogg Art Museum. On the way, Jim asked Guido what he was working on. Hemoglobin, Guido said. "Why?" Jim asked. Guido was hired anyway.

Installed on the "dingy, dark, dirty" fourth floor, Guidotti became accustomed to violin music occasionally coming from the office of Mark Ptashne, a postdoctoral fellow; the "war room" appearance of the office of Matt Meselson, who had embarked on a lifelong crusade to ban chemical and biological warfare; and "major conversations" that John Edsall (who died in 2002 at the age of 99) was having with himself in the hall. Occasionally, Guidotti would hear "this hurricane of invective coming down the hall, and it would be Jim Watson with something to unload," exclaiming that someone was stupid.

Some evenings, Jim, like Guidotti a bachelor then, dropped by Guido's lab for a chat, occasionally urging him to drop hemoglobin and work on something interesting, like RNA polymerase. Although Guido stuck to hemoglobin, he was promoted anyway. Soon after this, Jim dropped by to invite Guido to celebrate his promotion over dinner at a local restaurant called Henri IV, then one of the best in Harvard Square. Guidotti was "highly touched by this kind move." In the early 1970s, he and Jim taught a new undergraduate course on molecular biology and biochemistry, each one teaching for three weeks and then having three weeks off. A special feature was that students received full notes the day after each lecture so that they could get more out of the course, and so that Jim would have them for future writing. Guidotti's work thrived with the help of students "who were very good but couldn't get into Jim and Wally's lab." They came to Guido and "I had a great career."[109]

REVERSE GEAR

Forward-gear functioning of living systems, making proteins when they are needed, is only half the story of genetic control. Equally important is the ability to prevent the manufacture of proteins when they are not needed. This is important for bacteria, which must be fast on their feet, but is even more important in the process of specialization among the more complex "higher" cells of animals. In animals, including human beings, all cells except sperm and egg cells have the same endowment of DNA. But much of that hereditary information is shut down during the process called development, when cells specialize for particular functions, such as making hemoglobin for transporting oxygen into the body and carbon dioxide out. The escape of just a few genes from this shutdown process was and is thought to be a major factor in cancer. In the early 1960s, Jacob and Monod and Lwoff at the Pasteur Institute, doing studies of both bacteria and bacteriophage lambda, postulated the existence of the trains of genes they called operons. They also postulated "repressor" proteins that would bind to the DNA to prevent it from being copied into mRNA. Like a tape pasted over a start button, the repressor would block the action of RNA polymerase. It was not until a few years later that it seemed possible to isolate repressors and see where they functioned. "The repressor problem" excited scientists such as Mark Ptashne, who after graduating in chemistry from Reed College in Oregon in 1961 was drawn into molecular biology by the repressor. He recalled, "Jacob and Monod and Lwoff provided the first [genetic] evidence for the existence of molecules whose sole role is to regulate expression of other genes." They had the major insight of seeing "parallels between the two disparate systems."[110]

In 1966 and 1967, Gilbert and Ptashne, now also at Harvard, each isolated a repressor and proved that repressors bound to DNA. Jim encouraged rivalry between them, but, Ptashne recalled, "there was never a hint of self interest in his case, other than the goal of something important happening." The rivals talked often. "Almost every day," Mark recalled, "I would talk to Wally at great length about what both of us were doing. That was very helpful. On reflection, he was sure that "if any one of us had been working alone, we would not have been able to sustain the kind of effort required to do the repressor isolation." Mark was amazed at Wally's "stupendous ability in the face of failure to keep working . . . that much harder." He admired Wally's "incredible strength of personality." By contrast, Mark "got depressed sometimes. I'd just feel like my gut had been kicked out after really expecting something and then again and again failing."[111]

Gilbert, working with a German postdoc, Benno Müller-Hill (later of the University of Cologne), had gone after the repressor for the set of bacterial genes for making beta-galactosidase. This was the sugar-digesting protein on which Jacob and Monod had built their theories of genetic control. With the help of Nancy Hopkins, then a technician but later a graduate student, Ptashne sought the repressor of the bacteriophage lambda (of particular interest to Lwoff), which allowed the virus to turn its genes off after burrowing into a bacterium and to remain silent there for many generations.[112] Both teams had demonstrated "the first clear mechanism of gene control in molecular terms."[113]

The problem of finding the Jacob-Monod repressors began engrossing both Gilbert and Ptashne in 1965. By then Gilbert, who had left physics for good and moved over to the Watson lab full time, was on the prowl for a new challenge. Ptashne, having just finished his Ph.D., had been made a Junior Fellow at Harvard, which freed him from specific duties. Both had the leisure to attack problems that were important, and risky to get into.

They had "both scientific and psychological" difficulties in common. They didn't have a biochemical assay, a means for detecting the presence of repressors, and couldn't devise one because they didn't know how and where repressors worked. They suspected that a given cell would have only a few repressor molecules at one time, and they found that this was only too true. They did not know where to look: among proteins or nucleic acids or something else? They would have to "surmise" some property of the repressor that would lead to its capture. They did not know if their experiments would turn out to be relevant, or even have any result.

> It was possible that we would search but not find anything. Such a failure would not prove the negative (that repressors did not exist) but would simply mean that the question was still open. Even if we could isolate the product of the repressor gene, we might still fail to understand how that product functioned.[114]

Gilbert and Müller-Hill exploited the idea that the lactose repressor would bind to DNA when the enzyme wasn't needed, and fall off when it was. A factor called an inducer would bind to the repressor and deform it slightly so that the repressor would release its grip on DNA. The inducer was thought to be not the lactose itself but rather a related molecule produced by modification of lactose. Müller-Hill and Gilbert worked with a radioactive lactose-like molecule called ITPG that they hoped would grab the repressor, and allow them to separate out the molecules bound to their hot "label." After some nine months of stumbling, they succeeded. The search was arduous, and kept Gilbert at the lab constantly. His young children, John and Kate, who knew he was looking for something, would sometimes call up to their father from the lawn in the Bio Labs courtyard, "Did you find it yet, Daddy?"[115]

Down the hall, Ptashne and Hopkins worked the lambda phage and devised a system whereby irradiated cells would manufacture little else but lambda repressor. Ptashne recalled, "Even though the cells made very small quantities of lambda repressor, we could radioactively label it preferentially, a trick that made it possible for us to show right away that lambda repressor bowed to the lambda operators." By this different approach, they too could isolate repressor. Both the lactose and lambda repressors turned out to be proteins.[116]

Gilbert, Ptashne, and their colleagues were by no means done with the repressors. Ptashne recalled, "If you really are going to be completely hard-headed about it, it didn't really tell you much. . . . It didn't solve any biological problem. The real issues were, what does it do? Does it bind to DNA?" Was the DNA stretch they blocked indeed Jacob and Monod's "operator?" With the help of tricks to get even more repressor produced, they found that the answer to the questions was yes. They could move on to other significant problems such as how the repressors recognize specific sequences in DNA, how gene activation works and how these processes help explain gene regulation in higher organisms. They could explore the architecture of their repressors—four subunits of identical weights in the lactose repressor and two identical subunits, lighter than those in lactose, in the lambda system. Working out the exact sequence of amino acids in each protein was next.[117]

In the fall of 1966, Gilbert said in a lecture that the repressor discoveries had made "the dream of Monod and Jacob come true." It was in the spirit of the times in what was becoming the Watson-Gilbert lab. An excited graduate student who was there recalled, "That's a pretty impressive statement to hear."[118] A year later, Gilbert had the chance to say the same thing in a seminar that Monod himself attended—and to confidently answer a series of tough questions from him.[119]

By the mid–1970s, Gilbert and Ptashne were working out the sequence of DNA letters that their repressors recognized.[120] It was an early excursion into the world of genome sequencing (see chapter 12). Gilbert, with Allan Maxam, developed a widely used DNA sequencing method, and for that he shared the 1980 Nobel Prize for Chemistry with Fred Sanger of Cambridge, England, who worked out the method on which the Human Genome Project was largely based. Gilbert, a principal contestant in the race to use bacteria to produce human insulin, became a founder of one of the first biotechnology companies, Biogen, based in Cambridge, Massachusetts. Ptashne was a founder of Genetics Institute, also with headquarters in Cambridge.

In 1998, when a plaque honoring Watson was unveiled in the lobby of the Bio Labs, Watson said he hoped that someday there would be another plaque, honoring Wally and Mark.[121]

LIZ

When asked long afterward why he married so late, Watson fidgeted and said, "It's too complicated."[122]

Throughout the nervous years at Harvard, people noticed Watson's practice of hiring young women who were students at Harvard's sister college, Radcliffe, as lab assistants and secretaries for himself and others working in his lab. One of them recalled, "The fact that he had this whole batch of 'Cliffies' working there did not escape comment. People would run into him at Jolly-Ups, mixer things, and come back with these stories of this guy who was a Harvard professor. It was Jim. He didn't make the impression I guess he was looking for." But still, "he liked to have a bunch of Cliffies around the lab."[123]

As Gros had observed, Watson was a romantic. The best-selling memoirist Susanna Kaysen records that Jim came to see her during her stay at McLean Hospital, a psychiatric hospital in Belmont. They may have known each other because her father was on the Harvard faculty. When she went out to see the visitor, "He was standing at a window in the living room, looking out: giraffe-tall, with slumpy academic shoulders, wrists sticking out of his jacket, and pale hair that shot out from his head in a corona." He "drifted" toward her. She always had liked his way of drifting, wobbling, and fading out "while he was supposed to be talking to people." Asking what "they" did to her at the hospital, Watson looked around at the over-stuffed vinyl armchairs in the living room and whispered, "It's terrible here." He beckoned Susanna over to a window, pointed out at his sports car, and suggested that they speed out and head for England, where she could work as a governess. After a few seconds, she said, "I'm here now, Jim. I think I've got to stay here." Without seeming miffed, he said, "Okay" and left. Kaysen watched from the window as Jim's sports car puffed down the drive.[124]

Jim's bachelor status thrust him into some unusual roles. He had met a beautiful young chemist from Iran, Nasrine Chahidzadeh, in Geneva in 1966 during his sabbatical in Tissières' lab (mentioned above). She worked there but she wanted to go to the United States. Watson took her to dinner, where she described her chemistry studies in Zurich and asked about the Kennedy family—she had just met Senator Edward Kennedy. Jim told Nasrine that she could get a job at Harvard.

When she came to the Bio Labs, she and Jim had some "quiet suppers." She also dined with Senator Kennedy. But after a short time, she disclosed that she would soon marry a Swiss chemist who was working in the New York office of his family's perfume company. She had a surprising request. Because her father could not leave Iran, would Jim "give her away." He said yes, went off to give some lectures in Turkey, and on the way back to Harvard stopped off in Geneva, where he called on the woman in whose house Nasrine had lived.

The night before the wedding, the pre-wedding dinner was given at the Locke-Ober restaurant by the groom's father whom Jim assured that Nasrine wouldn't have had a job at Harvard if she weren't a good chemist. Next day, after the ceremony, there was a reception at the Copley Plaza Hotel, where telegrams from absent friends were read. One came from Nasrine's landlady in Geneva. It expressed the hope that "Jim Watson finds a wife as beautiful as Nasrine."[125]

Adding spice to this good wish was the fact that Jim had recently met Elizabeth Vickery Lewis of Providence, Rhode Island, a beautiful Radcliffe sophomore with "blue eyes and full cheeks," who had begun helping Watson as a secretary several afternoons a week. Watson began to "anticipate" her visits. He began to regard Liz as "the needed beautiful girl." She accepted a last-minute invitation to accompany him to a faculty cocktail party "that I felt would be dull without her." Leaving the party they drove in Jim's MG to Boston to see a movie, and afterward they walked slowly around the grassy courtyard surrounded by Radcliffe dormitories.[126] Soon afterwards, Jim went to the Gilberts' house to tell them. Celia recalled, "Jim came into our kitchen . . . and sat down. He sighed, he hissed through his teeth, and he said, 'I have just met the woman I am going to marry, and she is wonderful.' And that was Liz."[127]

After Liz returned from a summer job in Montana, from which she sent an affectionate postcard, their acquaintance blossomed. Jim and his father dined with Liz at a little restaurant called Piccadilly, a few blocks' walk from Jim's house on Appian Way. Their first real date was at the 1967 Christmas party for the newly formed Harvard Department of Molecular Biology and Biochemistry. Paul Doty spotted them together and afterward told Jim he had found "a peach of a girl."[128]

Soon Jim and Liz began holding hands in private, and began to wish they could do so in public. Planning for their secret wedding began. The circumstances were complicated, what with Jim's decision to take over direction of the Cold Spring Harbor Laboratory, and the explosion of notoriety over the publication of *The Double Helix*. The first opportunity for their wedding was late in March 1968, during Harvard's spring vacation. Jim was scheduled to lecture at the annual American Cancer Society science reporters' briefing, held that year at the Salk Institute in La Jolla. With the help of "the highly literate polymath" Jacob Bronowski of the Salk and his secretary—the only ones in the know except Liz's parents—a small wedding in a Presbyterian church in La Jolla was arranged to take place after the reporters' briefing. Liz would fly out to meet Jim there, and so he found himself nervously waiting at Lindbergh Airport in San Diego to greet her. There is a story that she had to get permission from a supervisor at the lab for a few days off to get married. The supervisor said the approval of "Dr. Watson" was needed. Liz replied brightly that she didn't think that would pose a problem, as it was Dr. Watson she was marrying.

Just before the wedding, Jim and Liz went to the Bronowskis' house to be photographed and filmed. A clip from the scene was used in the final episode of Bronowski's famous television series, *The Ascent of Man*. After the wedding there was a small reception at the Salk Institute; but one friend, Leslie Orgel, did not believe that Watson and Liz had married— he thought it was a practical joke. During a brief honeymoon at La Jolla's famous La Valencia Hotel, Jim sent Paul Doty a postcard on which he wrote, "Nineteen year old now mine." It was the start of a lifelong partnership in which Liz's role was crucial. Friends observed that they were devoted to each other, and never said a harsh word to each other or about each other. Jim could count on Liz when things went badly, and she smoothed over many awkward situations. One friend said, "She was the most stabilizing thing in his life."[129]

8

"FRESH, ARROGANT, CATTY, BRATTY, AND FUNNY"

*Francis is one of the few people who
have nothing to be modest about.*
JAMES D. WATSON, CENTURY CLUB,
NEW YORK, 15 FEBRUARY 1968

WATSON: "I CAN'T IMAGINE FRANCIS BEING HARMED"

In 1968, the same year that he got married and became director of Cold Spring Harbor Laboratory, Jim Watson vaulted into his own unique niche in literature when he published the most indiscreet memoir in the history of science. The romantic legend of DNA and of Jim Watson would be brightened immeasurably by the telling, and the special picture he painted of himself as a brat-genius would stick.

On 15 February 1968, the *New York Times* disclosed on its front page that the previous year, Harvard's president, Nathan Pusey, had forbidden the university's press to publish Watson's *The Double Helix*. Pusey thought that this personal memoir of the 1953 discovery of the structure of DNA, which Watson had spent the year 1965 writing, would thrust Harvard into "a dispute among scientists." For the first time in anyone's memory, Pusey and the other six members of Harvard's Corporation had twice set aside favorable votes of the Syndics, the governing board of the press, which had okayed the book's publication twice both before and after more than a year of softening the book's sharper edges. Pusey was reacting to objections to the book from Francis Crick and Maurice Wilkins, who hinted at legal action after Jim showed the manuscript to them and others such as Linus Pauling and Lawrence Bragg. Both had sent Pusey copies of their letters blasting the book as trivial, adolescent, gossipy, unrepresentative of science, and, above all, a gross invasion of their privacy. Crick's letter was addressed to Watson, and Wilkins had written to Thomas J. Wilson, director of the Harvard University Press, who had been first told about the book by Ernst Mayr.

The attention this contretemps attracted was dramatically augmented by an unrelated advance in molecular biology that hit front pages across the United States on 15 December 1967, just two weeks before Jim's book began appearing in a serialized version in *The Atlantic*.[1] The Stanford University laboratory of Arthur Kornberg, the 1959 winner of a Nobel Prize for Physiology or Medicine, had succeeded in creating fully infectious artificial virus DNA in the test tube. The news was considered so big that reporters were sent releases from Stanford, the University of Chicago, and the National Institutes of Health, and page proofs from *Proceedings of the National Academy of Sciences*. In those dark days of the Vietnam War, President Lyndon Johnson was eager for any good news, and featured the discovery in a talk in Washington that occurred just as a press conference began at Stanford to discuss the achievement. The president hyperbolically praised the Stanford work as "an awesome accomplishment . . . [one of] the most important stories you ever read, or your daddy read, or your granddaddy ever read." Headlines and television reports blazed. DNA was big, big news. Ironically, the din of publicity caused heartburn in Jim's group at Harvard, which considered the recent isolation of gene repressors a more fundamental advance in applying molecular biology to cancer than the Kornberg synthesis.[2]

Gossip about Pusey's veto of Jim's book had been bouncing around the university and the world for months. The news that Harvard had turned the book down, after much effort by Jim and his editors to meet scientists' criticisms, had been covered in at least one news story.[3] During the protests, editing, and negotiations, Tom Wilson was director of the Harvard University Press. After he had moved to the start-up publishing house Atheneum, which became Jim's publisher, Wilson recalled that Sir Lawrence Bragg, who wrote the book's foreword, spent a day at Harvard in 1967 talking with Jim, John Edsall, and him. Bragg asked Jim "to tone down this and that." He didn't worry about references to him, but couldn't Jim treat Maurice and Francis "a little less violently?" Neither Wilkins nor Crick was mollified, but neither wrote *The Atlantic* or Atheneum to block publication. Evidently, they focused not on stopping the book altogether but on preventing it from carrying Harvard's imprimatur.[4]

The full story broke in February 1968 in the student newspaper, the *Harvard Crimson*.[5] It was the month when the United States decided to commit no more troops in Vietnam, and when the second installment of Jim's book appeared in *The Atlantic*. A *Crimson* editorial accused Pusey of being "less interested in diversity of viewpoint than [in] bland tranquillity."[6] The *New York Times* picked up the story and put it on its front page, guaranteeing vast attention.[7] Watson hastened to write Max Perutz to deny that he or his publisher had leaked the story to hype the book.[8]

At lunchtime on the day the *Times* story appeared, a number of journalists, including me, gathered at a big round table in the Century Club in New York for a lunch sponsored by Atheneum to launch *The Double Helix*. From the episodes in *The Atlantic*, the reporters already knew that the book, as Nan Robertson of the *Times* wrote years later, was "fresh, arrogant, catty, bratty, and funny."[9] The host of the lunch was Simon Michael Bessie, publisher at Atheneum, which was starting off with more éclat than he had dreamed of. Also there was Wilson, who had joined Atheneum after a dispute with Harvard over mandatory retirement, and had signed his contract with the new house four months before Pusey's veto.

Watson, as usual almost inaudible, started off with a joke. He said he had rejected the title *Golden Helix* (actually, this was the name Crick had given to his house on Portugal Place in Cambridge), because it "sounded like Chinese pornography." The book was intended for 16-year-olds just going into science and wondering what it was like. "Basically the only book that told *me* about it was *Arrowsmith*," said Watson. Having kept no diary, Watson found his best source in his letters home. These, he said, were "understated"; he told his parents that "we may have done something interesting."

Jim whispered that he didn't think that it was unusual that he hungered for a Nobel Prize when he was young, or that he thought, when in his early twenties, that "anybody over forty was impossible." He said he had felt little influence from the scientist C. P. Snow's novel, *The Search*. He had read either Kendrew's or Crick's copy and found it "pretty implausible." Asked why he hadn't written his book together with Wilkins and Crick, Jim said they were different people, as different as Eisenhower and Montgomery, whose relations before and after D-Day were notoriously strained. Would he give up science for being a celebrated writer? "It would be a bore. I have one story to tell in my life and I've told it."

Reporters were told that almost everyone mentioned in the text had seen the manuscript, but none of the changes suggested by lawyers was adopted. Although the fight was really over "tone," Watson said, "I never tried to change the mood." Jim denied that he was trying to correct any image of science. "I was just trying to tell the story as it happened." Given his personality, it couldn't have been told any other way.

The notably aggressive science editor of the *New York Herald-Tribune*, Earl Ubell, demanded of Watson how he could object to Crick's seeking to block publication by Harvard. After all, the book's first sentence was "I have never seen Francis Crick in a modest mood." Referring to both

Wilkins and Crick, Watson answered, "I really didn't think the book would harm them in the slightest." He added, "I can't imagine Francis being harmed. I want to make a distinction between being harmed and being annoyed. . . . The great good luck of my life was to meet Francis. Francis is one of the few people who have nothing to be modest about."

As they would for decades ahead, the reporters asked Jim about the possibility of changing people's DNA. He called the idea "very premature." In 1968, he said, the story was: DNA makes RNA makes protein. Nothing had been discovered so far that opened the possibility of altering the DNA of people. To be sure, he wouldn't object. "We're suffering from the consequences of bad DNA." To him it was not a legal but a moral question.[10]

A few months later Watson told a reporter that biology was in its "infancy, like nuclear physics before the applied atomic energy revolution." The next step would be trying to understand more complex biological structures and how they are formed. Along the way, biologists also would try to find out something about the origin of RNA and DNA. "But I don't think this is going to enable us to improve a person genetically. . . . I am afraid at the moment there is no mechanism around to reproduce another Francis Crick."[11]

If Pusey wanted to take Watson down a peg, he failed. His censorship certainly widened the audience for the book, which remains in print in the twenty-first century after sales of more than a million copies. By demanding that Harvard not publish it, Crick and Wilkins had forced it into the hands of a feisty new publishing house, primed to sell to a much wider public than the Harvard University Press might have reached.

CRICK: "A VIOLATION OF FRIENDSHIP"

The Double Helix had plenty of passages that raised eyebrows.

On Crick: "At that time (1951) he was 35, yet almost totally unknown. Although some of his closest colleagues realized the value of his quick, penetrating mind and frequently sought his advice, he was often not appreciated, and most people thought he talked too much."[12]

On Wilkins: "Maurice had noticed that my sister was very pretty and soon they were eating lunch together. I was immensely pleased. For years I had sullenly watched Elizabeth being pursued by a series of dull nitwits."[13]

On Rosalind Franklin: "By choice she did not emphasize her feminine qualities. Though her features were strong, she was not unattractive and might have been quite stunning had she taken even a mild interest in clothes."[14]

On Sir Lawrence Bragg: "The thought never occurred to me then that later on I would have contact with this apparent curiosity out of the past."[15]

In the tangled maneuvering around publication, Wilkins and Crick protested frequently. Wilkins asked, "Concerning Rosalind, is there any mention in your book that she died?" He thought that Franklin was more than just "presentable," as Watson's manuscript said, and suggested the word "stunning," which Watson adopted. A main defect of the account, according to Maurice, was that comments that Jim billed as "impressions" read more like statements of fact. He pointed out that after Franklin moved in 1953 to Bernal's laboratory at Birkbeck College, "she was able to show her true scientific potential."

Wilkins held to his position that the book should not be published. To Tom Wilson he wrote, "The book is extremely badly written, juvenile, and in bad taste." While he admired Watson's "scientific brilliance, maturity of scientific judgment and distinction as a research leader," he thought the book displayed "what are probably his greatest and most personal weaknesses," so the press of Jim's own university should not publish it. *The Double Helix* was "unfair to me, Dr. Crick, and to almost everyone mentioned except Professor Watson himself." The manuscript, wrote Wilkins, should be turned over to the scientific historian Robert Olby, who had already begun his history, *The Path to the Double Helix* (1974).[16]

Crick was more savage. He dismissed Watson's book as "a fragment of your autobiography," not a history. "Should you persist in regarding your book as history . . . it shows such a naïve and egotistical view of the subject as to be scarcely credible." The book did not "illuminate the process of scientific discovery. It grossly distorts it." Personal details buried "the thread of the argument." Gossip, much of it irrelevant to a history of DNA, was preferred to science. "Anything with any intellectual content, including matters which were of central importance to us at the time, is skipped over or omitted." It was bad taste to write about your friends, at least when they were living. Scientific details, such as the complex field of crystallography, were "referred to rather than described." In science, Crick wrote, the point is "*what* is discovered, not *how* [or] by whom. It is the *results* which need to be brought home to the public." By focusing on competition, Jim ignored the fact that "the major motive was to understand." Jim had neglected such historical questions as "the advantages and disadvantages of collaboration or when the structure would have been solved if we had not solved it." And there was no attempt to document anything, even though Watson's memory was "faulty." Jim hadn't even mentioned the crucial documents, his letters to his mother.

The book would surely lead to a picture of Jim's character which was "not only unfavorable but misleadingly so." The whole exercise, Crick wrote, "grossly invades my privacy." It was "a violation of friendship."[17]

Years later, however, after rereading *The Double Helix*, Francis ruefully admitted that Jim had been "clever" to get so much science across in a popular book which was "very suspenseful . . . told in such a way that you can't put it down."[18] Delightfully, as a result of the book a number of people had invited him to dinner parties he would otherwise have missed.[19] In a mellow mood nearly 30 years later, Crick suavely observed, "Apart from this curious opening sentence, I come out rather well in the book."[20] Harvard eventually became "cross with me because the book was a best seller and would have made them a lot of money!" By then, he was telling people that he didn't have to seek publicity. With Jim, he had the best publicist in the world.

LWOFF: "MAY GOD PROTECT US FROM SUCH FRIENDS!"

According to Peter Medawar, *The Double Helix*, like "all good memoirs," had "not been emasculated by considerations of good taste."[21] Jim's irreverent account of one of the greatest of scientific discoveries elicited a torrent of reviews, many almost as disapproving as the earlier letters from Wilkins and Crick. Jim had trod on bunions and reviewers, including at least two Nobel Prize winners, had a grand time, denouncing or approving the book's accuracy. The reviews so underlined the incandescent effect of Watson's book that a selection of them, including Medawar's, was printed in a Norton Critical Edition in 1980.[22]

The reaction amounted to a public debate about the nature of competition among scientists, the type of popularization that science should receive, and the place of science in popular imagination. Was the book a healthy demythologizing of science, showing that scientists were human beings and not priests chanting deep within the temple of the Sun God at Luxor? Would it lower the greatly exaggerated public expectations of benefit and fears of risk from science, or induce disgust among the general public so great as to dry up government funding for research? Was the discovery of the double helix typical of how science proceeds, or highly unusual? Did other scientists now have a duty to step forward with their own memoirs, lest the full range of creative instincts that drove them remain undescribed? Would young people thinking of a scientific career be appalled and turn away, or glow with fervor to take part in the biological revolution? What are the obligations of a memoirist toward living associates? Did the avowed aim of writing in the tone of youthful arrogance excuse Jim's descriptions of the late Rosalind Franklin as an aggressive shrew? How could a man of 40, of such transcendent achievement and influence, fail to exhibit mature judgment and kindness in describing the lightning strike of 1953? Would Jim ever grow up, and put aside the brutality of the young?

Reviewers had a field day, and sales rocketed. Some delighted in the book's irreverence while disapproving of its immaturity. One reviewer remarked that Watson seemed to think that life in the lab was "some work, much idle talk, and considerable back-biting."[23] The book might be "breezy and gossipy," giving "vivid details" of feuds, a *New York Times* reporter wrote, but then he quoted an unnamed veteran of the events leading to the double helix: "You pretend to yourself that you don't do such things at all. And if you do, you try to forget them as soon as possible."[24] The famous English crystallographer J. D. Bernal had a simpler point of view. Soon after the book appeared, he wrote Jim that it was more exciting than *Arrowsmith*.[25]

An expectably bitter review was written by Erwin Chargaff, deriding Crick's and Watson's "quick climb up Mount Olympus." He lamented, "I know of no other document in which the degradation of present-day science to a spectator sport is so clearly brought out," and added, "I believe it is only recently that such terms as the stunt or the scoop have entered the vocabulary of scientists." Denouncing the book's "merciless persiflage" about Rosalind Franklin, Chargaff wrote that Jim had turned into a gossip columnist, "a sort of molecular Cholly Knickerbocker."[26]

The review was so waspish that the editor of the Norton Critical Edition, Gunther Stent, singled it out for rebuttal. (Chargaff had refused permission to reprint his review in Stent's book.) Stent accused Chargaff of being "a very intense sort of sore loser."[27]

Like some other journalists, the science editor of *The Saturday Review*, John Lear, was scandalized by Watson's revelations. Writing "a bleak recitation of bickering and personal ambition," Watson had given a "fragmentary and incomplete" account of the discovery. Lear was sure that "idealistic students considering a scientific career will be repelled."[28] They weren't.

The iconoclastic Harvard biologist Richard C. Lewontin reflected that perhaps scientists as a community should sue for libel.[29]

Conrad H. Waddington, animal geneticist at the University of Edinburgh, belittled the double-helix discovery. It was "only the very final stages in a scientific advance which had been put firmly on the rails long before he came on the scene." He said that "DNA plays a role in life rather like that played by the telephone directory on the social life of London." The race for the double helix was "not typical of most top-level science."[30]

To the virologist Robert Sinsheimer, head of the Biology Division at Caltech (who would tangle bitterly with Watson in the future), the private world Jim was describing was "unbelievably mean in spirit . . . a world of envy and intolerance, a world of scorn and derision." High-school students would "learn that [science] is a clawing climb up a slippery slope, impeded by the authority of fools, to be made with cadged data and a resolute avoidance of profound learning, with malice toward most and with charity for none."[31]

One of the most powerful critiques came from the magisterial André Lwoff. *The Double Helix,* he wrote in *Scientific American,* was "either a rape or an autopsy." To be sure, Jim told his story "with such absolute sincerity and innocence, and recorded his impressions with such candor, that he becomes transparent." But he had applied "cold objectivity" most particularly to people he liked, such as Francis Crick. "Very few are spared. May God protect us from such friends!" He added: "Jim appears to be ignorant of the fact that the naked truth can be a deadly weapon, even to those who are dead and have no way to forgive. He seems completely unaware of the injuries he inflicts, completely unaware of the harm he can do to his friends, to the friends of his friends, to say nothing of those he dislikes. His portrait of Rosalind Franklin is cruel."[32]

All this tut-tutting exasperated the editor of the British journal *Nature.* "If, of course, his picture is seriously awry, then other people are free to protest and even have a duty to do so. It is not enough simply to resolve never again to invite Professor Watson to tea and biscuits."[33]

MORRISON: "SEEKING ROOM AT THE TOP"

The approving reviewers, confident that Watson had told it like it was, tended to take a lighter tone, although not without bite. Sydney Brenner, one of the founders of molecular biology and a man of impish temper, found the aspirations in Watson's book similar to those of Holden Caulfield, the principal character in J. D. Salinger's *Catcher in the Rye.* Of Holden's wish to be the catcher in the rye, Brenner wrote many years later, "He knows it's crazy, but that's what he'd like to be." The 16-year-old "struggles to come to terms with the realities of the adult world. He finds most of the people in that world pompous and ridiculous, except for an older brother, D.B., who lives in Hollywood and has cars and girls."[34]

Equally forgiving was the scientist Jacob Bronowski of the Salk Institute (he later hosted the BBC documentary series *The Ascent of Man*), who helped arrange the wedding of James Watson and Elizabeth Lewis soon after his review appeared in *The Nation.* To him the book, despite bowdlerizing, remained "a classical fable about the charmed seventh sons, the antiheroes of folklore who stumble from one comic mishap to the next until inevitably they fall into the funniest adventure of all: they guess the magic riddle correctly."[35]

To the physicist Philip Morrison of MIT, the "cranky, intuitive" environment of the double-helix discovery resembled the story of the chemist August Kekulé, who dreamed of dancing atoms lining up "in pairs and threes and long chains" while dozing on a horse-drawn London bus. This, along with another dream of a snake biting its tail, is said to have led to the concept of six carbon atoms forming a benzene ring like those in A, T, G, and C. Morrison thought *The Double Helix* had "the air of a racy novel of one more young man seeking room at the top. Censored movies, smoked salmon, French girls, tailored blazers set the stage on which ambition, deft intrigue, and momentary cruelty play their roles. The story should kill the myth that great science must be cold, impersonal, or detached."[36]

Among the most positive assessments was Medawar's. He predicted, "It will be an enormous success, and deserves to be so—a classic in the sense that it will go on being read." Jim had blasted the notion of "the scientist who cranks a machine of discovery." He added, "In my opinion, the idea that scientists ought to be indifferent to matters of priority is simply humbug." Of the book's harsh tone, Medawar remarked, "Many of the things Watson says about the people in his story will offend them, but his own artless candor excuses him, for he betrays in himself faults graver than those he professes to discern in others."[37]

The sociologist and historian Robert Merton agreed. Intense fights over priority were normal in science, and had been far nastier in the days of Newton and Leibniz than in the 1950s. "Though it might surprise the outsider," Merton wrote, "this emphasis on competition in science will scarcely come as news to working scientists." Originality is crucial to the scientist, "a value central to the scientific enterprise," and that usually means having the insight first.[38] A year later, Merton made note of the competing scientist's "close awareness of the champion who must be defeated in this contest of minds." Watson had helpfully called attention to "alternating periods of intense thought and almost calculated idleness" as a factor in discovery, along with the false starts and wrong inferences, the fevered acquisition of needed new knowledge, the awareness of each other's strengths and weaknesses, scientists' "unfailing sense of the key problem, and an intuitive and stubbornly maintained imagery of the nature of the solution."

But Merton also saw the reverse of the coin. The competitors with big egos, treading on ground where they knew so little, also had to cooperate. A grapevine of fact and rumor ran between players in New York, Pasadena, Paris, Cambridge, and other places, all of them mindful of those who had gone before. Jim and Francis also had scientific "taste." They "knew that they had hold of a problem of the first magnitude."[39]

Some things, one may reflect, were left out of many of the reviews: Jim's desire for fame, his drive to recruit more soldiers for the crusade of modern biology by an account in the brash language of youth that made biology more exciting than other fields, exultation at having burst into the inner circle, getting his own back by writing the way he thought and talked in his early twenties.

Jim had no reason to feel burned by all the controversy. "The reception was better than I expected," he recalled exultantly in the 1980s. "I only published it because I thought it would be considered a serious work of literature, not just a magazine story. If I had thought it was not first class, I really would have suffered!"[40]

WHY ALL THE FUSS?

Enraging some and elating many, Jim's book unleashed a torrent of scientific histories and memoirs. None of them achieved the notoriety or sales of *The Double Helix* but they enriched the intellectual history of modern biology. They included Robert Olby's *The Path to the Double Helix* (1974); Chargaff's *Heraclitean Fire* (1978); Salvador Luria's *A Slot Machine, a Broken Test Tube* (1984); Francis Crick's *What Mad Pursuit* (1988); and Arthur Kornberg's *For the Love of Enzymes* (1992). A principal player in continuing this historical enterprise was the Cold Spring Harbor Laboratory Press, which later reprinted such books as Alfred Sturtevant's 1965 classic, *A History of Genetics*; a translation of François Jacob's *The Statue Within* (1995); and an expanded version of Horace Judson's *The Eighth Day of Creation* (1996), originally published in 1979.

When reread after more than three decades, *The Double Helix* has an effect much more like that of a light white wine than of a heavy red one. It is engagingly conversational, and its aura of playground roughhousing now seems a good deal less outrageous than it did when both characters in the narrative and reviewers first read the story and found so much to disturb them. The tone is that of an impudent young man—and it seems far more obvious now than it did in 1968 that impudence is required to look at things a new way so that new benefits can be achieved. That is as true in the microprocessor revolution as in biotechnology, in biology as in medicine.

So why did the book arouse such an outpouring of writing that now seems a bit overheated?

One factor is the book's apparent and actual revelations about Jim Watson's character. Several who commented on his memoir knew him quite well and had been repelled by him. The book seemed to make it crystal-clear why—there he was blatantly, even proudly, displaying to a vast and uncomprehending public the very motivations and qualities that appalled some of his acquaintances. What a dreadful representative of the sober and beautiful search for truth that scientists revered! Several of his critics knew that in the years since 1953, Watson had shown remarkable ability to attract some of the best young people in biology and to build an exciting Harvard lab, but the book seemed to underline how unlikely it was that Watson could do this. How could someone totally lacking in gravitas, someone who apparently believed only in acerbic direct assault on those who got in his way, someone who just wasn't "nice," get anywhere in a world where suavity, even indirection, is the usual course of success in any kind of politics?

Even worse, the book enthralled large audiences, including aspiring young scientists and the rapidly growing numbers of science journalists. By manifesting quite ordinary tendencies to rough play and a thirst for distinction, Watson succeeded, despite expectation, in making scientists more like the rest of us and therefore not quite as frightening as they had been portrayed in the years dominated by fear of the Bomb and novels like *On The Beach* and satiric films like *Dr. Strangelove.*

A second factor in the impact lies in the time of the book's publication. By 1968 the double-helix discovery of 1953 had reformulated the basic questions of biology, and the public was far more aware of these issues than before. Fundamental biology expanded dramatically against a background of revolutions in both medicine and agriculture that lengthened human life and increased food supplies. Between 1953 and 1968, one medical development after another had received enormous publicity. Among these were vaccination against polio, birth-control pills, the realization that smoking greatly increases the risk of lung cancer, the discovery of many connections between diet and heart disease, and the increasing use of open-heart surgery. Meanwhile, medical concern was shifting from infectious diseases toward the disorders that become more prominent when infection is beaten back and the average age of a population increases—disorders such as cancer, heart disease, stroke, and severe loss of memory. In those same years, the development of short-strawed fertilizer-responsive varieties of rice, wheat, and other grains was bringing about a "green revolution" in agriculture.

The Cold War was making it more credible that biological and chemical weapons, formally barred from military conflicts, might someday be used, particularly by small nations too poor to enter the big-power nuclear game. The fear of biowarfare was widespread.

In the years after the Soviet Union launched its first *Sputnik* in 1957, the United States and other nations began spending sharply more on science in general, and on the teaching of science. The number of graduate students exploded, notably in biology. Meanwhile the three letters DNA and the insights related to DNA were pushing their way into more and more textbooks.

For all these reasons, biology's promise and threat were a livelier presence in 1968 than before, and helped boost the attention for Jim's story. Undoubtedly, and contrary to many predictions, the excitement of the account helped draw very bright people into biology. It may be that young readers caught Watson's underlying idealism and love of learning beneath the surface bravado.

But the book may have had some unintended consequences as well. One was that, in public meetings forever after, Jim had to answer insistent questions about Rosalind Franklin. It was part of his public character that he had trashed her. People wanted to know why. The book may also have contributed to simmering unease about science. Only six years later, amid intense distrust of all authority that had been intensified by the Watergate break-in, consciousness of biology's place as a phenomenon as significant as atomic energy, and perhaps a lingering distaste for Watson's story, helped fuel the flames of the recombinant DNA hysteria.

9

A PASSION FOR BUILDING:
COLD SPRING HARBOR

The message was clear: whenever possible, we should design attractive buildings appropriate for the future of a great institution.

JAMES D. WATSON, COLD SPRING HARBOR
LABORATORY ANNUAL REPORT, 1993

ALWAYS EDGY

When Edward O. Wilson heard early in 1968 that Jim Watson was going to take over a barely surviving, ramshackle Cold Spring Harbor Laboratory (CSHL), he commented sourly to friends, "I wouldn't put him in charge of a lemonade stand." Thirty years later, in his memoir, *Naturalist,* Ed Wilson wrote, "He proved me wrong. In ten years he raised that noted institution to even greater heights by inspiration, fund-raising skills, and the ability to choose and attract the most gifted researchers."[1]

The whole performance stunned just about everybody who knew Jim. They didn't doubt that he would keep his sharp instinct for the next crucial problems in biology or his ability to charm, attract, push, promote, and exasperate young talent. But few would have predicted the other things he accomplished in 25 years of running Cold Spring Harbor Laboratory. In spite of his mercurial persona, he proved adept at exploiting fame. He could get away with declaring, right away, "Unless I can add $1 million a year to [the lab's] operating budget, I will have failed."[2] He reached this goal in a couple of years and more than doubled it in four. To get there, he focused the laboratory on cancer, the hot-button issue in medicine. He and others hoped that researching the underlying causes of cancer would speed up the transition of molecular biology—the field he helped define and advertise—from studying tiny, simple bacteria to tackling animal and plant cells many thousands of times more complex. To find the money for his lab, Watson somehow worked with the politicians and bureaucrats of Washington—the principal patrons of U.S. science—instead of alienating them as he had the president of Harvard.

Few appreciated at first that he had found in CSHL a power base far more satisfying than Harvard, even though it was many miles away from leading university centers of biology. Instead of uselessly punching the feather bed at Harvard, Watson went off to a scientific Elba, but unlike Napoleon, he stayed there. For the frontier problems he wanted to work on he needed to build research groups larger than universities could accommodate—as the unquestioned boss, he could do this at Cold Spring Harbor. With the laboratories at CSHL nearly empty, he had a clean slate to write on. To be sure, he could not offer young scientists the relative security of university jobs, but he could offer them the space and equipment to work on urgent problems and build a reputation fast.

Cold Spring Harbor's distance from major universities such as Rockefeller or Princeton or Yale—which had undertaken to make annual contributions in return for seats on the board of trustees—made Watson's attempt to create a major center of research there seem quixotic, almost like a defiance of gravity. Yet the new board chairman, Bentley Glass, quickly persuaded the affiliated institutions to pay up on their total financial commitment of $42,000 per year.[3] The CSHL board of trustees was made up of roughly equal numbers of scientific and business-oriented members. The scientist-trustees provided mild supervision of new appointments and promotions.

In 1969 a major cancer-project grant from the National Cancer Institute boosted and stabilized the income that the laboratory had to have for its expanding year-round research program. But Jim did not neglect the local constituency of wealthy neighbors who had rescued the lab more than once before. With Watson—and cancer—as magnets, the neighbors raised hundreds of thousands of dollars to renovate or rebuild labs and residences and to buy crucial adjoining land. Within five years, one neighbor, Edward Pulling, led Watson to Charles S. Robertson, who gave CSHL an endowed estate to be used for small conferences and, even more important, a make-or-break research endowment of $8 million. Ever after, this fund generated the money to help young scientists establish their careers. Young researchers—the ones Jim recruited year after year—were the most likely to come up with genuinely new things. Cold Spring Harbor Laboratory might not be rich, but it had a fountain of youth.[4]

With hitherto undetected fervor for architecture and horticulture, and with the help of his wife, Liz, Jim fanatically pushed the renovation of every building on the place, and the conversion of its heretofore jungly grounds into a park, to the delight of the neighbors. As the years passed, new buildings began going up, designed to blend into their sheltered surroundings.

For more than 30 years before Watson became director, the summer courses and conferences at Cold Spring Harbor had created a camplike, intellectually heady atmosphere for participants to catch up with the latest ideas and techniques, intensely discuss the next year's experiments, and announce significant findings. Watson built these activities up, using them to encourage new lines of work, to preserve Cold Spring Harbor as *the* place to try out new ideas and to add to a growing number of publications. These included conference proceedings, lab manuals, and monographs. These in turn not only boosted the reputations of the often-young researchers who organized the meetings and edited the books but also raised CSHL's intellectual influence even higher than it had been.

Surpluses from book sales provided so much money for renovation that CSHL could be regarded as the lab that books rebuilt. The surpluses strengthened CSHL's balance sheet, a document that riveted Jim's attention. Occasionally, he startled a potential recruit by displaying these numbers to tell a story of solidity. Jim had a healthy respect for money, and the people who made it and managed it, and he said so. This trait was no small help in managing a frequently changing board of trustees and the donors who were needed again and again as Watson kept adding activities over a quarter of a century. In 1990, as CSHL marked the centennial of the first lab on the site, Watson exclaimed, "I think about money all the time!"[5] A couple of years later he said, "I really like rich people."[6]

It was all so surprising. Jim was entering the world that many leading scientists disdain as "administration," in which he ran the risk of becoming overly concerned about procedural details[7] and struggling with the requirement for "instant shifts of attention."[8] As director, he would have allegiance to six constituencies: staff, trustees, the surrounding community, individual donors, institutional donors, and scientists elsewhere "whose interests overlap" Cold Spring Harbor's.[9] There was another difficulty, as one scientist lightly observed 25 years later: "He's not verbally adept. He could never sell cars."[10]

Watson certainly was aware that intellectuals, even geniuses, usually are not allowed too close to the cash box or the levers of power. They are regarded as otherworldly, not canny in the ways of finding money to finance their intellectual dreams, or, worse, idealistic spendthrifts. Even for Jim, who had a sustained vision of the key puzzles and of the atmosphere needed for tackling them, the role of builder could have been terrifying. Potential donors say no—so often—and nobody, least of all someone with a program to revolutionize biology, takes rejection at all philosophically. The likelihood of a no must be avoided if at all possible, and donors often must be courted for years. In the domain of philanthropy, one moves beyond logic into a forest of emotion. In order to follow the forest trails, one must have more than a little understanding of people who inherited a lot of money or made it themselves and want it to be used for a worthy purpose. Most of these potential donors have a strong sense of business, and more than a little experience in sizing up the horses they want to bet on. Often, their gifts are more to a person than to a cause.

The process is filled with uncertainties. Who could have predicted the behavior of the immensely rich Andrew Mellon, secretary of the Treasury for 11 years in the 1920s and 1930s under three presidents? Forced out of office because of a corruption case, he chose to found the National Gallery of Art as an expiatory gift to the nation. The famous British art dealer Joseph Duveen persuaded Mellon to buy an apartment full of paintings to start the gallery's collection.[11]

Woodrow Wilson, as president of Princeton in the early 1900s, invited Andrew Carnegie to visit the campus. Wilson hoped for major funding for such goals as graduate student residences, a school of law and government, a school of science, or a tutorial system. Carnegie was amiable, and showed interest in everything that was shown him, but, surprisingly, he focused on athletics, specifically football. Hating a game whose tactics seemed to be geared toward maiming opposing players, he told Wilson he knew just what Princeton needed. In great anticipation, Wilson asked what that would be. Carnegie replied, brightly, "It's a lake. Princeton should have a rowing crew to compete with Harvard, Yale, and Columbia. That will take young men's minds off football." He gave $400,000 for the lake and attended the dedication, irritating students by remarks that disparaged football.[12]

This was not Wilson's only travail with donors and their emotions and loyalties. He wanted the eagerly awaited new graduate school to be close to the undergraduates on the campus, but Dean West, more elitist, wanted it to be farther away on a golf course. The Gordian knot was cut when a donor died and left money for the graduate school—and specified it must be on the golf course. To make sure his wishes were followed, he made the dean one of the executors of his will.[13]

Donors can get fed up with the recipients of their largesse, as John D. Rockefeller did with William Rainey Harper of the University of Chicago. Building a great university, Harper always outran his resources and expected Rockefeller to continually cover the deficits. Rockefeller could not count on Harper's word. Having learned the lesson of not letting his benefactions grow too big too fast, Rockefeller doled out money much more carefully to what is now Rockefeller University.[14]

Watson, trustees remembered, always delivered on his promises.[15] Charles Stevens of the Salk Institute recalled, "Jim was wonderful with the trustees. . . . He was very respectful and he took [the relationship] extremely seriously. He always got what he wanted. He was very careful with them. He didn't upset them or rile them. . . . He showed that he appreciated them, always, and treasured their advice."[16] Harold Varmus, of the Memorial Sloan-Kettering Cancer Center, said that people might have dismissed him in advance as "that geek. Doesn't seem like a highly socialized character. But the fact is he loves it. He loves the interactions. He's good at both goading and beseeching people."[17] The institution grew despite Jim's impossible demands and outbursts of anger. Somehow, he could work with business managers and editors and accountants and librarians and cooks and grounds keepers and builders.

Helping Watson navigate this wild terrain were some of his own emotions. From the beginning, he had a house at Cold Spring Harbor, as he had long dreamed he would. He was defending and improving his home even before he moved into Airslie, the big director's house, surrounded by lawns sloping down toward the harbor. He was impatient and determined to be at the forward edge of molecular biology, and he had a grand place for accomplishing this.

Yet his collaborators, however hotly they fought with him or smarted from humiliations at his hands, were convinced that, somehow, this was not a game of self-aggrandizement. Something larger was at stake. Watson was "incredibly selfless in letting his own scientific aspirations melt away and get transferred to another generation." He was ready to accept the principle of, "Okay, we've done our science and now we're going to be effective leaders," said Varmus.[18] This is all too rarely seen among scientific leaders. Jim had an unusually strong sense of "ownership" of CSHL, and was alert to potential dangers and inventive in getting around them. Relentlessly optimistic, he hacked away at all the unexpected administrative problems that kept popping up, bringing to them the same enthusiasm he brought to campaigning against the vines that luxuriated in the moist air off of Long Island Sound. He had as many challenges as the chef-owner of a restaurant or the director of a summer music festival. His enthusiasm could never flag. With so much on his mind—in addition to scanning perpetually for what and who was scientifically hot—it was hardly surprising that he was perpetually visibly edgy. "He was always skating on thin ice," Norton Zinder observed.[19]

Once, a leading scientist at CSHL wondered, half humorously, if Watson was insane. "Is he certifiable?" The answer was "Closet normal."

"A LAB NEAR DEATH"

Back in the 1960s, the uniquely informal and intense environment of Cold Spring Harbor, which was associated with many of the most important discoveries in biology, had almost disappeared. But the enterprise was held together by an Australian scientist named John Cairns, who left something for Watson to rescue and revive. Watson, as a happy and frequent visitor and a trustee representing Harvard since the reorganization of CSHL in 1963, had seen the task coming, but he fought shy of it for a time before embracing it.

The cliff-hanger existence at Cold Spring Harbor, which was always a fabulous invalid, intensified around 1960. The rules of the Carnegie Institution of Washington (CIW), parent of the decades-old Genetics Department at Cold Spring Harbor, required Milislav Demerec to retire at 65. For 20 years Demerec had directed both the Carnegie Department of Genetics and the Biology Laboratory owned by the neighbors' association called LIBA (Long Island Biological Association). After Demerec stepped down and moved his research out to Brookhaven, farther east on Long Island's North Shore, the Bio Lab crept on under directors of short tenure, one of whom, to Watson's disgust, locked the library's door at 5 P.M.[20]

The Genetics Department included two great but low-paid scientists who had no intention of relocating, Barbara McClintock and Alfred Day Hershey (both later won Nobel Prizes). The president of CIW, Caryl Haskins, began looking around for a successor to Demerec, and after geneticists advised him against placing a protein chemist in that position turned to three noted young biologists: Robert Edgar; Sydney Brenner, the amazingly inventive associate of Francis Crick in deciphering the genetic code; and Norton Zinder of Rockefeller University, then a rival of Watson's on RNA viruses and later his long-term close adviser on many subjects. They all said no.

Zinder, for one, disapproved of the Carnegie policy of relying solely on income from its big endowment. It did not seek grants from outside agencies, which in those early post-Sputnik days were flowing freely. Zinder thought CIW was "rich but stingy." Haskins offered Zinder a budget of $200,000 a year to run the place. For that amount of money "I can't mow the lawn," Zinder told Haskins in a shouting match in the Cosmos Club in Washington. Buildings at Cold Spring Harbor needed at least $500,000 in repairs. Zinder's wife, Marilyn, told her 30-year-old husband that he would be going "out there for nothing. Why destroy your young career?"[21] Haskins decided to close the Genetics Department while continuing modest support for McClintock and Hershey.

At this point the neighbors stepped in, as they had earlier. Back in 1924, under the leadership of Arthur Page, they had formed LIBA to take over the foundering Biology Laboratory, which dated from 1890. When Demerec left, LIBA was unable to raise money for either an endowment or a new building. But it could help in other ways. In 1963, under the leadership of Arthur's banker son, Walter Hines Page (later chief executive of J.P. Morgan), a new institution embracing both the LIBA and Carnegie activities was chartered under New York State law, the Cold Spring Harbor Laboratory.[22] The new lab was launched with the help of several scientists, particularly Francis Ryan of Columbia.[23] Taking over as first director was Cairns, who had made important advances in molecular biology. During a sabbatical in Hershey's laboratory to work on DNA replication, Cairns had shown that the DNA of the virus called T2 was in a single molecule. Determined to understand how an entire genome would duplicate itself, he went on to find a way to take pictures of the entire chromosome of *E. coli* bacteria, which showed very few places for starting the DNA replication.[24] Cairns also had visited Delbrück's lab at Caltech, and Delbrück's opinion of him was that he was "as good a man as you will find anywhere."[25]

After a journey of 12,000 miles, Cairns, his wife, Elspeth, and their two sons and a daughter arrived at an institution that was world-famous yet "decrepit beyond belief." Like the Marine Biological Laboratory at Woods Hole and the Zoological Station at Naples, the newly reorganized Cold Spring Harbor Lab was hanging on by its fingernails. This was a desperate situation, Cairns thought, because summer courses and research at Cold Spring Harbor and elsewhere were vital to the change that biology required.[26] CSHL needed at least $750,000 worth of renovations. Despite the poor condition of the buildings and the finances, Cairns was able to recruit an Australian, Cedric Davern, and Joseph Speyer from New York University, both of whom received substantial grants. Cairns also had a grant for his own research from the National Science Foundation. His salary was covered for five years by the Rockefeller Foundation.[27]

Soon, Cairns was fuming. His board of trustees, headed by E. L. Tatum, from Rockefeller University, who had shared a Nobel Prize with George Beadle in 1958, seemed unable to raise any money. The new director was so "utterly taken up with the frustrating job of fund raising"[28] that he hadn't even unpacked his lab equipment for his research.[29] Still he reinvigorated the summer courses and conferences and their attendant publication programs; Watson later recalled that "a small but real cash surplus accumulated."[30]

The Cairnses, living in the director's house, kept open house for lab scientists and were popular with the neighbors. Watson recalled that "their home in Airslie always radiated gracious and good-natured concern for our staff and many visitors."[31] Other scientists were even more favorably impressed. Cairns was "sort of a father of the whole place." In those days, Airslie was "an open house." Cairns would "see you walking on the road, take you into the house, have a beer," Heiner Westphal recalled.[32] He displayed an amazing "innate capability of summarizing something" in a few sentences, so that people puzzled by a talk another scientist had given would understand it as explained by Cairns, even though he lacked Jim's "real outlook on where a field would move to what was going to be exciting in the future." Although Cairns was "contemplating great things," always looking for *the* next experiment, he would confess, "I can never think of the next experiment to do." Still, Cairns "really was a brilliant person . . . very wonderful to interact with . . . always thinking in complex ways."[33]

Still, by 1967, when Davern and Speyer both left for more secure jobs, the dominant mood was discouragement. Cairns offered his resignation and found the trustees, now including Watson as representative of Harvard, willing to accept it. One trustee, Zinder, who succeeded Tatum, was exasperated. He thought that while the lab was going downhill financially and physically, Cairns complained continually about a lack of help but preferred the role of "country squire." In Zinder's view, Cairns "went around bad-mouthing" the lab and the trustees.[34] Another director—more optimistic—would be needed to get beyond bare survival.

After considering more than one possibility, the board turned its thoughts to Watson, who had brains and a kind of glamour. At a board meeting, Norton Zinder recalled, Watson suggested that he come down from Harvard once a week or so and help. Zinder's view was that Cairns was excellent at bailing out the boat, but was also fond of punching more holes to do more bailing. The lab was stuck in a "decadent mode." Norton exclaimed to Jim, "If it's all that easy to do, why don't you just take over the job." Jim said, "I'll let you know tomorrow." Zinder thought, "He must have been waiting for this."[35] Jim said he'd come part-time; he could spend at CSHL the time that Harvard allowed its professors for consulting. Cairns reported to Delbrück, "The Board accepted Jim Watson without a murmur."[36] In Watson's view, Cairns had come in when "the outcome was in doubt," and put the lab "on an upward course"; he had taken over "a Lab near death and brought it back to life."[37]

BIG GOALS

Watson lost no time nailing his colors to the mast, telling the press in very personalized terms about his decision to focus CSHL on the role of viruses in cancer, which he described as "the next big hard problem" after DNA.[38] Late in March 1968, two days before his wedding, he pulled off his wrap-around sunglasses and told a *New York Times* reporter, "I'm looking for someone who would like to give $5 million to cancer and restoring old buildings." Fidgeting nervously and sipping from a can of Coca Cola, he stated the reason he'd taken the job: "I didn't want to see it [CSHL] disappear."[39] He showed another reporter who was visiting Cold Spring Harbor the buildings that would have to be torn down and replaced.[40] He told a third, "I came down here because no one else was going to take over this place and keep it the way I thought it should be kept." [41] He told yet another, "I am fond of the place. It has always for me stood for good science. . . . During the summers, well, it is the most interesting place in the world, if you're interested in biology."[42]

Even though Harvard was paying him a salary of $21,500 a year, and his association with the university would last another eight years, Watson was not shy in signaling that he wouldn't be at Harvard forever. Cold Spring Harbor offered a challenge that Harvard no longer did. As he ran his fingers through his wispy hair he told a reporter, "It's something new. Besides, my colleagues at Harvard are much brighter than I am." He could shape and direct Cold Spring Harbor Laboratory from scratch.[43] To be able to use research on cancer viruses for "working out a cell as if it were a Swiss watch,"[44] Watson was convinced that "the most effective labs are likely to be those in which large numbers of people with closely related interests work simultaneously." He scorned "dull factories" built around "superstars."[45]

In the eyes of Ray Gesteland, who had recently moved to CSHL from Harvard, Watson quickly showed himself to be a "bold" administrator, and not only in his tumor virus initiative, which built up quickly.[46] By May 1969, with help from Cairns and Joe Sambrook, newly recruited to run the tumor virus lab, he had secured the $1.6 million cancer project grant from the National Cancer Institute, which brought in more than $500,000 the first year. In 1972, the first $1 million installment of a five-year $5 million grant from NCI to establish a cancer center came in. Against fierce competition, the grant was renewed many times.[47]

FIXING THE PLACE UP

One of Jim's first steps was to declare an "emergency" need for building repairs. Zinder got Rockefeller University to contribute $5,000 and other institutions gave a total of $15,000 more.[48]

Jim had a strong focus on aesthetics and he was determined to move fast to "transform a dilapidated relic of early American science into a productive blend of the past and the very new." But the transformation must maintain the character of a whaling village of the nineteenth century, from which many lab structures dated. Construction activity grew so quickly that Watson was relieved to find a successful contractor, Jack Richards, who was willing to join the lab full-time as the "in-house contractor."[49] Over decades, working mostly with the notable postmodern architect Charles Moore and his partner, William Grover, Richards oversaw not only the construction and reconstruction of buildings but also the extremely touchy problems of sprucing up the grounds.

Not everyone was pleased by the war against the vines and the relentless campaign to turn the laboratory grounds into a park. Barbara McClintock, an inveterate walker, complained bitterly to the younger scientists who delighted in walking with her about the droppings from the many geese attracted to the expanded lawns. She slipped on them and muttered that the only proper place for a goose was on a Christmas platter.[50] She was remembered as "very spry," busy after snowstorms shoveling her sidewalk or driveway, and always available for conversation as she walked around, "looking at Queen Anne's lace and watch[ing] the horseshoe crabs mate."[51] Joe Sambrook, who resided on the grounds for 16 years, approved of most of the changes but missed the tangle of trees and vines and their association with "intense evenings." He wrote in 1984, at the end of his year as acting director:

> The grounds were disorderly and delightful with cover, vibrant with birds and animals. For four months of the year, Bungtown Road was little more than a leafless tunnel for the bitter north wind, but in early spring it softened into a narrow culvert whose flowered walls were a mass of head-high brambles and rhododendrons. In late evening, with the light slanting at a certain angle, it was easy to believe that this was the perfect setting for a Wagnerian opera, so overpowering was the combination of color and perfume to the senses.[52]

Watson moved aggressively to make buildings usable year-round for research, to find housing for more scientists, and to provide more meals at more times of day, especially the conversation-promoting coffee breaks and beer parties. To help, LIBA raised $250,000 by 1973. In 1976–77 LIBA raised an additional $225,000 so that an old house dating from the whaling era, called Williams, could be torn down and replicated to create five apartments for scientists, each of which had views to the east over the harbor.[53] Additional full-time living space with picture windows was created in the Firehouse, a residence with a curious history. Up to the 1920s, it was the firehouse of the village of Cold Spring Harbor, but in 1926, no longer needed, the building was purchased for $50 and towed on a barge across the harbor to the lab.[54] A derelict nineteenth-century cottage called Wawepex was restored as a year-round guest dormitory, with money donated by Max Delbrück's wife, Manny.[55] With LIBA support, yet another winterizing occurred in the dining hall, Blackford, so that scientists could eat lunch there throughout the year.[56] Planning an environment where scientists would live minutes from their labs, Watson did not neglect recreation. In 1972, Watson and Manny Delbrück contributed money to build new tennis courts to replace those that had been torn up for a sewage-treatment facility.[57]

The pace of laboratory extension and renovation kept increasing. Part of LIBA's largesse went for additions to James, the center of the new work on tumor viruses. A new wing housed a vital seminar room, and offices for Jim and Joe Sambrook's growing group of tumor-virus researchers. Jim had shown his priorities by not moving into the nearby Nichols Administration Building: "I believe that the Director of a scientific institution should always be a scientist, not an administrator."[58] An old lab called Davenport, after Charles Davenport, the director of Cold Spring Harbor for the first three decades of the twentieth century, was renovated and later winterized for year-round research on yeast.[59] In 1974–75, the 80-year-old white clapboard Jones Laboratory was transformed, with Moore and Grover as architects. Four gleaming stainless steel "modules" for neurobiology were installed in the interior, while the wainscoting on walls and ceilings was preserved. This example of "adaptive reuse" received an Honor Award from the American Institute of Architects in 1981. Watson said he suspected "that we now possess the aesthetically most interesting scientific laboratory to be built anywhere."[60]

Another project was the complete rebuilding of a cottage called Osterhout as a home for Jim and Liz and, later, their boys, Rufus and Duncan, born soon after. The architect gave an estimate for a restoration and then remarked that he could build a new house, with a modern interior, air conditioning, and picture windows, for the same price. Feeling flush with royalties from *The Double Helix*, which had reached $20,000 the first year after publication, the Watsons said go ahead, donated the disguised new house to the lab, and in June 1969 moved in. Jim wrote proudly, "Though much of its external appearance retains a colonial flavor it is a thoroughly modern house with marvelous views out on the Harbor."[61]

Meanwhile, as soon as Jim became director, he applied for and obtained in 1969 an American Cancer Society professorship for Cairns, thus guaranteeing his salary. The Cairnses continued to live in the director's house until 1973, when John became director of an Imperial Cancer Research Fund laboratory at Mill Hill, outside London.[62]

In 1973, major renovations began at the director's house, Airslie. The changes were designed by Moore and Grover at an estimated cost of $200,000. Untouched since the lab bought it in 1943, the 6,600-square-foot old farmhouse, dating back to 1806, needed conversion "for twentieth-century living," including enlarged living and dining rooms. By this time Jim and Liz, with their sons nearing school age, had decided that they soon would not be able to maintain homes both in Cambridge and Cold Spring Harbor. Late in October 1974, the Watsons moved into Airslie, this new and grander home, a waterside version of "the house on the hill" that, he once had told Wally Gilbert, he had always coveted.[63]

There Liz meticulously planned a cascade of lunch and dinner parties for family members, visiting scientists, trustees, potential donors, old friends, and other visitors. The number often was 12. She kept careful lists of "who was there, what they were fed," to make sure that when "you came back . . . [she] hadn't served the same thing as last time. The consummate hostess."[64]

The search for more on-campus housing turned Jim's eyes to the two remaining properties along Bungtown Road that did not belong to the laboratory. One of these, a yellow frame house owned by a daughter of Charles Davenport, had already been rented for scientific staff housing and negotiations to buy it were under way. But the other, owned by the Takami family, went up for sale when a family member moved to Honolulu. The moment to buy it arrived in 1973, but, a scientist-trustee recalled, "the pantry was somewhat bare." The board and Watson had a contentious discussion, which he ended by saying, "Well, if you won't let me buy it, I quit." And he walked out. Trustees James Darnell and Norton Zinder had some fast talking to do. While Watson might be temperamental, they told the board, "he was seriously devoted to solving these kinds of issues." A board member, possibly Walter Page, went out to catch Jim. Darnell chuckled as he recalled, "Jim came back and they bought the house." It was done with LIBA donations and the house was renamed for the fund-drive chairman, Robert Olney.[65]

Housing was not the only bone of contention between Watson and his board. In 1972, a problem across the inner harbor grew more acute. A decade earlier, a 50-slip boat marina called Whaler's Cove had been established there, detracting from what Jim called the harbor's "essentially non-commercial character." Trustees, many living not far from the harbor, feared further development. In 1972, the developer of the marina in nearby Huntington Harbor, which was thickly dotted with boats, was looking to expand. He bought Whaler's Cove and adjacent land with a historic house, and planned to double the marina's size and convert the house into a clubhouse. Abutting neighbors tried legal challenges but failed. It was time to invoke the influence of powerful neighbors like Walter Page to try to stop the development. A deal was negotiated to buy Whaler's Cove and the development rights to the adjacent land for $300,000.

Watson thought the lab was lucky to do the deal before it became "prohibitively expensive." He argued that Cold Spring Harbor was the only North Shore harbor that preserved a sense of former days. The existing seclusion was "vital to the spirit of our many meetings and courses." Despite months of campaigning, including invocation of environmental concerns among local officials, however, Watson faced trustees who wondered where the money would come from, even though hard times appeared to be ending. Jim's presentation of the issue had been unemotional, and the trustees doubted that he cared deeply about it. They found out otherwise. He left the room while the trustees voted—no. Coming back and hearing that the trustees had turned down buying the marina, he exploded, "I give you three minutes to change your mind." Trustee Bruce Alberts recalled, "We wanted him more than we wanted three hundred thousand dollars."

Norton Zinder had another perspective on the incident. Having played the resignation card, Jim went stomping up and down Bungtown Road during a June Symposium, telling all and sundry that he was out. Zinder recalled that in those years, his jobs as a trustee were to oversee promotions of scientific staff and "to try to control Jim." That day Jim joined Liz and Zinder's wife, Marilyn, for lunch. Characteristically forthright, Marilyn told Watson, "Don't worry, Norton is not going to let you resign." Indeed, Norton recalled being delegated by the trustees to assure Watson that the money would be found, "and he unresigned." Zinder considered it "one of the funniest days of my life." The board had to "beg, borrow, and steal" the money the lab didn't have. He was convinced that the money was secretly lent by Walter Page.[66]

Clearly, Jim and the wealthy neighbors shared a passion for leaving the local environment as little changed as possible. Together they worked successfully to kill off the plan of New York City's transportation czar Robert Moses to run a parkway along the east shore of the harbor to link Caumsett State Park on nearby Lloyd Neck with his network of highways. Around 1930, the wealthy of Cold Spring Harbor had forced Moses to divert his Northern State Parkway southward, away from their estates. Watson also helped mobilize public opinion to scuttle a nuclear power plant planned for Lloyd Neck. The emotions stirred by the issues of environmental protection helped Watson get the backing to save and build and beautify CSHL.[67]

THE CRUCIAL GIFT

A critical aspect of Watson's success with the trustees was his ability to explain the connection between the basic research at Cold Spring Harbor and the attack on human disease. "The importance of genes in human disease," one scientist-trustee, Tom Maniatis of Harvard, observed, "was a theme of his whole career," and he kept his eye on that goal as he built the basic research. Jim explained the connection with medicine "in a particularly persuasive way," obtaining "the support of people who you would be surprised would do this." Gradually evolving from an eccentric scientist who was "on the surface rather awkward socially," Jim mastered the social interactions with friends of the laboratory with increasing skill.

His board meetings were "exceedingly well organized . . . really the most organized and efficient board meetings I've attended," said Maniatis.

Materials were handed out, including slides. "The people doing the most exciting science . . . would give these wonderful talks and get people excited about what was going on at the lab." In his annual director's report Jim would range from financial matters to the Human Genome Project and Congress. Having pondered in advance, "What would these people really like to hear?" Watson always seemed "right on tune," said Maniatis. Trustees "would leave the meeting just feeling wonderful."

Trustees were not only loyal, they were "functional," working hard at fund-raising and buildings. These were not just trustees put on the board after a big gift, who would "come and have tea and smile and listen and go home."[68]

The trustees came through most remarkably in June 1972. Jim and Liz had gone for a vacation to the Sea Ranch colony of country houses on the California coast north of San Francisco. At Sea Ranch, Charles Moore was the dominant architectural influence, and Jim and Liz took notice. Their vacation was interrupted by an urgent phone call from Edward Pulling, a neighbor and trustee, summoning them back to Cold Spring Harbor at once to talk to a neighbor who wanted to visit the laboratory. It was Charles Sammis Robertson.

Pulling was a handsome man of modest beginnings who had married Lucy Leffingwell, daughter of a J.P. Morgan partner. They had honeymooned aboard J.P. Morgan's yacht, *Corsair,* and went on to found the Millbrook School in Dutchess County, New York. The school's campus of white clapboard buildings resembles the most respectable of New England towns. Probably their most famous alumnus was William Buckley, the author of *God and Man at Yale* and founder of the *National Review,* a notable conservative periodical. After several decades of highly successful academic entrepreneurship, the Pullings retired to their family compound not far from CSHL, and Pulling had breathed new life into LIBA with the encouragement of Walter Page. The motto of Pulling's many years of fund-raising was, "Ask and ask, and thank and thank." Fond of sailing, riding, and poetry, Ed Pulling was a perfect agent for reeling in the big fish.

Charles Robertson's estate lay on 50 acres along Banbury Road in Lloyd Harbor, a village across from the lab. He and his second wife, Marie Hartford (the Hartford family owned the huge Atlantic and Pacific Tea Company chain of grocery stores), had worked together since their marriage in 1936 to put together the Banbury estate, piece by piece. They built a Georgian brick mansion, with tree-framed views to the west over Cold Spring Harbor. The emotion behind this effort was not casual. Ancestors on Robertson's mother's side had long before owned the 50 acres and much more. Robertson had been born in a house adjoining the property. Devoted to the management of his wife's fortune, Robertson

commuted daily to New York City. Charlie and Marie were strong supporters of Britain, particularly during World War II, and firmly committed to internationalism. After Marie cashed out her A&P stock holdings
in the late 1950s, she and Charlie astonished Princeton University, his
alma mater, with the huge gift of $35 million to the Woodrow Wilson
School (worth some $500 million in 2002), and insisted on anonymity.
An adopted son and daughter lived near them. Charlie was a Lab neighbor par excellence.

In the spring of 1972, Marie died of cancer, leaving Charlie Robertson
not only bereft but under pressure because of tax laws to relocate to a
low-tax state like Florida. He was determined to keep his and Marie's
beautiful estate together even after his death. Only an institution like
CSHL could take it and pledge not to break it up. Pulling heard of
Robertson's dilemma and began exploring whether the Lab could use the
Banbury land and buildings. He tracked Jim and Liz down to make what
Watson recalled as Pulling's "fateful" telephone call.

The next step was a tour of the lab by Robertson and his attorney, followed by one of Liz's best lunches at Osterhout, "the most important
lunch ever held at the Lab." Robertson could look east through a picture
window and see "the summer beauty of Cold Spring Harbor at its best."
The handsome little house breathed institutional solidity, as did the new
wing of James, where Jim's office was located. This impression was important because Charlie knew that if he were to entrust his estate to the
lab, he not only would have to endow the estate so that it would not become a white elephant, but also would have to do something dramatic to
enhance and ensure CSHL's hard-won financial solidity. "He wisely realized," Watson recalled, "that unless the Laboratory were itself financially
secure it would not be able to intelligently use or lovingly care for his estate." So, he decided to give $1.5 million to endow the estate and $8 million to support research at the laboratory.[69]

At one point in the process, Jim judged it useful to bring two scientists
from the lab over to the brick mansion at Banbury to be introduced to
members of the Robertson family as examples of the kind of men carrying out the lab's work. The chosen were Robert Pollack and Ray Gesteland. Jack Richards, the construction chief, drove Pollack, Gesteland,
and Watson over to Banbury in a lab van. Watson's instructions, Pollack
recalled, were "You don't speak until you're spoken to." When they
reached the front steps, "Jim very carefully bends down and unties his
shoes and musses up his hair. Wow! If I hadn't seen it, I wouldn't believe
it." Watson went on to do what seemed to Pollack "the Nobel laureate,
mid-western white guy's version of Steppin Fetchit. He does a full number of goofy scientist and they eat it up."[70]

But Robertson wanted to be sure that Walter Page's down-to-earth financial acumen would be available in hammering out the deal. Page agreed to rejoin the lab's board and over the next year helped to set up the independent Robertson Research Fund, which was tapped for income at a conservative rate that guaranteed growth. It gave CSHL "for the first time a partial independence from unpredictable fluctuations in Federal monies for science." In the next 25 years, the fund grew more than tenfold and its yield soared from $250,000 a year to several million dollars, paying for people and equipment and courses and lab space.[71]

The Banbury estate was officially turned over to the lab in 1976, the year Watson left Harvard for good, and over the next year its big garage was converted into a conference center, named the Banbury Center. In 1977, Francis Crick gave the dedication speech. His message, according to Watson, was that scientists must meet frequently with peers who have the same interests. "Going to each other's labs is generally much too time-consuming and expensive, and the best way to sort out the meanings of new ideas and experiments is to come together at small conferences" like those planned for Banbury.[72]

No longer would CSHL live hand to mouth. To Watson, Robertson's was "an offer too wonderful to believe." Watson reflected that gifts like these usually go to big universities. Never underestimating the importance of environments, he was convinced that the architectural excellence of the James wing and Osterhout had been a factor in Charlie's "great philanthropy." Watson asserted, "The message was clear: whenever possible, we should design attractive buildings appropriate for the future of a great institution." The crisis atmosphere had evaporated, scientists could concentrate on their research, and CSHL looked "very alive."[73]

"DNA TOWN"

Despite such signal success, for the rest of his directorship, Jim Watson was faced with a constant ratcheting effect. Fund-raising targets went up from hundreds of thousands to millions of dollars. New fields of research needed new space and equipment for ever advancing technologies—including advanced electron microscopy, X-ray crystallography, the making of monoclonal antibodies—and ever more computers. The computers had to be upgraded and linked usefully.[74]

Watson's dogged insistence from the beginning on a commitment to neurobiology culminated in a $21.5 million Neuroscience Center, which he expected to become "the hub of the Laboratory's future year-round development." The ever-skeptical John Cairns was less enthusiastic: "How will he populate that huge building?"[75] After it was dedicated in 1991, Watson moved his own office there. To build the center, Jim had to attract funds from many sources; ultimate donors included the scientific-instrument magnate Arnold Beckman and the Howard Hughes Medical

Institute. A bell tower with a helical staircase and a one-ton brass bell, donated by one of the trustees, Lita Annenberg Hazen, was ornamented with the letters A, G, T, and C, one on each face of the tower.[76]

A laboratory building that had been converted from a facility for lab animals and named for Barbara McClintock was remodeled again.[77] Wings were added to buildings, two of them to Demerec Lab, as part of a five-year collaboration with Exxon Research and Engineering.[78] Greenhouses that were turned into extra offices for the growing number of scientists were named for Alfred Hershey.[79] The old Davenport Lab was doubled in size by the addition of a south wing and the whole building was named for Max Delbrück. Later, a wing to the north, for plant research, was named for Walter Page and his father, Arthur. To make room for it, the Firehouse was winched 50 yards northward.[80]

The need for more housing never let up. The success of the summer courses and conferences created a constant demand for more guest rooms, such as the 60-room dormitory next to the Neuroscience Center. Up the hill from it, 11 new cabins with four bedrooms each were built with gifts from the scientists Mark Ptashne and Herbert Boyer and others.[81] Those who could remember the primitive accommodations of the 1960s and '70s reveled in the new comfort. Harold Varmus, a frequent visitor, said, "It's awfully nice to have a reasonable place to stay." Later, he said, "It's hard to remember how miserable some of those hot nights were."[82]

The need for housing extended to the Banbury estate. The conference room there accommodated 35, but the mansion, now called Robertson House, where conference participants slept and ate, could house only about 20. Trustee Betty Schneider led Jim to the foundation that gave the money to build the dormitory named Sammis Hall, which was completed in 1981, just across the lawn from the conference center. Designed by Moore and Grover in the style of Andrea Palladio's Villa Poiana in northern Italy, the guest house has 16 bedrooms arranged on two floors around an atrium.[83]

Helping to meet the demand for housing was a highly visible restoration of the white Late Victorian Charles Davenport House along Route 25A that for years had provided decaying quarters for staff scientists and their families and had been called "Carnegie dorm." Liz Watson had been trained in architectural history, and she went to work. The house was drastically renovated inside to provide eight dormitory rooms, one apartment, new bathrooms, a laundry room, a telephone room, and public rooms. Using existing pieces of trim as templates, a local carpenter, Frank Parizzi, recreated many pieces of Victorian trim that had been lost.

One of Liz's former Columbia graduate-school classmates in historic preservation, Frank Matero, painstakingly identified the colors originally used on the house's exterior. In its heyday, Liz wrote, the house had a "body painted in a rich deep golden yellow, deepening to pumpkin on the gables, its trim executed mostly in deep green, but accented in certain areas with a bright yellowy green, and its window sashes painted dark maroon." Despite Ed Pulling's protests at abandoning white, Pulling and several other neighbors raised the money for the new paint, and the Davenport house reblossomed with the Late Victorian colors. The cost, however, ran beyond the budget, and the trustees and Jack Richards were nervous. The trustees balked at paying the extra cost. Liz was crestfallen. Jim, sticking up for Liz as usual, responded by arranging to give lectures across the country and brought Liz the honorarium checks to close the gap. A piano recital in the music room celebrated the project's completion in 1980, and Liz and Jim were photographed there soon after for a newspaper story.[84]

North Shore neighbors were continually recruited as contributors and board members. Oliver and Lorraine Grace and 400 members of LIBA funded a new and larger auditorium, named for the Graces, and completed in 1986. The growing activities demanded more parking, including many spaces hidden beneath the Neuroscience Center. The blizzard of projects continued.[85]

LIMITS

In the years after he became full-time director of what he called "DNA Town," Watson pushed out in many directions. One place he failed was in his hope to acquire an estate called Fort Hill, an antiquated great stone pile atop a bluff on Lloyd Neck with a dramatic view of Long Island Sound, a good 20 minutes' drive from the laboratory. The lab was using it while the owning family debated how to dispose of it. As trustee Charles Stevens remembered, "It was beautiful. He loved it. Jim was dying for that house, but he couldn't take it unless it was endowed."[86] He tried many expedients, including the establishment of a center for cancer epidemiology to be headed by John Cairns or Julian Peto from England or Malcolm Pike from California. They all fought shy of the idea. Jim planted one of his young scientists, Ronald McKay, his wife, Jill, and their small daughter, Ramah, in a lonely apartment in the Fort Hill mansion.

In the summer of 1979, he installed a team of scientists, including Martin Raff and Bruce Alberts, at Fort Hill for six weeks so that they could draft *Molecular Biology of the Cell* without the distractions of family. Alberts remembered it as a tense "summer from hell." Watson had had the vision to launch the project but dreamed it could be done in a single summer (it took years). Jim's draft of chapter 1 was well written, but his colleagues decided it was far too long. Jim stopped writing. Alberts's wife and children were housed in a "shack" without air conditioning in the village of Cold Spring Harbor. "It was the hardest thing I ever

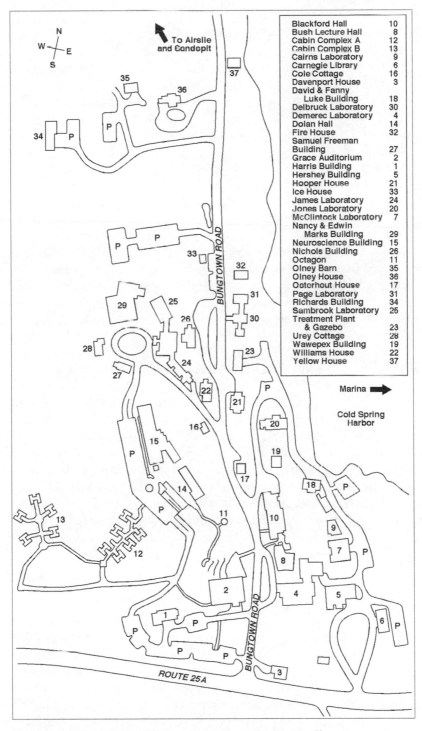

Site map of Cold Spring Harbor Laboratory, by Jim Duffy.

tried. . . . Hellishly hot, 150 hours a week. . . . It took a couple of months to recover . . . That's never happened before or since."[87]

Jim used Fort Hill as overflow housing for conferences at the Banbury estate. He tried to interest the Council for Religion in International Affairs, endowed by Andrew Carnegie, whose director was his brother-in-law, Bob Myers, in sharing expenses with CSHL. The family that owned Fort Hill did not have the money to endow the place, the trustees stood firm about not acquiring it without an endowment, and the property was soon subdivided.[88]

Another estate that drew Watson's attention was much closer to the lab. It was called Uplands Farm, and belonged to Walter Page's mother-in-law, Mrs. George Nichols, who was the daughter of J. P. "Jack" Morgan, son of the financial titan J. Pierpont Morgan. Here, with help from Page, Watson was more successful. Most of the property, just across Lawrence Hill Road from the Pages' own residence, came under the control of the Nature Conservancy at Mrs. Nichols's death in 1981. There was discussion of locating what became the Whitehead Institute at Uplands Farm, but the institute's designated director, David Baltimore of MIT, was reluctant to leave the Boston area, and anyway, Watson was said not to want the institute with Baltimore as director. With a loan, CSHL was able to acquire from the Conservancy a 10-acre piece of Uplands Farm with some outbuildings for $700,000. Now, with the new acreage, the lab could restart its plant genetics research using recombinant DNA techniques. These techniques made possible a fresh approach to plant genetics, a subject of vast importance to agriculture. CSHL had been famous for the early work of George Shull on hybrid corn and of Barbara McClintock on "jumping genes" in corn. In the 1980s, to enhance the plant genetics work, the lab formed a five-year, $2.5 million alliance with the leading seed company, Pioneer HiBred, whose longtime owner, Henry Wallace, was one of the first commercial users of Shull's techniques. The labs for the plant scientists using Uplands Farm were installed in the new building named for Walter and Arthur Page on Bungtown Road.[89]

MANAGEMENT STYLE

From 1978 to 1982 I reported to Watson in my position as director of the Banbury Center. My interactions with Jim can be considered typical: respect for my autonomy accompanied by explosions. At Banbury, Watson operated in his bold, even reckless way. Not content to stage small meetings and courses at the frontier of molecular biology or neurobiology, he decided that there should be a satellite CSHL that held conferences followed rapidly by publication of reports on environmental sources of cancer. It was quite an experiment, a leap in the dark, to get something going at Banbury. To fill the directorship, Watson attracted me from the *New York Times,* where I had been the technology reporter. The subjects were controversial and Watson needed an organizer-arbitrator who would be seen as not favoring a particular viewpoint, even if that person were devoid of experience in organizing conferences or raising money. Involving me in the process, Watson obtained grants from the Sloan and Klingenstein foundations to cover operating expenses for two years, including the salary of the director and editors for the reports. As principal adviser he recruited John Cairns, hoping to attract him back to CSHL. Cairns' interests had shifted strongly to the epidemiology of cancer.

Because environmental cancer risks related closely to hopes of cancer prevention, Watson expected substantial funding from the National Cancer Institute. But the response from NCI, where Watson was not universally popular, was long delayed and in any event was insufficient to cover the costs of the conferences; NCI support took the form of book purchases. It turned out that NCI was a lot more interested in treatment than in prevention, and probably was nervous about getting into a particularly contentious zone of public policy. There was more success in obtaining funds for such activities as workshops for congressional staffs and reporters and editors for *Newsday* and Time Inc. publications.

The trustees found themselves funding Watson's new project on faith while the director learned his job. In new territory himself, Watson met frequently with me, often at gossipy lunches at a pub in Huntington called Finnegan's. These meetings usually resulted in multi-item to-do lists covering everything from recruitment of speakers to finding alternate sources of money. Watson was fertile in suggesting new options, despite the immense range of other things on his mind. His gossip included remarks on his attempts to acquire additional estates and to start a company. There were sudden, agonizing, but infrequent descents upon me in his well-known style of management by annoyance—I derided it as "hobgoblin management." On occasion, Jim and Liz entertained me and my wife at Airslie, and we reciprocated at our house in Lloyd Harbor. Ever attentive to the trustees and to the reality that he was taking a flyer on me, he saw to it that I got to know them.

Slowly, the program began to go well: Conferences on the key subjects were attended by the key researchers; books appeared about six months later and sold modestly; the workshops for journalists and congressional staff, supported by a fresh grant from the Sloan Foundation, went smoothly. Many key people on the laboratory staff, themselves involved in frequent Watson turbulence, helped again and again, including Joe Sambrook, who could pull together a list of invitees in an hour; William Udry, the administrative director; Jack Richards, head of buildings and grounds; and Nancy Ford, the chief editor. But the deficit, several hundreds of thousands of dollars, exceeded the "overhead" that Banbury was expected to yield for the lab.

The culmination came, appropriately, on Halloween, 1980, when the CSHL treasurer, Robert Cummings, said at a board meeting I was attending that the deficits were not sustainable. In a sad tone, Harry Eagle, the great pioneer of tissue culture, who was chairman, turned to me for a rebuttal. I said I thought we had a system that worked—but it was their money. A trustee, Boris Magasanik of MIT, asked if he had heard aright, that the program was successful but should be closed. Watson sat in silence and I assumed I was being thrown to the wolves. But it may be that he planned what happened. After the formal session, trustee Alex Tomlinson of the First Boston Financial came up to offer help in contacting industry for funding, and Ed Pulling went out of his way, in Watson's presence, to praise my remarks.

Money did roll in, but the array of expectations was too broad. My contract was year-to-year, and my staff was being cut. Knocking the carcinogen of the week on the head was less interesting to me than news reporting. I was living someone else's dream. So I went off to another entrepreneurial assignment: starting a program of fellowships for science journalists at MIT.

The Watsons threw a farewell party for everybody who had helped the Banbury program, and took my wife and me to dinner at the Piping Rock Club. Watson said that a conference on patenting life forms that I staged in 1981 had opened the way to the $7.5 million research cooperation between CSHL and Exxon. Over the years at MIT, I reflected that I would never have survived there but for Watson's boot-camp training.

SUCCESSION

Even for Watson, running Cold Spring Harbor Laboratory got to be too much. The same problems kept coming around again but on a larger scale. The constant watchfulness with trustees and donors and government agencies and staff was wearing. With so many other concerns claiming attention, it was hard to keep up with the science. It was a challenge to find time for batting around ideas in late-night lab visits and morning coffees and strolling up and down Bungtown Road. In the early 1980s Watson began to think about yielding the directorship. He might leave altogether, or, if he stayed, he would be free of the daily grind and yet be consulted on the big issues.

In 1983–84 Watson took a sabbatical in England and his principal deputy, Joe Sambrook, became acting director for the year. During that year, one of the young scientists at CSHL, Douglas Hanahan, who was a favorite of Watson's, visited him in London. Turn the reins over to Joe, Hanahan boldly advised. He felt Jim was spending far too much time on raising money and landscaping, and not enough on "recognizing important, accessible problems and the right people to do them." He should regain the attention of young graduate students by becoming "a senior thought leader." Watson should bail out of being director and spend more time using his intellect "to be scientifically influential."

Although Sambrook showed energy and competence, he may have reached a little too hard for the steering wheel. As Hanahan saw it, Sambrook "was ready for a leadership position . . . and Jim wasn't ready."[90] Joe's candidacy faded and not long afterward, after being elected a Fellow of the Royal Society in London, he left to become head of the Biochemistry Department of the University of Texas Southwest Medical Center in Dallas. Watson praised Sambrook's "forceful, innovative mind that consistently and wisely worked for the good of this institution." Sambrook later became Director of Research of the Peter McCallum Cancer Institute in Melbourne, Australia.

Watson asked trustee Norton Zinder, his trusted troubleshooter, to convene a committee to make a discreet search for a successor. The problem, Norton recalled, was "how we were to convince any candidate that Jim would really give up total control." The immediate problem was a job description, a definition of the director's powers, but Watson hesitated. He would keep fund-raising and public relations, but day to day, the staff would report to the director. He was ambivalent about stepping aside. Zinder recalled, "He wanted out but he didn't want out."

At first the assumption was that Watson would leave for a job in England. But now a new factor supervened: the emerging mental illness of his son Rufus, now a teenager. So Watson abruptly changed his mind about England and focused on stepping up to the ill-defined role of president. After Bruce Alberts declined the position of director, attention focused on David Botstein, then of MIT and later of Genentech and Stanford, a leader in the molecular side of human genetics and a longtime instructor in the CSHL summer course in bacterial genetics. He was very smart— and the opposite of a shrinking violet. Even if Watson stayed, Zinder thought, Botstein "was sufficiently strong in himself that he could live with Jim. There aren't many people like that in the world."[91]

At first Botstein thought he would be alone at the top and began to think of what major focus he would choose as director, and he decided that it should be the nascent human genome idea, not the difficult-to-define field of neurobiology that Watson favored. Watson and Botstein met at Botstein's MIT office, and Jim told him he would not be going to England. Botstein replied, "Oh, okay, so the deal's off?" Watson: "Oh, no, no, no. I want you to come down anyway and be the director of the Lab and I'll be president." Botstein recalled asking, "What happens if we don't agree on something fundamental? Not whether we're going to cut down another tree, accept another estate, but something substantive, like, you know, scientific stuff?" Watson: "Oh, that won't be any problem. We can just work it out." But Botstein knew well Joe Sambrook's travails over the succession. In a spirit of compromise Botstein suggested, "Look, let's do it the way you say, but if we have a disagreement . . . we'll take it to the board and let them decide."

Jim looked at David, looked around, looked up and down, looked at his untied shoelaces—silent. He left to catch the shuttle to New York.[92] He telephoned Zinder the next day to tell him that the search was canceled. For now Jim would stay as director.

But the problem was only postponed. It came back into focus when Watson agreed to take the part-time job of starting the Human Genome Project at the National Institutes of Health. He needed to be sure of competent leaders at both ends of his New York–Washington shuttle trips. At Cold Spring Harbor, Terri Grodzicker and Richard Roberts became assistant directors, as did the Australian Bruce Stillman. Stillman had arrived at the lab as a postdoc in 1979 and risen to head a large research group focused on DNA replication. He had been co-organizer of the 1991 Symposium, on the cell cycle. In 1990, Jim announced that he had given Bruce a role in the recruiting effort, which, Watson said, "I consider the most important job in the institution. I wouldn't do it lightly."[93] It was becoming clear that Roberts, who considered that the cutthroat spirit of the 1970s had made the Lab "a terrible place to work,"[94] would not be the successor. After 20 years, he left to join New England BioLabs—just before sharing the 1993 Nobel Prize with Phillip Sharp of MIT for the discovery of split genes. Terri Grodzicker, now editor of the Lab's first journal, *Genes and Development*, stayed at CSHL.

A committee of trustees searched for a successor. Tom Maniatis of Harvard, chairman of the committee, recalled, "It became clear that we weren't going to be able to recruit a really internationally famous figure as long as Jim was calling the shots." Sambrook, another member of the committee conveyed this to Jim, whereupon he called Maniatis and said angrily, "You know, it sounds like you feel that the only solution to this director problem is for me to die."

Eventually the committee settled on Stillman. Despite skepticism about a young director operating under Watson's thumb, Watson knew that Stillman would be successful. The transfer became official at the end of 1993, soon after Watson's noisy departure from the successfully launched Human Genome Project; Watson remained on the scene as CSHL's president. Over the next few years, Jim "gave Bruce more responsibility and public exposure, but he didn't do it all at once," Maniatis recalled. "He didn't just put him out there and say, 'Go to it.' He groomed him. He worked with him." The process was "graceful and systematic."[95] Watson's role as president gradually became less day-to-day, and Stillman stepped more and more into the spotlight.

James Dewey Watson, age 10, all dressed up and posing a bit stiffly, perhaps impatiently.

Watson's grandmother Elizabeth Mitchell, whom her family called Nana, looking forward intently, in a pose Watson also assumed.

Watson's mother, Jean Mitchell Watson (1900–1957), organizer at the office and the Chicago South Side home. She made sure her son and daughter applied for scholarships.

A summertime expedition. James Watson flanked by his son, Jim, about 11, and his daughter, Elizabeth (Betty), about 9.

The canonical photo of Francis Crick and Jim in Cambridge, England, aspiring, walking, and talking.

Rosalind Franklin, crystallographer (1921–1958). Masterful, determined, a loner. A "superb experimentalist" for whom the structure of DNA was a preoccupation for only two years. She regarded Watson as an "amateur."

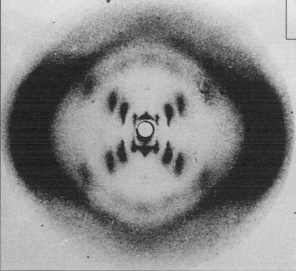

Franklin's May 1952 X-ray image of the wetter, extended, B form of DNA, shown to Watson by Maurice Wilkins in January 1953. Watson was struck by the "beautiful helix," evident from the pronounced X pattern of dots.

Soon after Watson heard that he'd won a Nobel Prize, on 18 October 1962, Watson's regular Harvard biology class was interrupted by news photographers, who demanded he put on a white lab coat (clean, starched, never used) and hold a pipette. *Rick Stafford photo. Courtesy, Harvard University Archives*

Elizabeth Lewis, 19, and James Watson, 39, on their wedding day in La Jolla on 26 March 1968.

The Watsons' sons, Duncan and Rufus, in the mid-1970s.

At the fortieth anniversary of the double helix in 1993. Watson puts his arms around two leaders of molecular biology, Sydney Brenner of Cambridge, England, and François Jacob of the Pasteur Institute in Paris. *Ed Campodonico photo.*

On his way from Cambridge to the Salk Institute in June 1997, Francis Crick attends the annual Cold Spring Harbor Symposium, at which newly discovered "split genes" are the hot topic.

Partners in rebuilding Cold Spring Harbor Laboratory, Jim and Liz stand in front of the new Neuroscience Center in a picture taken in the mid-1990s. The bell tower behind them shows the letter G for guanine. The other three faces bear the initials for adenine, cytosine, and thymine.

Watson and Bruce Stillman, his successor as
director of Cold Spring Harbor Laboratory,
flank a model of DNA in 1993.

President Bill Clinton has just presented Watson with
the National Medal of Science. Five years later, in June
2002, Watson received an honorary British knighthood.

Two of Watson's earliest colleagues in ribosome research at Harvard, David Schlessinger and Alfred Tissières, in 1982. *Herb Parsons photo.*

Nancy Hopkins of MIT. Recruited into science by Jim, veteran of the 1960s work on the lambda repressor, she went on to viruses and then zebrafish, a main organism for studying mechanisms of development.

Paul Doty, who brought Watson to Harvard, and Mark Ptashne, whom Watson encouraged in his isolation of the lambda repressor in the 1960s, in 1993.

Masayasu Nomura, a visitor in Jim's Harvard lab in the late 1950s and a lifelong student of the structure of ribosomes. Nomura served a term as a scientist-trustee of Cold Spring Harbor Laboratory.

Edward Pulling, founder of the Millbrook School, neighbor, and organizer of local donations to the lab, including the most significant—from Charles S. Robertson.

Walter Page, hard-headed trustee, mainstay of Watson's directorship—and head of the Morgan bank.

Jack Richards, CSHL's durable and efficient "in-house contractor," whom Watson respected and eulogized at his death in 2000. *Marlena Emmons photo.*

Norton Zinder of Rockefeller University, CSHL trustee and Watson's right arm in starting the Human Genome Project in 1988—as usual seeing the humor of something.

Ray Gesteland, early leader of cell biology at Cold Spring Harbor, later at the University of Utah.

Joe Sambrook, feisty emperor of tumor-virus work at Cold Spring Harbor, who served as acting director in 1983-1984. In 1985, when he left to head a department at Southwest Medical Center in Dallas, a new laboratory building was named for him.

Two veterans of the high-intensity 1970s: Tom Maniatis of Harvard, later a scientist-trustee, and Michael Botchan, later chairman of the biology department at the University of California, Berkeley.

Richard Roberts, a 20-year veteran of CSHL, gives the 1995 annual lecture to the lab's local supporters. Two years earlier, he and Phillip Sharp shared the Nobel Prize in medicine for split genes.

All together in 1979 as the renovated greenhouses of CSHL are named for Alfred Hershey. From left: Watson, Franklin Stahl, James Ebert, and the three winners of the 1969 Nobel Prize in medicine, Hershey, Max Delbrück, and Salvador Luria. All had been influential in Watson's career.

Two key advisers, pictured in 1994, for Watson's determined drive to build up the study of neuroscience at Cold Spring Harbor: Charles Stevens of the Salk Institute (left) and Eric Kandel of Columbia University (a Nobel Prize winner in 1999).

Ron McKay, later of MIT and the National Institutes of Health, pictured at the 1983 Symposium on molecular neurobiology that he co-organized with Watson.

Phillip Sharp of MIT, born in Kentucky on D-Day in 1944, collaborated with Joe Sambrook on adenoviruses at Cold Spring Harbor. A codiscoverer of split genes, Sharp shared a 1993 Nobel Prize. Noted RNA researcher, and successively head of MIT's cancer center and Biology Department, he became in 2000 founding director of MIT's new McGovern Institute of Brain Research.

David Baltimore, 1990, at the twentieth anniversary of the discovery of the reverse transcriptase enzyme, later vital in the fight against AIDS. He announced the discovery at Cold Spring Harbor and shared the 1975 Nobel Prize with Howard Temin and Renato Dulbecco. Inaugural director of the Whitehead Institute at MIT, later president of Rockefeller University and then Caltech. He received a National Medal of Science in 1999.

Douglas Hanahan, later of the University of California, San Francisco, one of many younger scientists encouraged by Watson, a sometime tennis partner.

Michael Wigler, one of the first scientists to clone cancer genes, in 1981, who later became the first CSHL scientist to be granted tenure.

Terri Grodzicker, a mainstay at CSHL and longtime editor of the lab's first journal, *Genes and Development,* sits with Bob Tjian of the University of California, Berkeley, at the 1985 dedication of Sambrook Laboratory.

In September 1985 at Cold Spring Harbor, Renato Dulbecco urges that a human genome project be started.

In June 1986, Walter Gilbert of Harvard jots down on the Grace Auditorium blackboard the anticipated $3 billion cost of the genome project—and runs into a storm of criticism.

The advisory committee of the Human Genome Project, established in 1988, gathers at CSHL's Banbury small-conference center in summer 1989 to make detailed plans. Third from left, top row, is Francis Collins, who in 1993 succeeded Watson as director of the project. Top row, at right, is David Botstein of Stanford, pioneer in genetic mapping and a prominent early skeptic about the program.

10

"HIGHER CELLS": SCIENCE AT COLD SPRING HARBOR

I quickly lose interest in a scientist if I discover that he lacks virtual monomaniacal interest in his work.
JAMES D. WATSON, 26 MAY 1982

THE PRESSURE TO KEEP IN TOUCH

One afternoon in early June 1968, Jim and Liz strolled south on Bungtown Road in Cold Spring Harbor, away from the Sandspit Beach and toward the traditional Symposium picnic. Watson wore shorts and a long-sleeved shirt, and his trademark protection from the sun, a flat Irish cap. In stunning contrast, Liz wore a bikini. The newlyweds appeared delighted with each other, basking in Watson's new task of directing Cold Spring Harbor Laboratory and the soaring sales of *The Double Helix*. The funeral of the assassinated Robert Kennedy in St. Patrick's Cathedral in New York seemed far away.

The topic of the thirty-third Symposium was the replication of DNA in microorganisms. The meeting came after dizzying years of discoveries: messenger RNA, the genetic code, details of protein synthesis, and the first genetic control mechanisms. Just ahead lay the plunge into normal and cancerous animal cells, and the surprises of the 1970s: restriction enzymes, a special mechanism for RNA viruses to reproduce, the transfer of genes from animal cells into bacteria, the development of DNA sequencing, the discovery that genes are in pieces, and the birth of the biotechnology industry.[1] Some of the surprises were achieved or announced at Cold Spring Harbor, and others were exploited there feverishly.

DNA replication remained a puzzle. Matt Meselson, who opened the 1968 Symposium, commented: "It's extraordinarily easy to describe what we would like to do, but extraordinarily hard to do it." Hunches abounded, including Walter Gilbert's "rolling circle" model, or Meselson's "pinwheel" theory of DNA duplication.[2] Irreverent questions peppered the Symposium speakers, amid much laughter. There even was mock hissing when a speaker showed a slide full of unreadable numbers and said, "As you can see . . ." In the crowded and imperfectly air-conditioned Bush Auditorium, slide after slide of sucrose gradient centrifugation results flashed on the screen, producing drowsiness and an impression that "when you've seen one sucrose gradient slide, you've seen them all." But in the traditional Cold Spring Harbor way, the largely youthful audience could escape outside to eat barbecued chicken on the grass, drink 25-cent glasses of beer in the basement of Blackford, play volleyball, or swim or canoe or sail. Much of the time, the participants talked about the unending puzzles of their field.[3]

Like many others in molecular genetics, Watson sensed that after 25 years of intensive study, the field of bacteria and their viruses was peaking as a way to illuminate how life works. Now he wanted to push on to far more complex living things—cells. Now, viruses, the familiar tools of molecular biology, would be exploited for their ability to cause cancers in cells they invaded. It was to be able to do this without inhibition that Watson needed to escape from the straightjacket of Harvard into a small, free-standing laboratory. At Harvard, he was convinced, "there was no way to do it within the confines of our building." At Cold Spring Harbor, "one could think somewhat bigger." From the start, he reflected, "I saw that the task of understanding cancer would require large team efforts involving many senior scientists with highly different backgrounds."[4] These teams, some focusing principally on viruses and DNA and others probing the architecture of normal and cancerous cells, sometimes competing and sometimes collaborating, would constantly be renewed by sending off successful workers on mature topics and bringing in young replacements for the new fields.

On these young tumor virologists and cell biologists Watson operated somewhat like the famously choleric conductor Arturo Toscanini. An uncompromising rationalist, interested more in the science than in the people who did it (although people interested him, too), he dealt like a denying father with extremely ambitious, iconoclastic researchers, ignoring what bored him and responding excitedly to what seemed interesting, occasionally exploding at seminars, exasperating the young scientists into doing their best. He advertised them to visitors at the summer conferences and courses and at frequent seminars all winter long. He made sure they were featured at his own meetings, and communicated their papers to such journals as the *Proceedings of the National Academy of Sciences*. He drafted his nascent stars to organize large and small meetings at Cold Spring Harbor, and then to edit the proceedings.

This astonishingly successful process took constant scrolling on dozens of screens. By reading and endless gossip on the telephone and during visits, Watson kept hearing of new developments from people eager to tell him something interesting. He would decide what looked important and exciting, even if it didn't pan out, and lobby fiercely with his researchers to get going or keep going on what he saw as the hot topics. More interested in moving forward than in being right, he was fascinated by surprises. He was determined that Cold Spring Harbor would be at the forefront of the most exciting fields—and the development of the scientific techniques and instruments on which those fields rested. These included restriction enzymes, fluorescent antibody staining of cells, two-dimensional gels, cloning and sequencing genes, and monoclonal antibodies—some invented at Cold Spring Harbor, and the others pushed hard there.

Organizing the summer courses and then major conferences at appropriate times was a principal means of establishing Cold Spring Harbor as an energizing center for important topics, such as the brain, the immune system, tumor viruses, the organization of the chromosome, and the humble and highly useful yeast. Watson had to make sure that key figures, including Nobel Prize winners, showed up to present papers and chair sessions, and had to make equally sure that the sharpest younger scientists were there.

His gossip network and personal encounters guided his efforts to attract talent, from places like Harvard and Berkeley and Cambridge across the sea. Once at Cold Spring Harbor, researchers found little to distract them. No teaching, no committees, and usually no commuting. A constant flow of coffee in the Blackford dining hall helped keep people awake as they habitually worked late, and stimulated conversation during the day.

Given his own highly competitive nature, it seemed natural for Watson to foster competition between the tumor virus workers who gradually filled James Laboratory and the cell biologists grouped on the other side of Bungtown Road in Demerec Laboratory. Watson also favored competition among researchers within the same buildings. To maintain a sense of competition with other labs, he dive-bombed the researchers constantly with things he had heard or read. Never barging into researchers' labs, and never putting his name on their papers, he still kept up constant pressure. This policy ran the risk of overclaiming on results, not only hindering useful collaborations between workers in Demerec and James but also fostering backbiting among scientists living at close quarters. Inevitable contests over credit for discoveries, including split genes—the greatest discovery of Watson's 25-year directorship—were messy to palliate, let alone resolve.

Another domain of anxiety was the task of attracting and keeping research grants from government, industry, foundations, and individuals.

Although he largely succeeded in shielding Cold Spring Harbor scientists from the details of fund-raising—the money just seemed to appear—Watson still insisted fanatically that grant proposals hit the right note, and that presentations for the "site visits" by review committees be rehearsed meticulously. When it came time for someone to leave, Watson worked hard on making the recommendations that would land the departing scientist a good spot, from Berkeley to the Medical Research Council lab in Cambridge, England. From the beginning, Watson was determined to "keep the average age of our staff low," and to "provide an atmosphere for the importantly unexpected." As James Feramisco, who spent 10 years as a cancer researcher at CSHL, observed, "It is not a place to grow old."[5]

The pressure to keep in touch intellectually was unrelenting. It ran alongside all the other pressures connected with restoring and building up the laboratory's physical plant and environment, and finding donors to give the institution stability, a floor to build on.

"THE NEXT BIG HARD PROBLEM"

Watson had been intrigued by viruses as causes of cancer since his graduate-student days with Luria, and he became even more interested in the late 1950s.[6] Even then, however, the field did not look amenable to experiments. But in the late 1950s and 1960s, biologists such as Harry Eagle at the National Institutes of Health and later at Albert Einstein Medical College had learned how to grow cells routinely in laboratory dishes, including cells infected by viruses. Now Watson saw a chance of understanding tumor viruses at the molecular level and then "working out a cell as if it were a Swiss watch, figuring out everything in it."[7] "There is no point to wait 10–20 years to accumulate more basic knowledge," he wrote. CSHL must not look back. "The future is what counts."[8] He could count on plenty of recruits to the new field because of what looked in retrospect like "a morale panic," a fear that cells with nuclei, or "eukaryotes," would be "fundamentally different" from the "prokaryotes," nonnucleated bacteria, on which modern biology had rested for so long.[9]

Building on the work of Ludwig Gross, Sarah Stewart, and Bernice Eddy, Renato Dulbecco at the Salk Institute in La Jolla became the "founder" of precise techniques for manipulating animal viruses.[10] The induction of cancer in animal cells by viruses had become a practical subject of research. Now, not one but two questions could be explored: How do cells become cancerous, and how do normal cells regulate the "reading" of their DNA? One could ask a specific question about how viruses transform cells, and, at the same time, go much deeper into a question of basic knowledge: the molecular genetics of cells with nuclei.[11] One could probe not only the origins of cancer but also the fundamental biology of higher cells.[12] Even though the exact nature of viruses was in doubt,[13] one could hope to find cancer genes.[14] Cancer began to seem like a more "solvable" problem than the study of development from egg to embryo to adult. Best of all, the basic questions would endure in spite of roadblocks (and there were plenty) with any particular experimental system.

Principal tools would be cells cultivated in laboratory dishes and animal viruses, following the analogy of "the great age of phage and bacterial genetics."[15] Then, the focus had been one-cell nonnucleated prokaryotes. The leap into complex, nucleated, eukaryotes, or "higher cells," could be a little less daunting than it had seemed. To the ebullient Watson, it seemed "a very small jump,"[16] although he admitted it was "the next big hard problem" after DNA.[17] According to Heiner Westphal, who had trained as a medical doctor in Germany and moved from Dulbecco's lab to Watson's, "The Dulbecco lab had shown that there was a science to be reckoned with—just emerging—so Jim Watson didn't take so much of a gamble."[18] Watson reflected, "Viruses gave us a logical handle to use our brains, as opposed to our emotions, in deciding what to do next." He later wrote, "These tiny packages of proteins and genes somehow contained the necessary information to generate cancer."[19]

Major aims of the twin gamble on cells and viruses were to uncover genes for cancer-causing proteins, later called "oncogenes," and "get at the essence of cancer at the molecular level."[20] This hope was realized in the late 1970s and early 1980s by Harold Varmus and Michael Bishop at the University of California at San Francisco, who won the Nobel Prize in Medicine for their work on oncogenes, and Robert Weinberg at MIT, and Michael Wigler at Cold Spring Harbor, who first cloned cancer genes.[21] Wigler, who told one journalist, "Actually, I don't do anything but work," became a member of the National Academy of Sciences and the first scientist at Cold Spring Harbor to be granted life tenure. In 1985 he was named an American Cancer Society professor and received a seven-year, $8 million Outstanding Investigator grant from the National Cancer Institute.[22] Further achievements on cancer genes at CSHL included Earl Ruley's finding in 1983 that the oncogene called RAS needed a second oncogene to work, which Watson called "a major coup."[23] In 1988, Ed Harlow, later of Massachusetts General Hospital, succeeded in making a connection between genes promoting and retarding cancer, oncogenes and anti-oncogenes, which was "a watershed observation" in Watson's view.[24]

As usual with molecular biologists, as Francis Crick remarked, this strategy of focusing on the simplest possible phenomena would take advantage of the fact that biological mechanisms, such as amino acids or DNA bases or cell membranes, are widely distributed in the living world. Hence, something learned in a simple system could apply to humans.[25]

Dulbecco's lab started with a tiny DNA virus—one whose genes are embodied in DNA—called polyoma and showed that the viruses could become "proviruses." In this situation, the DNA of the cell-invading virus formed permanent bonds with the cell's DNA. The genes for normal virus growth (and destruction of the cell) were repressed. Instead the virus transformed the cell into an uncontrollably multiplying tumor cell. Apparently this happened because the host cell was able to turn off some virus genes. Thus the study of proviruses went beyond the confines of cancer research and "became a tool for studying regulation of DNA transcription in animal cells."[26]

Thus, cancer research, hitherto so focused on treating existing illness, would extend into "pure science," Watson reflected. "The solution to cancer lies in a fuller understanding of what happens after viruses enter cells."[27] In the drive to make Cold Spring Harbor a "major center of basic cancer research and training,"[28] the lab's summer course on "higher cells" would be divided into three courses: animal-cell culture, cell transformation by tumor viruses, and animal viruses. There would be year-round research on the molecular biology of tumor viruses in James Laboratory and a companion effort in cell biology.[29] The swelling flow of federal money to fight cancer would nourish a "tiny frontier encampment full of upstarts willing to take big risks."[30]

Among Dulbecco's colleagues on the provirus experiments at the Salk Institute were Joe Sambrook, a wiry, feisty, even flamboyant scientist from Britain; the medically trained Heiner Westphal; and Carel Mulder, a microbiologist who trained at Leiden in the Netherlands and afterward worked in Paul Doty's lab at Harvard. In the summer 1968 animal-virus course at CSHL, Sambrook was the lecturer who excited Watson most.[31] Watson, "always curious about what's going on," soon "dropped by" the Salk, a center of the boom in "eukaryotic anything,"[32] to recruit Sambrook and his colleagues Westphal and Mulder. It was a challenge, because Sambrook was "one of the younger stars" in Dulbecco's lab at the well-equipped, "almost new" Salk Institute.[33] But Sambrook said yes. To him, the animal viruses looked like "the safest of a bad bunch" for confirming that "the truths previously discovered in prokaryotes really were universal."[34] The plan was for Dulbecco to keep going on polyoma and for the CSHL group to use another DNA virus, the very tiny simian virus 40 (SV40). According to Mulder, Dulbecco was not resentful. "That's not the way Dulbecco is. . . . He was willing to extend every help." The more people working on tumor viruses, the better.[35]

Sambrook's strong talents for organizing and writing were put to work at once, and he established himself early as central to Watson's hopes for Cold Spring Harbor. Even before leaving La Jolla in 1969, Sambrook wrote the crucial proposal for a $1.6 million grant from the National

Cancer Institute, and once at Cold Spring Harbor he became for many years a dominant force. To be sure, he as well as Watson knew that repeated budget increases for cancer research were generating "far more federal money . . . than could be used effectively by scientists working on this disease." Biochemists who had shunned cancer as a graveyard of scientific reputations could afford to get into it.[36] According to Watson, Sambrook was taking "virtually no financial risk" in coming to Cold Spring Harbor, even though the Laboratory's endowment was only $20,000 in 1969. Joe found it possible to attract other scientists with more interest in "important science [than] long-term job security."[37]

Sambrook had a will to power and made James Laboratory into his own empire. He was not shy about standing up to Watson, with whom he spent 17 "fairly tumultuous" years.[38] Every now and then, Jim would "fire" Joe and then unfire him, and Joe repeatedly considered leaving, and then didn't. When Jim imposed bright young researchers from Harvard on Joe, he ignored them—effectively leaving them free to work on what they wanted. On one occasion, when Sambrook thought he was on the point of joining the Imperial Cancer Research Fund in London, he informed Keith Burridge, a new recruit, about this, warning him that he would be on his own.[38]

To Joe and his associates, the work was expensive and slow,[40] almost a leap into molasses. Westphal, for example, found the months and months of setting up the work all over again at Cold Spring Harbor were not easy. An annoying problem was contamination by molds.[41] To be sure, the research used extremely simple viruses like SV40, with just a few genes,[42] but scientists wanted to get at individual genes. New tools were needed to get free of dependence on whole viruses.[43] And these soon appeared. One of the big surprises of the 1970s was the class of enzymes that could recognize particular short sequences in DNA, four to eight base-pairs long, and chop the DNA up into more manageable pieces whose sequences could be worked out. For the discovery of these "restriction enzymes," Werner Arber of Basel and Daniel Nathans and Hamilton Smith of Johns Hopkins shared the 1978 Nobel Prize in Medicine.

Soon, the restriction enzymes received their first intensive use, in studying a virus easier to work with than SV40. It was adenovirus, which causes common colds as well as cancer. A scientist visiting CSHL, Ulf Pettersson of the Uppsala laboratory of Lennart Philipson, brought adenovirus samples with him and offered them for the tumor virus research. Fewer labs were focusing on adenovirus than on SV40. Besides, Westphal recalled, "Adeno was so simple. . . . You could get decent quantities of it, and quantities mattered in those days very much. So it was a godsend."[44] To be sure, adenovirus had about seven times as much DNA as SV40, but the restriction enzymes could cut it up into pieces little bigger than SV40, and one could try to decipher the short sequences that the enzymes recognized.[45] CSHL jumped into adenovirus with both feet.

Colleagues at Cold Spring Harbor held views of Joe that balanced admiration with criticism. Douglas Hanahan, at the lab in the late 1970s and 1980s, said Sambrook "is the great devil's advocate and loves to tease people scientifically."[46] According to Robert Crouch, a postdoc at CSHL in 1969–71, "Joe is a provocateur. And he does it intentionally just to make people think, [to] get [them] to do things." He "drove people pretty hard."[47]

Joe was "a very passionate and emotional guy," according to Tom Maniatis, who worked at Cold Spring Harbor in the mid-1970s. He was "a very clear thinker" who enjoyed writing "enormously," but "his real strength is organizational. He gets things done." Although he "thought about things," he really was "an action person." The "flamboyant" Sambrook could be "very nasty," with first responses to an idea or a situation that were "quite violent." Underneath, though, he was "perfectly reasonable and kind and understanding." Joe succeeded in putting together "a very effective DNA tumor virus group that really dominated the field at that time."[48]

In Michael Botchan's recollection, Sambrook had limitations. "Joe had great strengths, but he was not the sort of person who understood other people. . . . He didn't have the sense that Watson had for who was good and who wasn't." Furthermore, although Sambrook was "great at solving a puzzle," he "lacked a certain kind of boldness . . . a certain vision for big problems."

Mike Botchan had been recruited from Berkeley in 1972 by a typical telephone call from Watson. "Do whatever you want," Watson said before demanding, "When are you coming? You got to come. We need you now." Botchan was flattered that "this almost mythological figure" had called him. He later concluded that Watson "just wanted a critical mass," and had the intuition that Botchan "was brazen enough and young enough . . . to fill in gaps that were missing," and was ready to "duke it out."

Sambrook's increasing dominance irritated Mulder and Westphal, who, Mike said, "saw themselves as equals [to Sambrook]. . . . They didn't see this as Joe's lab at all." By contrast, Botchan enjoyed what he saw as the ambiguity about who was boss; he thrived on "no boundaries, shared facilities, different personalities," with Watson as "this overarching figure . . . in the shadows."

He stayed for seven years, returning to Berkeley in 1979 because "I saw myself as a professor. . . . Much as I loved Cold Spring Harbor, there was something missing there. . . . [There was] a certain narrowness. . . . I like to think about lots of things in an unfocused way." At Berkeley, if he felt like it, he could drop into a class on medieval history.[49]

At Cold Spring Harbor, Botchan did not begin either on a Sambrook project or on DNA-cutting restriction enzymes, the prevailing enthusiasm in Demerec Laboratory. He regarded the enzymes as "an enabling technology," although he did use them in the work he chose: the mechanism of transformation of cells into tumor cells by viruses that integrated their DNA into the cells' DNA. He asked himself: "How did the virus integrate? . . . What did the integration patterns look like? What was the arrangement of viral genes in the [cell's] chromosome?" Eventually, exploiting a new blotting technique invented in Edinburgh by Ed Southern, Mike found that SV40's way of getting into the cell was not "very fancy . . . basically a very sloppy mechanism." Most of the virus's DNA was scattered around the host cell's chromosomes—except for its "early genes," which carried the blueprint for taking over.

Despite his different focus, Botchan admired Sambrook and collaborated with him through most of the 1970s. In 1978, he and Joe went to the Imperial Cancer Research Fund in England to do an experiment that was restricted in the United States as a result of rules regulating recombinant-DNA research (discussed in the next chapter). Botchan recalled that they were the first to get clones—and hence large quantities—of viral DNA integrated with that of its host cell. Sambrook showed "technical brilliance" in the development of an experimental method called "single copy blotting" into a tool for obtaining maps of cellular DNA. Courses at Cold Spring Harbor, attended by scientists from both CSHL and other institutions, helped spread this method far and wide.[50] An example was the cloning course Jim recruited Tom Maniatis to teach in 1980. Very soon afterward, in 1982, Sambrook, Maniatis, and Edward Fritsch produced the first edition of the manual *Molecular Cloning*, for which there were 5,000 advance orders. Sales totaled 18,000 in the first year and eventually reached 60,000. Thus, in the courses, new technologies were inculcated at CSHL and elsewhere, and then, in Maniatis's view, the books were used as "the most efficient way to disseminate the methods widely. This sequence of courses leading to books was a subtle but pervasive aspect of Watson's intellectual leadership and influence.[51]

One of Joe's collaborators in 1971–74 was Phillip Sharp, who in 1993 shared a Nobel Prize in Medicine with Roberts. Sharp had come to Cold Spring Harbor to work on what seemed a naïve question: "What was the structure of genes in a vertebrate cell?" Sharp was a graduate of Union College in his native Kentucky and the University of Illinois. He had then gone to Caltech as a postdoc in Norman Davidson's lab, but now he needed another postdoctoral fellowship. Jerome Vinograd at Caltech told Sharp, "You could learn a lot from Joe Sambrook." Sharp recalled, "It was when cell biology was opening and viral approaches and tumor biology were the ways to do it." Joe, a more senior scientist with much knowledge of virology and cell biology, struck Phil as "a very stimulating, interesting guy," and the two collaborated on projects that could use Sharp's complementary skills in electron microscopy and in hybridizing nucleic acid strands.[52] Because of Sharp's interest in how transcription of DNA into RNA copies started and stopped, he and Sambrook worked together making physical maps of the distribution of genes of adenoviruses as they became the lab's virus of choice.[53] With the help of restriction enzymes, Sharp, William Sugden, and Sambrook combined the use of staining with ethidium bromide and electrophoresis with agarose gels to create a "powerful new way" to separate different-size DNA fragments.[54]

RESTRICTION ENZYMES

The center of cell biology at Cold Spring Harbor was Demerec Laboratory, which was in a kind of rivalry with the "DNA racists" in James. For nearly a decade, Ray Gesteland was the leading researcher in Demerec. Formerly a Harvard graduate student, Gesteland had moved to Cold Spring Harbor in the late sixties, at the end of John Cairns's directorship. His main focus was the synthesis of protein in the test tube; he labeled the amino acids of the proteins with radioactive sulfur to study the translation of nucleic acids into proteins.[55]

Robert Crouch, who had trained at the University of Illinois, had come to Cold Spring Harbor when Gesteland asked him a simple question at a conference in Norway (Crouch was a lecturer at the University of Bergen): "Why don't you come to Cold Spring Harbor? Jim's taking over." Crouch said yes. Gesteland told Crouch he could work on what he wanted, and Crouch chose a protein involved in stopping the transcription of DNA into RNA, called by the Greek letter *rho*. Later, Sambrook's taunts led Crouch to work with a protein that degraded RNA that had formed hybrid links to DNA—ribonuclease H, and this problem dominated his later career. It was a heady experience and there were others during Crouch's stint at CSHL. Like the other young scientists at Cold Spring Harbor—"Here I was, just a post-doc"—Crouch was put to work on meetings, helping to recruit session chairmen for the annual bacteriophage meeting.[56]

Gesteland was quite ready to stand up to Watson. On one occasion Watson, concerned about the renewal two years hence of the grant from the National Cancer Institute, took Mulder aside at a party. Because the restriction enzymes Mulder had been working on were from bacteria, Watson couldn't see "how we can fit that into our cancer grant." This was before Mulder had used one of them to cut up cancer-causing adenovirus. Shocked, Mulder did not know whether Watson had missed "the vision of what these things could do" in cancer research or whether "he was trying to challenge me." When Mulder told Gesteland, the latter recalled a similar injunction to Dick Burgess at Harvard to stop working on an enzyme. Gesteland said, "Oh, Carel, it's happening again. . . . Don't pay attention to him."[57]

Demerec Lab was a hotbed of work on the new restriction enzymes. Soon after they were discovered, one of the discoverers, Daniel Nathans, lectured at Harvard Medical School about how these enzymes could chop up SV40. One who heard him was Richard Roberts, who had received his doctorate in organic chemistry from the University of Sheffield in England in 1969 and then joined the Harvard laboratory of Jack Strominger, a longtime student of the action of penicillin. At Harvard Roberts concentrated on the transfer RNAs involved in synthesizing the membranes of cells. While he was working at the Harvard lab he visited Fred Sanger's lab in Cambridge, England, to learn techniques for sequencing bases in RNA. Roberts later described his enthusiasm for sequencing: "When studying a new compound, the first and most important thing is to determine its detailed molecular structure. For a molecular biologist, that usually means determining some DNA sequence, since an accurate knowledge of sequence will then allow the proper design of experiments to examine [protein] function." The potential of restriction enzymes had "an immense impact" on Roberts. Soon, at Watson's behest, Roberts moved to Cold Spring Harbor with the aim of sequencing tumor viruses. At CSHL he found Joe Sambrook and others using restriction enzymes to study adenovirus. They encouraged Roberts to find more of the enzymes.[58]

Installed in Demerec late in 1972, Roberts soon clashed with Watson's eagerness for competition. To Roberts, the goal was to find noncompetitive fields where entry by others would be difficult. Roberts joined others in suggesting that instead of trying to sequence SV40 in competition with at least two other scientists, which struck him as "foolish," Cold Spring Harbor should work on another virus, and adenovirus was available.

Most attractive to Roberts was the idea—which he carried out—of building the biggest inventory of existing and newly discovered restriction enzymes—even if Watson and others at CSHL thought of this as, said Roberts, "a non-intellectual activity." Indeed, the disdain went further. When in 1974 Roberts proposed setting up a small company to sell restriction enzymes, Watson was dismissive. In Roberts's words, the line was, "Why would anybody want to do that? We're a research institution here. We have no business in the commercial world." To Roberts, "This was stupid." Watson later acknowledged that he had made a big mistake.

Roberts then tried to get NIH to set up such a facility to distribute restriction enzymes. This failed, but Roberts heard of a company, New England BioLabs, that was getting into the business. He became the company's principal consultant and in 1992 he left Cold Spring Harbor and went to BioLabs full-time as research director. A few months later, Watson haled Roberts before the CSHL trustees to be rebuked for his private-industry affiliation. After the meeting, Walter Page quietly told Roberts not to "take it too much to heart."[59]

In after years, Watson had to acknowledge how important Roberts' initiative had been, with a bow to Mulder.

> By 1972 an important component of this work and in fact of most of the research at the Laboratory was the newly discovered restriction enzymes that cut DNA molecules at very specific base sequences.
>
> Carel Mulder was the first staff scientist to recognize their importance, but it was Richard Roberts working in his new facility in Demerec Laboratory who screened diverse groups of bacteria for new enzymes and made his reputation through their isolation and characterization. By 1974 a total of 17 new enzymes were isolated there, 11 of which showed specificities never before seen. Soon more than half of all the known restriction enzymes had been discovered at Cold Spring Harbor and these were available for use not only by Laboratory scientists, but also by a steady stream of visitors from all over the United States and Europe.[60]

RETROVIRUSES

When adenoviruses were emerging as attractive tools of tumor virus research, a scientific explosion occurred that might have diverted Cold Spring Harbor in another direction. This was the independent discovery in 1970 by Howard Temin of the University of Wisconsin and David Baltimore of MIT that when they infected cells, some RNA viruses—whose genes are carried in RNA—carried with them an enzyme for making DNA copies of the RNA, so that the newly made complementary DNA could take over the cell. Appropriately, because the process involved the reverse of the usual doctrine of DNA makes RNA makes protein, the enzyme acquired the name of reverse transcriptase. Viruses that possessed the enzyme, like the human immunodeficiency viruses (HIVs) that were discovered in the 1980s, were called retroviruses.

Temin announced his finding at a medical conference in Houston, and two weeks later Baltimore reported on reverse transcriptase at the 1970 Cold Spring Harbor Symposium, whose subject was the transcription of nucleic acids. In Watson's view, reverse transcriptase broke open "the previously very murky world of RNA viruses."[61] The discovery aroused immediate hopes that interfering with reverse transcriptase might be useful for fighting cancer. Soon afterward, six more viruses with the enzyme were reported, and by the end of the year the total had reached 20. Another study looked for reverse transcriptases in 40 viruses that did not cause cancer, and found it in only one.[62] As Watson later remarked, these hopes of medical applications were not realized,[63] but reverse transcriptase did become a scientific workhorse of the newly opening era of restriction enzymes and recombinant DNA and sequencing. Medical utility came dramatically in the 1980s and 1990s, when drugs attacking the enzyme became a principal tool in controlling AIDS in HIV-infected patients.

In the early 1970s, Watson may have been tempted by the hot new field of retroviruses, but he made the strategic calculation to stick with DNA viruses like adeno. Westphal recalled that "RNA viruses were very hard to handle."[64] DNA still looked like "the central theme of understanding biology." It was before biologists learned to clone genes, and many others were jumping into retroviruses.[65] So adeno became the primary focus for asking the basic questions of how normal cells and cancer cells work, questions that would lead to future achievements like cloning cancer genes—and surprises like the discovery of genes in pieces.

JIM DROPS IN

As at Harvard, Watson found his scientists by a seemingly haphazard process; in this he was aided by his brightening aura. To energize his new venture at Cold Spring Harbor, he brought many researchers down from Harvard, sometimes by heavy pressure. Prompted by direct contact and word from others he trusted, he would scout other labs, such as the Salk, and pull people across the country. To recruit Westphal from Dulbecco's Salk Institute lab, Watson got him to Cold Spring Harbor for the summer of 1969, where other scientists chimed in singing the recruiting chorus. The barn where Westphal was staying had a grand piano in the old hayloft, and Max Delbrück came to join in the music by playing his recorder. Westphal told him he had attractive offers to return to Germany. Should he go back? Delbrück: "Well, that depends on whether you want to do science or not." Westphal reflected that there was nothing in Europe like the tumor-virus work, and stayed in America for the rest of his career.[66]

Watson's recruiting of Phil Sharp was typical. Knowing that Sharp would soon finish at Caltech, Watson telephoned him and said, "I got to know in a week. I'm going to Europe to try to recruit someone else if you don't take the job." Phil recalled that Jim, painting from "a very big palette" and having the "broadest perspective of people," wanted to have around him personalities "who are gregarious, articulate, bright, energetic [and] have a sense of style," people eager for "center stage" [and] "aggressive in terms of intellectual activities." Such researchers "move fast. They know what's going on. They keep abreast of the latest and the new." People came to the "high-visibility celebrity culture" of Cold Spring Harbor just because of who Watson was "and how visible he was." Jim was "one of the absolute doyens of science."[67]

The impulsive and informal atmosphere continued after the recruits showed up at CSHL and went to work, usually being left to their own devices. There was, for example, the "court" in Blackford dining hall around 10 A.M. most days, where scientists were expected to interact with each other and with Jim.[68] Another favorite Watson technique was to drop in, often late in the evening, at scientists' offices—many of them along the corridor leading to his own office in James. While Sharp collaborated with Sambrook on adenovirus, "Jim would come in from time to time and ask me what was going on." Watson loved to gossip. "It's the intellectual forefront of science he's interested in," said Sharp, "but if there's a little scandal around the sides . . . [it] makes a better story."

Around nine o'clock in the evenings, after dinner in his campus apartment with his wife and children, Sharp would come back to his office. Perhaps once a week, Watson would walk down the hall and stop at Sharp's open door. "He'd look at you for a while and you'd feel compelled to strike up conversations. The only thing you had in common was what you were doing." If Watson found this interesting, "he'd talk about it for a while. Occasionally he would have a bit of news about a nice scientific advance from someone else."[69]

Westphal recalled that Watson "would just pop up in my little office, which actually was a very pretty office." The visit was not casual. Watson was always in a hurry. "You could see it in his eyes. Because he would rarely look into your eyes. He had a point. He wanted this or that, or he had an idea. . . There was always a purpose behind it." Westphal found Watson "very supportive," even though he was just an M.D. who had done one successful experiment at the Salk Institute and was struggling to get his tumor-virus experiments going at CSHL. "Jim impressed me . . . by his directness, which is proverbial, and his very sharp insight into things. Very brief, straight to the point, and always something for me to think about."[70]

Botchan thought that "descending" was a good word for Watson's unscheduled drop-ins. He recalled, "I always felt his presence. . . . Jim's expectations were always high. People rise to the expectations of those around them who respect them and who they respect. He knew how to do that, in a very subtle way." Botchan reflected, "I actually love the guy. I love his personality. I love his idiosyncrasies. He was cruel to me at times. He fired me a couple of times. He said nasty things to me—but I always deserved them. Even at the time, I realized that he was just being honest with me." Watson was always pushing Botchan "to solve this problem and do it in a creative, elegant way. . . . He didn't expect me to be a good boy."

When Botchan arrived at Cold Spring Harbor, he recalled, "I was a very undisciplined kid." He was impressed by how hard Jim worked; Watson's office was just down the hall, and there he spent "hours and hours . . . writing and reading, talking on the phone. He would get there very early and he would have an amazing ability to concentrate and focus." Watching Watson, Botchan "picked up some discipline." When something wasn't working, Watson would tell him not to "throw good money after bad. . . . 'Forget about it.' He would make one feel that you could do important things just by paying attention."[71]

Keith Burridge came to CSHL from Sydney Brenner's lab in Cambridge, England. Sydney and Jim were frequently in touch. Burridge was impressed that Watson, a leader with so many administrative cares, "could be so excited at the simple sort of data." But Burridge found that Watson "didn't make small talk very easily. . . . We often would stand and both look at our feet and there would be long pauses. I found him to be a difficult person to have a casual conversation with." But Watson kept up with the literature and could talk about "the big picture," and would ask, "Have you seen this?" Impatient for results, he would ask, "What's new today?"

Watson's traditional sink-or-swim policy was in full force. During his six years at CSHL, Burridge recalled, "Jim gave me enormous freedom and in some ways very little advice . . . [treating me] almost like a senior investigator," when he was just a postdoc. "I probably did need more advice and guidance. I think sometimes I floundered." Watson's suggestions often were very general: "We've got to get more pictures." There seemed to be "no particular agenda" for his work, Burridge recalled. Jim was keen for pictures of the proteins that form the cell's "skeleton," Burridge recalled, "but he wasn't going to really direct me." Aware that a scientist could be a star one minute and out of favor the next, Burridge took care to leave CSHL when his work was going particularly well.[72]

The cell biologist Robert Pollack, who came to CSHL in 1971 and set up "a deli on cells" for the cancer work, also found that it was "never easy" to start a conversation with Watson. "Sort of like deciding to have your teeth cleaned." He added that both he and Watson were "thinking all the time. When he had to see me or I had to see him, it was an agenda item, which took away all pleasure."

Watson was "a student of everyone," in Pollack's opinion. "Whatever weirdness happened, and there was plenty of it, he was always a student, always ready to hear a new idea, always ready to get the picture, always ready to be excited. . . . The people who think of him as a complete goof do not realize they're missing the point. . . . The goofiness of the guy is the downside of his absolute enthusiasm for ideas, his absolute blind sincerity. He has no tact, no grace, but he is not a hypocrite—about anything."[73]

Taking temporary refuge at Cold Spring Harbor during the recombinant-DNA fight of the 1970s, Maniatis felt that his unpredictable conversations with Watson taught him that "having a broad view of things was very important and could lead to insights in your own work in a very unexpected way." But Maniatis also recalled that he was "intimidated. . . . I always wondered whether I was, you know, reacting in the right way or saying the right things, or looking stupid, or whatever." Watson's usual reaction to something he thought boring or wrong was nonverbal. "He usually just turns away." One could never tell if he disagreed, or had something more urgent to do. Still, "The fact that he continued to come back and engage in conversation led to the conclusion that it must not have been so bad."

Maniatis was impressed that Watson, always dominated by the theme of genes and human disease, turned out to be "a master manager" with a rare ability to deal with "these quirky personalities of scientists, and egos and so on." It helped that Watson, by very hard work, honed his "incredibly good taste" in science. "He gets to know who to believe and who not to believe, who to respect, who knows what's interesting." Watson was ready to take risks on relatively untried people, and able to get the "very technology-oriented" Rich Roberts and the "virologist and tumor biologist" Sambrook to complement each other.[74]

Roberts's view was less favorable. He objected to Watson's fostering of competition, even though it influenced his own greatest discovery—split genes. "Jim thinks competition is good, inherently good." But Roberts recalled "the downside of competition. . . . [It] is just too high a price to pay for the good things that come out of it." Watson's criticisms, Roberts claimed, "tend to be very destructive. . . . When I criticize people, I always try to do exactly the opposite [of Watson]." When Jim heard that "you're going to do something that he doesn't want you to do," he did his utmost to "make that sound so unattractive that even after a while you begin to think to yourself you don't want to do it."[75] Despite his many discontents, Roberts worked at Cold Spring Harbor for 20 years.

LIGHTING UP THE "CYTOSKELETON"

Many at Cold Spring Harbor noticed the highly visual nature of Jim's appreciation for research results. He wanted to "see" the transformation of cells into tumors, through pictures showing changes in the contents and architecture of cells. Nowhere was this more evident than in his reaction to the method known as immunofluorescence that was developed in the mid-1970s by the researcher Klaus Weber from Harvard and Elias

Lazarides, a colleague who came down to CSHL with him. In this method, fluorescent antibodies were made to bind to specific proteins that make up the superstructure, which can be thought of as the muscles and bones of the cell, the so-called "cytoskeleton." The method unexpectedly showed that muscle proteins like actin were also present in many other types of cells. With the fluorescent antibodies, the cytoskeleton lighted up and Jim could see it.[76]

Watson was equally intrigued when Lan Bo Chen, who received his Ph.D. from MIT in 1975 and spent the next two years at CSHL, used such techniques to find a protein called fibronectin on the outer membrane of cells. Fibronectin promotes cells' sticking to each other. Cell "adhesion" was important. Normal cells stick to each other and keep each other from multiplying. But cancer cells don't stick together and multiply uncontrollably. They also are deficient in fibronectin, and Watson excitedly thought that the "glue" that sticks cells together had been found.[77]

He thought so much of immunofluorescence that when Weber and Lazarides left Cold Spring Harbor, consumed with anger at each other over who should get credit for the discovery, Watson insisted that the newly arrived Keith Burridge, who wanted to switch from cytoskeleton research into tumor viruses, stick with the cytoskeleton and transfer from the virus center in James down to the cell biology center in Demerec.

This excitement with the visual extended to the "two-dimensional gel" technique for gaining an overview of the population of proteins in a particular type of cell, to which Jim Garrels devoted himself for 17 years. Because the analysis of the gel sheets required heavy computer work from the beginning, and Garrels had the knack of finding grant money for "bioinformatics," Watson championed Garrels for many years against the carping of lab colleagues who thought the work was too "technological."[78]

By 1980, Watson's impact on CSHL was manifest. To a visiting reporter, Monte Davis of *Discover* magazine, Watson summed up his approach: "You only start something if it seems you can do it well." He made this comment, the reporter wrote, as he unfolded "from the driver's seat of his dusty green station wagon." Watson added, "You wouldn't want to hire someone unless you were sure they'd be doing something new." Asked how he spotted talent, he replied, "I don't know how I would describe it. . . . You read things in the journals, hear things at meetings. . . . You get a sense of who's stirring things up."

Davis gave this description of Jim Watson: "His movements suggest a caricature of an Oxford don: he walks loose-kneed, toes out, and his arm seems to have extra joints as he points to the site of a new lab annex." His style "is that of a commando operation: small, aggressive, narrowly focused, quick, and opportunistic." Many in biology steered their best students to Watson, one unnamed observer at the Lab said. They "are delighted to help him as long as they don't have to have anything to do with him." Researchers at CSHL, who usually stayed only three years, spoke of their unusual autonomy. One said, "Jim doesn't know the details of my experiments, but he knows exactly why I'm doing them, and he'll know if *I* don't. You can't get far on hot air around here."[79] Another visiting journalist was told that Watson's philosophy was, "Get them young, give them no responsibility [for committees, teaching, or administration], and either work them until they drop or you drop them." One advantage of working at Cold Spring Harbor, according to Tom Maniatis, was to "establish a direction" in one's research.[80]

The comment about friendliness at a distance had echoes among scientists at the lab. David Zipser, a molecular biologist and veteran of Watson's Harvard lab, who later became interested in the computational side of neurobiology and moved to the University of California at San Diego to study it, was at CSHL from 1970 to 1982. During that time, his wife, Birgit, began her work in neurobiology at CSHL.

Looking back 20 years later, Zipser remembered Watson coldly. He acknowledged that nastiness was not Jim's "distinguishing characteristic. . . . He could be generous or selfish . . . sort of nice or incredibly nasty." Jim, in David's eyes, "is a genius, and like most geniuses his special gift was restricted to a certain limited sphere. In his case the sphere is self-promotion. He is a genius at making himself an important person. And what's more, he is extremely highly motivated, and interested in his topic." Jim's focus on "power and self-promotion" made him very different from Francis Crick. "When Crick talked about science, the issue was science, and when Jim talked about science the issue was Jim." Zipser wondered at Watson's persuasiveness as an administrator. "He was able to convince the people he had to convince, whether [it was] the head of Morgan . . . or whether it was the maid that cleaned the rooms, that they should do what . . . he wanted them to do [which] was something . . . that would promote Jim Watson. . . . I have no idea how he did it. I don't know why they gave him the time of day. . . . He understands power, he knows what it is and how to get it, how to use it and get more." In Zipser's opinion, Watson had left Harvard "because he realized he wasn't going to put it over on those guys."[81]

"EVERYBODY IN THE WORLD CAME THROUGH"

Even at the beginning of Watson's directorship, scientists at Cold Spring Harbor found the summers "almost overwhelming," with meetings that seemed to go on 24 hours a day, seven days a week. "Everybody would come to you. . . . You were on display. . . . I could hear the latest and just be on top of everything."[82] "You could go over and hear the latest, greatest thing that was going on," to the point of saturation. The year-round researchers could run into the visiting conference attendees on the lawn having lunch or drinking cocktails, or over a beer at night. The visitors would come over to the year-rounders' labs. "You got to know everyone [and] everyone gets to know you."[83] All this made Cold Spring Harbor a "really very stimulating, fruitful place. . . . It was Grand Central Station."[84] The meetings were so valuable as a way to find out what was happening that the "privileged" scientists working at Cold Spring Harbor "didn't need the phone."[85]

Watson was not always polite to visitors. He would, as was his wont, take a seat at the back of the lecture hall and scan the *New York Times* while listening. "If something either struck him as particularly interesting or if somebody raised his anger," Westphal recalled, "he would just burst out and say something."[86] On one occasion he interrupted a scientist named French Anderson, later a famous advocate of gene therapy. "French? When are you ever going to do anything significant?" Anderson, evidently shaken, replied that he and Watson had had this discussion before and didn't need to continue it during his lecture. Robert Crouch thought it was "one of most rude things I've ever heard in my life."[87]

Francis Crick was equally rude. At the 1977 Symposium, on the structure of the chromosome, Crick, relaxing in a neighboring room where he could see the proceedings on closed-circuit television, was watching a young researcher present a model. Crick and others with him snickered. The 61-year-old Crick strode over to the back of the lecture hall and interrupted the talk, exclaiming that amateurs should leave this sort of model building to the professionals. The speaker was "devastated." The next evening Crick got his comeuppance. Charles Weissman, a sober Swiss scientist who usually wore business attire when lecturing, came up to the podium wearing an apparently blank white T-shirt. A few minutes into his talk, without any comment, Weissman turned to write on the blackboard behind him. The back of his T-shirt carried an inscription that Maniatis remembered as "huge." It read: AMATEUR.[88]

At some point, however, saturation with the nonstop cascade of new research results would set in, and "you would just have to get up and leave," said Crouch.[89] Robert Tjian, who worked at Cold Spring Harbor about ten years later, recalled, "I loved it the first summer . . . because it was like a kid in a candy store. . . . But by the second summer I was a veteran. I knew this was just going to eat my time up, and if I wanted to get something done, I better stay in the lab. . . . My brain is too small to have more information than I need."[90] For Westphal, there was "a real conflict. You were constantly torn between running your experiment or going to an interesting lecture." Adding to the pressure were the clusters of scientists from all over the world "all talking about the same experiment competitively." A researcher who thought his work contained something "glorious," would often learn that "nine other people" had also done it.[91]

Richard Gelinas, a participant in the 1977 split-genes discovery, put it this way: "The intellectual atmosphere around here during the summer is almost too intense. You want to do your work but here are these world experts who've just gotten off the plane. You have a bunch of highly motivated people, who've been waiting for a year to disgorge their latest results. . . . It can leave you emotionally exhausted."[92]

The annual Cold Spring Harbor Symposium on Quantitative Biology, which had been started back in 1933, became Watson's special care between 1968 and 1993. The meetings focused on a different theme each year, returning at intervals to such topics as the brain, the immune system, tumor viruses, and the organization and replication of genetic material. Watson was determined to make each Symposium as important and up-to-date as possible, to cover "an exploding phase of biology" and bring together "most of the key practitioners."[93] He sought lists of participants from leading researchers on the Symposium topic, including many past and future Nobel Prize winners, and used CSHL scientists as editors of the resulting volumes, which seemed to grow thicker each year. In his prefaces to these, Watson often wrote of "spectacular advances," sudden changes in methods, discoveries, and insights, giving hope that "fog" could be dispelled.[94] There were last-minute additions to the list of presentations, such as the report in 1970 of the just-made discovery of reverse transcriptase by Baltimore and Temin. To open and close these meetings, Watson strove to recruit eminent, not necessarily senior, people in a particular field. Baltimore, the star of the 1970 Symposium, was the closing speaker who summarized the 1974 Symposium, on tumor viruses, just a year before he received the Nobel Prize with Temin and Dulbecco.

The Symposium and a growing number of other conferences each summer provided Jim with an education. They fed his ability "to find out what's happening," so that he would know "what questions to ask and to whom," and be up to date. Maniatis recalled: "He would come around the first time at the meeting, talk to who he thought were the key people, get a sense for what they thought were the most interesting things happening at the meeting, and make sure that he was at the session when the interesting papers were presented." He popped in, stood at the back, heard the key talk, and then left, often to talk to participants taking a breather outside. He seemed to be a fast learner. During a meeting on neurobiology, in which he had little background, Watson sat at lunch with several scientists and, said Maniatis, "basically summarized what he thought were the most important issues. Very clearly."[95]

CLOSE QUARTERS

Many Cold Spring Harbor scientists lived in buildings dotted around the campus, close to their labs and close to each other. Watson, creating a

DNA village, encouraged this so that his rebels could work long hours. He made it clear to the science historian Horace Judson that "wives, children, houses, regular hours, are the bane of committed laboratory research."[96] In addition, the scientists found it difficult to afford housing in the surrounding affluent suburbs. Even though their school-age children went to local schools, they had little to do with the neighbors who commuted into New York. Phil Sharp recalled, "You entertained together. You worked together. There was no way to shield personal problems from . . . one family to another. . . . It had all the bad aspects and all the good aspects of a commune."

It was not easy for Sharp's wife. Having "bought into my continuing to be a postdoc at a measly salary," she and Sharp and one of their daughters drove across the country from Pasadena, and arrived at the Blackford Dining Hall. There, Sharp recalls, they "were told they didn't have a place for us, except in a dorm—and she blew up." If there was no proper housing, they'd leave at once. Space was found for them, and six months later they moved into a house called Hooper, right above the Walter Kellers and beside the Sambrooks. The three families got to know each other "extraordinarily well."[97]

Although "scientifically it was an incredibly good time," Mulder recalled, "socially the atmosphere at Cold Spring Harbor was rather unpleasant. . . . Your scientific life and your professional life and your social life [are] all together. So it is a bit incestuous."[98]

The scientists and their families saw each other a lot, "sitting on their doorsteps, right and left of Bungtown Road." They all "knew each other too well. They definitely stepped on each other's toes all the time. You would know what was frying . . . in your neighbor's pan at night," Westphal recalled. "During my time I saw two marriages falling apart."[99]

Beneath the veneer of a successful laboratory whose members' papers were cited more often than those of independent research institutes such as the Salk or the Whitehead or of Rockefeller University[100] was "the competitive nature" of life at Cold Spring Harbor. Steven Blose recalled, "Most of us worked seven days a week, eighteen hours a day, straight for years, with no time off." Although he met his wife while at CSHL and moved on, "a lot of personal lives were thrown on the rocks from that kind of a lifestyle."[101]

The scientists recognized the toll their commitments took on their families. "I was like in scientist heaven," Bob Tjian recalled. "My home was across the street [from the lab] at Hooper House. I could roll out of bed and be in James in seconds. You know, my wife didn't like that too much, but it was great for me."[102]

Robert Crouch and his wife lived in the ramshackle Victorian mansion then called the Carnegie dorm and later Davenport, just off the heavily traveled slope of Highway 25A. Soon the Westphals moved in upstairs. The two families became acquainted when the Westphals' daughter fell downstairs into the Crouches' kitchen. Later, the Crouches moved to an apartment in the renovated Greenhouse. He recalled growing "some snow peas in there that were really quite good." When they left, the Sharps moved in.

Liz Watson was remembered as a great help in lowering the pressure. "She had this ability to interact with people in a much gentler way" than Jim, Crouch reflected. She saw to it that a color television set was purchased for the families in the Carnegie dorm, so that they could watch football and baseball games. When the Crouches' son was born, Liz attended a baby shower and visited his wife in the hospital. The Watsons not only set up a swing set for their two boys but also had one constructed "for the rest of the people."[103]

The tight-little-island atmosphere occasionally extended to political matters. In 1974, as events ground on to the resignation of President Nixon, Bob Pollack found an essay by Philip Roth in the *New York Review of Books* in which Roth imagined Nixon giving a speech in which he declared he was not resigning—and also declared martial law. Pollack thought Roth had captured Nixon's cadences so well that he posted the clipping on his door. "Jim came by, and he said, 'You've got to take it down, you know, we have donors that would be offended.' To my everlasting shame, I took it down." The incident crystallized Pollack's determination to leave.[104]

After six months in a rented house in Huntington, Mike Botchan and his wife, Ruth, became the first occupants of a ground-floor apartment in the "wonderful" newly renamed house called Olney, whose purchase was spurred by one of Watson's threats of resignation. Their daughter, Rachel, had just been born in Syosset Hospital. Among the friends helping them move was Phil Sharp. Botchan, who worked in James, a few minutes' walk down Bungtown Road, recalled Olney as "a beautiful place to raise a young child." They parked "Showboat," the big American station wagon they had driven across the continent from Berkeley, in the driveway, alongside the used Volkswagen hatchback they bought when Showboat began breaking down. Ever concerned with the appearance of the grounds, Jim was annoyed. He told Mike, "You get rid of those cars. Get yourself a car." Botchan, who was later impressed by Watson's concern for the beauty of the place, was struck at the time by something else: "that he had time to even notice."

When the boss wasn't around, the scientists blew off steam like adolescents, "childish, fraternity-like things." Botchan and a colleague, Ashley Dunn, would jump on the "incredibly strong" Yakov Gluzman and try to beat him up. "You know, he would take the both of us." A scientist visiting from Switzerland, Walter Schaffner, would stage competitions for who could leap up the most stairs in a single bound. Botchan proved that he could hold his breath longer than anyone else.[105]

PRICE OF FAME

As Watson built up Cold Spring Harbor Laboratory, he became more of a public figure, participating in big controversies, and he gradually wound down his involvement with Harvard.

His fame led to invitations like one in 1974 from an organization in Palm Coast, Florida, to join other headliners in a two-day "great dialogue on the human condition." The meeting was arranged for Florida college presidents and deans, as well as Palm Coast residents who assiduously collected signatures on the famous panelists' photographs in their program booklets. Among the participants were Arthur M. Schlesinger, Jr., Saul Bellow, William F. Buckley, Gunnar Myrdal, Truman Capote, and Vernon E. Jordan. The sessions covered human nature and human destiny, uplifting the underprivileged, the role and clientele of universities, and popular and elite culture.

The panelists on human nature were Watson, Gunnar Myrdal, author of *An American Dilemma,* and Saul Bellow, who was to win the 1976 Nobel Prize for Literature. They were pessimistic about the chances for people to control their own destinies. According to Watson, hard science was still "so primitive" that it could do little to uncover the underlying chemical basis of human behavior. He reiterated his belief that living and nonliving objects alike were governed by the same laws of nature: "It would be nice to think that there is something distantly wonderful that keeps everything in line, but there is no reason for it."[106]

A month later, Watson's fame contributed to even more prominent coverage of a fraud in his group at Harvard. Working in the lab of David Dressler, a student named Steven Rosenberg had such initial success with an immunology experiment that two reports were published in the *Proceedings of the National Academy of Sciences,* and one in the *Annals of Internal Medicine.* Watson was so enthusiastic about the early results that he invited Rosenberg and Dressler to his house to pop open a bottle of champagne.

But other scientists, including the later Nobel Prize winner Baruj Benaceraff, an immunologist at Harvard Medical School, could not replicate the results. Neither could the student, who, it turned out, had not only faked his results but also had forged letters of recommendation for his entrance to graduate school and medical school, for election to the Phi Beta Kappa honor society, and for a scholarship. The published papers were retracted. "One's mind never considers fraud as a possibility," Watson told a reporter. "I was getting increasingly worried over the summer; the last successful preparation [experiment] was in March."[107]

"'HE'LL GIVE YOU A PROJECT"

For a time Watson served as chairman of Harvard's Molecular Biology Department which had been formed in 1967 to hold him at the University. But one day, during a departmental meeting, Matt Meselson recalled, "Jim became infuriated about something, lost his patience. I was sitting next to him. He got out of the chair and said, 'Here, Matt, you sit here. You be chairman.'" Matt was next in line to serve as chairman because he was next in chronological age.[108]

While Watson's formal responsibilities for teaching and supervision of graduate students at Harvard waned, he remained influential. Robert Horvitz was formally Matt Meselson's student, but Meselson was frequently away pursuing his campaign against chemical and biological warfare. What should Horvitz work on? Watson suggested that he come down to the annual Cold Spring Harbor Symposium. At the end of the meeting, Horvitz went to Osterhout, where the Watsons were still living, and asked, "Now what?" Watson replied with the instruction to talk to Klaus Weber in his lab at Harvard, "and he'll give you a project." The focus at Harvard was still the purification of RNA polymerase, the enzyme for copying DNA into RNA, on which Andrew Travers and Richard Burgess had been working (see chapter 7).

Horvitz was impressed by the "incredibly complementary and synergistic talents" of Watson, Weber, and Wally Gilbert, whose responsibility for the Harvard lab kept increasing. Going "almost blindly," Watson had an "unparalleled" intuition for science. Gilbert, "as smart as they come," was brilliant at designing approaches and critiquing entire projects, and had an instinct for whether a piece of data—including the fake numbers from the student forger—was right or wrong. Weber had "magic fingers . . . the ability to make anything work and to innovate experimentally."

When the frequently absent Watson was no longer allowed to have graduate students, Horvitz switched from Meselson to Gilbert, but continued to work on the project Watson had steered him to. The experiences with the thrice-weekly group meetings, now bearing the tongue-in-cheek title "Stalking the Secret of Life," were as daunting as many others had found them. Horvitz was fearful because "I saw some people sink who were actually pretty good." And he saw the "Jim look" when another graduate student, Bill Haseltine, trying to mix scintillation liquids in bulk, rolled a barrel up and down the corridor. Haseltine didn't do that again.

Jim wasn't shy about involving himself in trivia. Bob Horvitz once was summoned to Watson's office because of a tiff over a typewriter he was using in David Dressler's lab to type his thesis. Horvitz was afraid of what Watson would say. Dressler had objected to what he thought was a transgression, and Watson had gotten into a fight with Gilbert, who had taken Dressler's side. Watson merely said, "This is silly." But Horvitz had seen "people have their heads taken off by Jim, and it is not pleasant." To Horvitz's relief, Watson was in a peacemaking mode. Watson said Horvitz was in the right, but he should apologize anyway to the aggrieved Dressler. "The situation is very bad. . . . It's not worth it. The repercussions are too great."

When it came time for Horvitz to move on, Watson disagreed with Bob's wish to get into neurobiology. Watson's idea was for Horvitz to move up from bacteriophages to animal cells in Paul Berg's lab at Stanford. When Bob balked, Jim touted the alternative of the Stanford lab of David Hogness, where he could work on the nervous system of the fruit fly, *Drosophila*. Horvitz had another idea, to work on the tiny worm *C. elegans* with the emperor of the subject, Sydney Brenner, of the Medical Research Council lab in Cambridge, England. Weber had told him about this work. Watson called this "a silly decision," pointing out that Horvitz could have no idea whether the experiments would work, and, besides, it could be hard for him in those tight-money days to come back from England to an American academic job. When Horvitz persisted, Watson asked him whether he knew anything about neurobiology. Bob said no. Watson replied, "I've signed you up for all three neurobiology courses in Cold Spring Harbor. You'll spend the entire summer there." When Horvitz arrived for the courses he was handed a hefty bill; he went to see Watson and told him, "I don't have this kind of money." Jim said, "Give me the bill." Meanwhile Watson had persuaded Brenner to take Horvitz on. Brenner had at first written to Horvitz, perhaps ironically, that "I know nothing about neurobiology," but later accepted him."[109] In 2002, Horvitz shared a Nobel Prize with Brenner.

"JIM WAS REALLY HEARTBROKEN"
A major irritant for Watson at Harvard was Jack Strominger, an expert in the biosynthesis of cell walls[110] who had come to Harvard from the University of Wisconsin.

Strominger ran not only a large lab in the Harvard Molecular Biology Department but also another at the Dana Farber Cancer Center in Boston. Watson, who was still coming up to Cambridge to lecture once or twice a week, thought that Strominger was overextended and that this wasn't good for the department. He denied the analogy with his dual responsibilities at Harvard and Cold Spring Harbor, saying that at CSHL he didn't run a lab directly. A sharper issue arose when Strominger began taking Harvard graduate students over to his Dana Farber lab, just as Harvard students were gravitating to Cold Spring Harbor. Eventually, Watson persuaded his colleagues to pass a rule against such dual responsibilities, and then immediately said that the rule applied to him. He would go full time to Cold Spring Harbor.

Soon after, Watson went to Cambridge for a farewell interview with Harvard's president, Derek Bok, and his deputy, Henry Rosovsky, dean of the Faculty of Arts and Sciences. It turned out to be a "most unceremonious" occasion. Late that night, after flying back to Long Island, he came into the office where Tom Maniatis was working. "Jim came bursting into the room. He was dressed very nicely. Had a suit and tie on, and was obviously kind of panting. Obviously had just returned from the airport. And he literally cried in my office about how he was treated by the dean. . . . Jim was really heartbroken that somehow the university did not convey in any way their appreciation for what he did there."[111]

Two years later Harvard conferred an honorary degree on Watson, but the university's ambivalence toward him continued. Twenty years later, when Watson turned 70, his Harvard colleagues planned a party, and several of them who had done well in the biotechnology industry pledged about half the cost of endowing a professorship in Watson's name. One was Mark Ptashne, who had left Harvard a few months earlier to work at the Memorial Sloan-Kettering Cancer Center in New York. Ptashne was designated to bring the idea up with Rosovsky, who was now a member of Harvard's seven-member governing Corporation and was already irritated at his colleagues' reluctance to spend any of the billions in Harvard's swelling endowment. Rosovsky sent Ptashne to another Corporation member, Robert Stone, who said he'd "grease the skids" with colleagues and would speak to Harvard's president, Neil Rudenstine. But when Ptashne went in to see Rudenstine, the latter said no, alluding vaguely to competing priorities for the money Harvard itself would need to allocate to finish endowing the chair. Ptashne found it "quite interesting" when, shortly thereafter, a new professorship in education was announced.[112]

At the seventieth-birthday party, which was hosted by Paul Doty, Watson spoke sharply but not sourly. He remarked that Harvard leaders have "got to think that you're more important than the French department or any of these other departments which have great histories, but which, if they vanished, wouldn't make any difference. . . . DNA is more important than the hockey team." He hoped that "Harvard's most famous ex-freshman," Bill Gates of Microsoft, would consider a major gift to strengthen molecular biology at Harvard for "the really difficult science which lies ahead." He added, "It would be great someday if there was actually a scientist who made real decisions at Harvard."[113]

He was already lobbying with Jeremy Knowles, the chemist then serving as dean of the Faculty of Arts and Sciences, for what became in the next year a $200 million commitment by Harvard for a genome center. In September 1999, he was invited to give the main talk at the center's opening. The room in the Science Center where he spoke was jammed—

every illegal seat on all the stairways was filled, and more students clustered on the floor just in front of the lectern. Neighboring rooms with closed-circuit television were also filled. Knowles and Rudenstine were both present. Watson said he'd left Harvard because "you didn't make me president." At this, Rudenstine rose briefly from his front-row seat and lightheartedly turned around to the audience to show who the president of the moment was.[114]

Harvard was not the only university where Watson aspired to be in charge. A particular sore point was Rockefeller University, headed by a succession of Nobel Prize–winning biologists: Joshua Lederberg, David Baltimore, and Torsten Wiesel. The story goes that he was being looked at for the top job at Rockefeller in 1978, but he gave a talk that was so insulting that his candidacy vaporized. To many observers, he was passed over because he was too indiscreet. In spite of Watson's epochal success at Cold Spring Harbor, leaders of major universities were "all petrified that Jim would get up and say something or do something that would deeply embarrass their institutions."[115]

"WE WERE HIS YEAST GUYS"
While he pushed on into the study of higher cells, Watson was determined not to neglect other pathways for understanding the workings of such processes as the replication of DNA.[116] Baker's yeast, *Saccharomyces cerevisiae,* was a tempting target. The simplest cell with a nucleus, yeast has only about three times as much genetic information (15 million base pairs) as *E. coli* bacteria (4.5 million).[117] Much more closely related to human cells than to those of bacteria, yeast proved "as easy to manipulate genetically as bacteria."[118] It did not require special equipment, as the animal viruses did.[119] Yeast turned out to have many proteins that are also found in humans.[120]

At the start of his CSHL directorship, Watson was confident that yeast would turn out to have a basic biology similar to that of higher organisms, as he had just argued in *Molecular Biology of the Gene.*[121] He obtained an expanded government training grant for CSHL in order to fund a summer course on the molecular biology and genetics of yeast; it began in 1970, with Fred Sherman of the University of Rochester and Gerald Fink of Cornell as instructors.[122]

In Jim's scheme of things, such a course focusing on techniques often led on to the production of laboratory manuals such as *Molecular Cloning* and annual meetings of researchers in the field, and even year-round research. Young participants, including those who shied away from "those nasty animal viruses," could latch on to lore not available in existing books, and in later years encourage other colleagues to attend the course. The idea was to make yeast "the *E. coli* of eukaryotes."

Teaching year after year, Fink and Sherman not only helped build up yeast as "a practical reality," but also led in founding a "church," with a culture of intellectual openness analogous to that for bacteriophages. Behind the openness with information was "a deeper idealism" held by Delbrück and Luria, that science was not personal property but belonged to "all of civilization."[123]

One participant in the course's second year, 1971, was David Botstein, a geneticist who gravitated into Jim's orbit at Harvard before moving on to Michigan and MIT, and a faithful participant each year in the CSHL phage meeting. He was thinking of the best way for him to get into eukaryotes. He didn't like animal viruses, or plant viruses either. "I guess I have a black thumb or something, and I didn't get interested in the plants." So his choice fell on yeast. Three years later, Fink and Botstein inaugurated the annual yeast meeting at CSHL. Botstein had already suggested to Fink and John Roth of the University of California that they spend a sabbatical year together, working on yeast. When the Salk Institute turned the idea down because of fears that yeast might contaminate their animal cell cultures, Watson suggested they come instead to CSHL for what became, in Jim's phrase, "nonstop experimentation and conversation," based in the newly winterized Davenport Laboratory.[124] Botstein later thought of the work there as "possibly the most important year of my career."

Fink, Roth, and Botstein did more conversing than experimenting. They looked generally at the future, considering what they, and others, were doing, and the relative power of genetic and biochemical and other tools of molecular biology to answer basic questions. With a hotbed of animal virus work just across the street, they could spend weeks pondering how to apply tools being adapted by Botchan and others in Joe Sambrook's group to studies of yeast. They learned how CSHL pioneers were using restriction enzymes. With Ray Gesteland, they established that the "suppressors" in yeast were molecules of transfer RNA. They discussed how chunks of DNA, called transposons, would move around in bacteria, an idea that might be useful in yeast. And they concluded that yeast viruses might not be necessary for the research they contemplated.[125]

Fink helped finance his CSHL stay with a Guggenheim fellowship, for which Watson wrote a recommendation. During the year, he and his family lived in Osterhout, which had been vacated when the Watsons moved to Airslie. Fink's wife and Watson's became friends. Watson, who "popped in all the time to talk," also instructed the three not to distract themselves by attending CSHL fund-raising events.[126] Botstein recalled Jim as being "very supportive, but of course in his uniquely weird way. You didn't know when you were being supported or . . . being attacked much of the time. You had to have a thick skin, no doubt about that . . . [but] for better or worse, we were his yeast guys."[127]

Although Roth did not continue in yeast research, Fink and Botstein did. About two years later, using recombinant DNA techniques (described in chapter 11), Fink and such colleagues as James B. Hicks achieved a heritable transformation in yeast by injecting DNA. They concluded that yeast might be superior to bacteria for growing large quantities of animal genes of interest, such as that for human insulin.

Watson had already heard of Hicks. With Jeffrey Strathern, Hicks had proposed a "cassette" model of how a group of silent yeast genes become active when transferred to a new site on the yeast chromosome.[128] The two had been working in the University of Oregon lab of Ira Herskowitz, who had attended the 1971 yeast course with Botstein. Soon they were pulled into year-round research at Cold Spring Harbor along with several colleagues.[129] To make room for it, the Davenport laboratory had to double in size. In 1980, "movable genetic elements" were the theme of the annual Symposium, which was organized by Hicks. Among the session chairmen were Fink, Botstein, and Herskowitz.[130]

SPLIT GENES

An article of faith in the 1960s and 1970s was that the code sequence in DNA directly matched—was "colinear" with—the messenger RNA that specified the makeup of proteins. Suddenly, in the spring of 1977, scientists including Richard Roberts at Cold Spring Harbor and Phillip Sharp at MIT found that this idea of an uninterrupted sequence would have to be modified. The sequence was still colinear—but interrupted. "Expressed" stretches of DNA were interspersed with often-long stretches that were not expressed. Of special insterest were particular adenovirus genes, those coding for a group of "late" proteins, including one labeled "hexon," mostly for the protein coats of newly forming viruses. To everyone's amazement, both groups found that the genes were not discrete but were scattered in separated segments of the virus genome. Active DNA sequences were mixed in with stretches of "nonsense DNA with no discernible function."[131] This discovery of split genes demonstrated that the raw original RNA copy of the DNA was edited in the nucleus into a shorter working copy that had nothing but the "expressed" codons following one after the other and that would move out to the cytoplasm and direct the synthesis of particular proteins. This cutting out of unessential intervening stretches was called "RNA splicing." The finding was "unexpected," as Roberts said. "No one had predicted it. It went totally against the grain."[132]

The discovery was soon repeated in the genes for rabbit globin, chicken ovalbumin, SV40, and polyoma. The message for the string of amino acids in proteins was not continuous, but was broken up by stretches of "silent" DNA, which some liked to call "junk." The organization of genes in prokaryotes, cells without nuclei, and eukaryotes, cells with nuclei, was strikingly different. It began to look as if the eukaryotes, instead of evolving from the simpler prokaryotes, had kept a "genetic plasticity" that bacteria had lost.[133] According to Thomas Broker and Louise Chow of Cold Spring Harbor, two electron microscopists who participated in the discovery, "This startling fact has abruptly changed our concept of what a eukaryotic gene is and how its expression is controlled."[134] Broker and Chow, like Sharp, had worked in Norman Davidson's lab at Caltech. Because the discovery arose from Cold Spring Harbor's drive to sequence adenovirus, it redounded to the credit of Watson's original gamble on tumor viruses.

At first highly skeptical of the discovery, Watson became enchanted with the idea. He was sure the discovery would "revolutionize our ideas of genes and the proteins for which they code." The previous guess, he said, was that messenger RNA would only be processed by snipping off either end of it. Of the Cold Spring Harbor and MIT experiments, he wrote, "The counter proposal that processing occurs by removal of multiple internal sections would have been thought irresponsible speculation had not our experiments been so foolproof." Now it looked as if this kind of splicing happened in all eukaryotes. He went on to forecast "that a multiplicity of proteins will be generated from 'single' genes," an observation that was confirmed and gained force after the human genome sequence reached completion 25 years later. "Splicing is likely to be the most important observation in molecular genetics since the 1960 discovery of messenger RNA."[135] An outpouring of analyses followed, mentioning the implications for both evolution and the regulation of the cell's life.[136]

Some months later, in a famous one-page article, Walter Gilbert proposed the concept of "genes in pieces," whereby the usually long, unexpressed DNA, or "introns," divided sometimes numerous, shorter expressed sequences that he dubbed "exons." Each gene had an average of up to 40 exons. He thought of the introns as "frozen remnants of history . . . the sites of future evolution." With their genes organized differently from the prokaryotes', higher cells might be more adept at evolution by making new proteins out of fragments of old ones.[137] Phil Sharp was a little sorry he hadn't thought of the term "genes in pieces," commenting, "He saw the opportunity to do it and I didn't."[138] Commenting on this total surprise, Crick called it "a mini-revolution . . . [that] has given our ideas a good shake."[139] In a magazine interview, he said the discovery was "almost as surprising" as finding "a rather garbled version of an advertisement for a deodorant" in the middle of a Jane Austen novel.[140] The fourth edition of *Molecular Biology of the Gene* said that the concepts of gene expression no longer applied "automatically" to higher plants and animals.

This discovery of "interruptions in eukaryotic genes" was "a complete shock to the scientific world." Data from several labs "forced a conclusion no one could have foreseen."[141] Watson later called it a "historic discovery," and a "bombshell" that left him "overwhelmed and dazzled."[142] Soon afterward, scientists began discovering diseases, including one form of anemia and one of leukemia, in which foul-ups in splicing were a factor.[143] And Sidney Altman of Yale and Tom Cech of the University of Colorado promptly built on insights from the split genes discovery to find that RNA itself could be a catalyst for chemical reactions in the cell.[144]

The road to split genes began with the adenovirus that Ulf Pettersson had brought from Uppsala, which Phil Sharp and Joe Sambrook used in their collaboration and Rich Roberts sought to sequence with the help of his growing inventory of restriction enzymes. It was a "wide open" field, and the first task was to put the genes in order in a map. Sharp went off to MIT in 1974 to continue his studies of how much adenovirus RNA was produced in cells infected with the virus.[145] At Cold Spring Harbor in 1975, Roberts and his associate Richard Gelinas tackled the sequences of the 20 or so messenger RNAs they expected to find in the adenovirus genome. To their puzzlement, Gelinas found just one, a string of 11 nucleotides. Roberts repeated the experiments and got the same result.[146] Late in 1976 they "began to suspect the existence" of split genes.[147] Meanwhile, up at MIT, Phil Sharp and his associate Sue Berget were also scratching their heads over the "small tail at the end of the five-prime end of purified hexon message," and wondering, "What the hell is that about?"[148]

In March 1977, the competitive pressure built up between Cold Spring Harbor and MIT. Phil Sharp made one of his frequent calls to Ray Gesteland down at CSHL and told him he had done something that would "revolutionize molecular biology." Without saying what it was, he conveyed the idea of "a massive discovery." Ray immediately wondered what Phil Sharp had. Sharp hadn't left CSHL all that long ago, and before he left his subject had been adenovirus. Sensing that "something must be up," Cold Spring Harbor went into what Rich Roberts called "heightened intellectual activity"—red alert. Roberts suspected that Sharp's call had to do with the 11-nucleotide string, although Sharp later said it only had to do with mapping. The MIT-CSHL competition was so intense that scientists at each place suspected that their competitors were learning each other's facts and insights by the back door.

Evidence pointing to split genes came from many people at CSHL, including Gelinas, Sayeeda Zain, John Hassell, Daniel Klessig, Carl Anderson, and John Lewis. Under the pressure of Sharp's call, early one Saturday morning Roberts conceived an electron microscope (EM) experiment. Because, Roberts recalled, neither he nor Gelinas was an electron microscopist, he proposed the idea to Louise Chow and Tom Broker.

Refining the concept, Louise and Tom took electron microscope pictures that showed an additional major surprise: loops of RNA that would not "hybridize" with the relevant strip of DNA. It was RNA being spliced out of the sequence. Roberts recalled, "They did the experiment two days later, and we had splicing." But Broker and Chow remembered it differently. They recalled that the pictures actually disproved Roberts' main theory. In Sharp's MIT lab, the postdoc Sue Berget also was taking similar electron microscope pictures. Suddenly a phenomenon that had seemed absurd, RNA splicing, now looked real.[149]

Word got out quickly. Roberts threw in the startling results just at the end of a previously arranged talk at Harvard Medical School on "the adenovirus terminal protein." He had brought one or two slides about splicing. He recalled, "This was really hot off the press, and there was no one in the audience who even had an inkling about anything like this." He surmised that because "we'd done all of the biochemistry and the electron microscopy . . . capped it off . . . no one got up and said, 'Oh, this is shit. . . . How can this be right?' Everybody accepted it, essentially immediately. A number of people left that seminar, got on the phone, called around to their friends. Within days, the same phenomenon had been discovered in several labs." Not long after, when Roberts was asked to discuss splicing at the first session of Cold Spring Harbor's regular tumor virus meeting, Michael Bishop of the University of California at San Francisco showed slides that had puzzled him but that he could now interpret in terms of splicing. "Now, things moved very quickly. Everybody realized this was a big discovery. How could you not?"

Because Watson was "a very visual person," he had not believed the biochemical results Roberts had brought to him more than once. "He thought it was another contaminant," Roberts recalled, and shared the skepticism of the people up in James laboratory, who may have disliked the idea of a major discovery coming from Demerec Lab. Jim kept saying, "Don't bother me with this until you've got a decent result, until it's absolutely clear." Even with the electron micrographs by Chow and Broker, Jim had to ask for a second viewing. Coming from Roberts, with whom he had several major disputes, notably over the idea of a Cold Spring Harbor company to market restriction enzymes and about the general computerization of CSHL, the discovery "was not something he had found particularly likeable." And when he became "keen" about it, Roberts recalled, "he wanted it to be a Cold Spring Harbor discovery and not a Roberts discovery. . . . Everyone was talking about Nobel Prizes except for Jim."

The maneuvering over credit for discovering split genes began at once, with a tussle at Cold Spring Harbor over who would present the bombshell at the June 1977 Symposium. "Everybody wanted to present it," Roberts recalled, and Jim's choice fell on Tom Broker. In the fight over credit, which went on for many years, Jim stuck to the position that it was a collective, not an individual, discovery. He said this to people who visited during fund-raising efforts and to Roberts himself. "It was a team effort and they don't give [Nobel] prizes for team efforts." But Roberts felt that Jim "did not want to know what had actually happened."[150]

Meanwhile, Sharp began to be nominated for a Nobel, perhaps as early as 1978, and received a shower of other major awards. Watson's stand left both Roberts and Chow bitter, along with more than half a dozen collaborators at CSHL, most of them graduate students and post-docs. One of these, Sayeeda Zain, was not given credit in any of four papers submitted as a group to the journal *Cell*. Chow and Broker carried their bitterness with them when they went first to the University of Rochester and then to the University of Alabama Medical School in Birmingham.

The issue was resolved in the wake of the award of the 1989 Nobel Prize for Physiology or Medicine to Sidney Altman and Tom Cech for their discovery in 1981 and 1982 that RNA could edit itself—"RNA as an enzyme," as Roberts termed it. Watson received a telephone call from Susumu Tonegawa, a 1987 Nobel winner, saying, in effect, that Sharp was waiting, and that Jim should decide whom he would nominate.[151]

This caused an immediate dilemma for Jim. If Louise Chow was included along with Rich Roberts, the question of including Sue Berget arose—and the limit for a Nobel Prize was three. (This was the same issue that would have arisen in 1962 if Rosalind Franklin had lived.) Watson asked Roberts to write a 15-page "intellectual history" assigning responsibility. This history brought a detailed rebuttal from Broker. After the prize was awarded in 1993 to Roberts and Sharp, both documents reached the press.[152] Davidson at Caltech was quoted as saying of Chow, "The evidence she discovered formed an important part of the total creative insight that splicing was taking place. Only she could have interpreted those data [the pictures]." Chow may have been slighted, in Davidson's opinion, because "she's a woman, an Asian woman who's a little quiet. Sometimes they get ignored."[153] Watson told another reporter, "Louise did it and it's terrible that she didn't win."[154]

GAZING "UPWARD AT THE BRAIN"

The sciences of the brain and nervous system have been one of the Holy Grails of biology for much of the last century. Over many decades Nobel Prizes have been awarded to neurobiologists, most recently, in 2000, to Eric Kandel of Columbia University and two other neurobiologists. In the late 1960s, as molecular biologists concluded that they must move on to the study of higher cells, neurobiology was felt to be one of the most important fields to explore. It was an established theme. Francis Crick, when he got into biology in the late 1940s, regarded the brain and the boundary between living and nonliving as the two major possible fields of study for him.

As soon as Watson became director of CSHL and announced that cancer would be the Lab's principal target, he also embraced the goal of establishing neurobiology there, and talked it up constantly. He hoped for "a small but talented group" at CSHL[155] who would work on "the ultimate frontier" and help bring in "many of the brightest and most courageous minds in science."[156]

As was so often the case, Watson's first step was to set up courses heavily emphasizing "the most modern techniques . . . [the] highest tech."[157] He won a five-year grant from the Alfred P. Sloan Foundation to help pay for courses on basic principles and lab techniques using the sea slug *Aplysia*, which began in 1971, and to renovate a building for the course work. Eric Kandel was an adviser from the start. At the end of 1972, Watson announced that the neurobiology courses would be expanded with the addition of one on "behavioral genetics of a nematode." This organism was the famous *C. elegans*, a favorite of molecular biologists for decades.

Still, Jim admitted, neurobiology was "a subject of unlimited intellectual challenges now largely hidden by the horrendous anatomical complexity of the vertebrate brain," so that setting up year-round research would come slowly.[158] Nonetheless, Jim kept mentioning neurobiology as a high priority[159] and used some of the first money from the Robertson endowment to renovate the white frame Jones Lab for the study of neuroscience.[160] A course on the neurobiology of the fruit fly *Drosophila*, the classic organism for genetics, was added in 1973, and one on the leech got going in 1974 with Robertson funds. One of the students attending the leech course was Birgit Zipser of Downstate Medical School in Brooklyn, who began year-round research at CSHL in 1978, using nerve cells of the leech, an organism with an extremely simple brain.[161] Other foundations joined the Sloan Foundation in supporting neurobiology at Cold Spring Harbor.

In 1975, Jim arranged an annual Symposium devoted to the synaptic connections between nerve cells. Ever the booster, he wrote that "the question of what synapses are and how they are made remains a mystery which currently is intriguing to an increasingly large number of the world's better scientists." The meeting was opened by Stephen Kuffler, one of the great pioneers of the field.[162] The following year, Watson announced that the first income from a special fund honoring the late Marie Robertson would pay for equipment in the renovated Jones Lab. At the same time he acknowledged that neurobiology still lacked the "path through the woods"—a picture of the problems to be solved and in what sequence—that molecular biology had opened for itself.[163]

By 1980, Ron McKay from Walter Bodmer's lab in Oxford had joined Zipser and developed highly specific monoclonal antibodies by injecting leech nerve chords into mice. The mouse spleen cells began producing antibodies against specific types of leech nerve cells. These antibody-making cells in turn were fused with rapidly dividing mouse myeloma (cancer) cells to form many lines of "hybridomas," cellular factories for producing large quantities of the antibodies. Watson wrote that the achievement "helped revolutionize neurobiology." The antibodies bound to specific "antigens" on neurons. Susan Hockfield, newly arrived from the University of California at San Francisco, began trying to identify what these antigens were.[164]

In the early 1980s the neurobiology courses were still operating on a hand-to-mouth basis, and required the lab's own money and weeks of fund-raising each year for "essential last-minute assistance."[165] Even so, Watson was still increasing the number of courses; in 1981 there were eight. In 1980 he brought in the Yale neurobiologist (and later scientist-trustee) Charles Stevens for the summer, putting his family up in Osterhout and paying for all their meals, so that Stevens could "go around and talk to all of the people who were doing neurobiology there," Stevens recalled, "and sort of encourage them and find out what was good and bad, and tell Jim." A related assignment for Stevens that summer was to learn all he could about molecular biology and how it could be applied to neuroscience.[166]

Following the typical Watson pattern for building a field of inquiry at Cold Spring Harbor, the laboratory published a text, *Monoclonal Antibodies to Neural Antigens*, edited by McKay and two others and based on a meeting at Banbury in 1980.[167] The next step was an annual Symposium on molecular neurobiology, with Watson and McKay as organizers and Kandel and Stevens as advisers. The tone was ebullient. McKay recalled the atmosphere at the meeting as "electric." Watson and McKay wrote of "a new mood among those molecular biologists who gaze upward at the brain." New techniques created "a fighting chance to understand the uniqueness of nerve cells" and how they form networks.[168] Joe Sambrook was more skeptical. He thought the 91 papers at the Symposium "were far more neurobiological in nature than molecular." Although the meeting showed that a molecular approach might help spell out the uniqueness of nerve cells, it "left unresolved" the role of molecular biology in understanding how neural networks are set up.[169]

But then, in 1984, Zipser, McKay, and Hockfield all left. Zipser went to NIH, McKay to MIT (and later to NIH), and Hockfield to the neurobiology department of the Yale Medical School. The summer courses were no longer backed up by a year-round program. Watson responded in the spirit of double or quits. The skeptical David Botstein, who never understood the "conventional wisdom" that molecular biology would accelerate neuroscience, thought that Watson had "got it into his head that if he built a magnificent building to house them, then he could get a critical mass of good guys." Despite his dissent, Botstein nevertheless helped in the courtship of Arnold Beckman, a major donor for the Neuroscience Center, which was completed in 1991. Thanks in part to his lobbying, the center included space for the field of structural biology. In later years, Botstein regarded the leadership and quality of the effort as not "as revolutionary as Sambrook was" with tumor viruses. But it had been a success.[170] A problem, according to Stevens, was to get neurobiologists to come to Cold Spring Harbor. "It's kind of away from things. It's kind of lonely out there, and there was not a big tradition."[171]

Nonetheless, the Neuroscience Center was dedicated with great flair on 3 May 1991. The local congressman was present, and a brass band played. As the bell on the Hazen Tower tolled, a cloud of colored balloons rose from the air shaft of the parking garage and into the sky.[172]

11

"ODD MAN OUT": RECOMBINANT DNA

Why should we be afraid of something that goes on in nature all the time?

JAMES D. WATSON, 17 APRIL 1978

GERM WARFARE? "RIDICULOUS"

Beyond the glare of publicity, Nobel Prizes create an occupational hazard: The winner gets pulled into the realm of public policy. For an already-famous Watson, the process began even before he won the Nobel Prize, during the Kennedy administration. He took his first small, by no means reverential, steps along a stormy path of being a public scientist, long before the controversy that exploded when scientists figured out in the 1970s how to move DNA freely from one species to another, the technique known as recombinant DNA. In the fall of 1961, the Harvard chemistry professor George Kistiakowsky (whose labs were just below Paul Doty's) invited him to help the President's Science Advisory Committee (PSAC) oversee the program to develop biological warfare, and soon after, Watson had a pass to the Executive Office Building just across a parking lot from the White House. He liked the idea of working for Kennedy, and hoped to get a chance to meet his wife.[1] A central question was, Did germ warfare make any sense in superpower conflicts? The answer: No, the germ warfare program was "ridiculous."[2]

He also worked on another national question: Could the cotton pest, the boll weevil, be eradicated? Again the answer was no. Of the boll-weevil inquiry, he recalled that the sharecroppers' shacks in the Mississippi, still numerous then, resembled the houses he had seen when he visited Cambodia in 1961. The cotton farmers, encouraged by huge roadside billboards, sprayed their fields with toxins 10 times a day. When genetically modified crops that required less pesticide began to appear decades later, Watson's reaction was "Great!"[3]

Watson also was recruited to join a government panel reviewing Rachel Carson's *Silent Spring*, which had been published just before he received the prize. Should Carson be taken seriously? The answer was yes.

Small potatoes. But Watson was going beyond a lifetime of omnivorous reading about everything from politics to show business. He had begun threading the maze of personalities and issues in Washington, gaining experience that paid off in public debates about human cloning, the overblown "War on Cancer" of the 1970s, the possibility of risks from recombinant DNA experiments, and, in the 1980s, the Human Genome Project.

The Kennedy administration's interest in biological warfare was strengthened by the hope, which later influenced the American plunge into Vietnam, of finding means of "flexible response" in the struggle against the Soviet empire. Could biological agents, particularly those that would make people sick but not kill them, be tools of "limited wars"? A principal assignment for Watson, during a three-year period when U.S. spending on chemical and biological warfare tripled, was searching for "a satisfactory incapacitating agent."

When his participation in the limited-war study became public in 1968, Watson maintained that his role had been largely that of a "devil's advocate." He concluded that "the questions we were being asked were primarily political, not scientific." He thought chemical and biological warfare programs should be discontinued. "They are not a good way of winning wars. . . . Militarily they're a waste of time." But he feared that the U.S. military, unsuccessful in Vietnam, might be tempted to grasp at the illusory promise of biological warfare.[4]

He gave a fuller account in 1976, when he was campaigning against ideas that recombinant DNA methods would allow creation of even more fearsome biological agents. At Cornell University, he ridiculed as "nonsense" the notion that normal bacteria could be equipped with genes for deadly botulinus toxin and dumped into Ithaca's water system, killing a lot of intellectuals.

Biological warfare was an assignment of PSAC's limited-war panel. On the panel's behalf Watson went off to Fort Detrick, Maryland, the heavily guarded enclave where the military tried to make biological weapons out of deadly pathogens, and soon found that, as Watson said, "there was nothing good to tell the President." The pathogens were useless in a superpower conflict between the United States and the Soviet Union, each with thousands of nuclear warheads. "There was no way that you could decisively use them to win a realistic war." Supposing you wanted to kill the people of Leningrad with toxins carried by rockets. "You never know for sure which way the wind will be blowing, and instead of killing the people of Leningrad, we might wipe out thirty percent of the Finns."

If your aim was just to incapacitate people, Watson concluded, "There's also nothing there, for all such agents will inevitably kill a small percentage of those made sick." He mentioned psychotropic drugs like LSD or viruses and bacteria designed to be "not too virulent." Biological weapons were useless for anything but "random mass murder." Ultimately, Watson's Harvard colleague Matthew Meselson helped convince President Nixon to stop the work and destroy supplies. The generals could see no extra military advantage in the toxins. Said Watson, "You can't imagine them banning anything they thought would work."[5]

Biological warfare was a powerful example of classical anxieties about biology that gained force in the face not only of avalanches of new forms of medical care and plant breeding but also of rapidly increasing knowledge of how life worked. The biologists were working not only on one-cell bacteria and yeast cells but also on flatworms, fruit flies, mice—and humans. Biologists of Watson's stripe were determined to carry molecular genetics on into "higher" cells, in order get at the mysteries of both cancer and normal human development. For this they were using cancer-causing viruses. They were beginning to understand how to transfer genes directly from one species to another, including genes for drug resistance that helped make some experiments more convenient. They began to confront the risks faced for decades by microbiologists working to understand many infectious diseases. They also confronted a distrust of all authority that the Vietnam War brought to a pitch, and the evergreen fear of eugenics, seen as a way a powerful few could control the fate of many.

CLONING

In 1971, as public attention to prenatal genetic screening picked up,[6] Watson got into the topic of cloning animals and humans. Largely an idea from the realm of novelists and film makers, it was now creeping toward reality. Cloning an animal involves taking a complete nucleus from a cell of an adult animal and inserting it into a female's egg from which the nucleus has been removed. An English researcher, John Gurdon, had done the trick with frogs years earlier—although the odds of success even with young tadpoles was 99 to 1.[7]

Later, Watson would be far more dismissive of the idea of cloning humans, even joking, "Can you imagine what it would be like to be the clone of Arthur Rubinstein? You might be a wonderful pianist, but you'd be stuck having to spend all your time trying to live up to the original."[8]

In 1971 his view was darker. Actual cloning of humans could seem remote, Watson argued, but already on the horizon was the technique of making test-tube babies. R. G. Edwards and P. S. Steptoe in England were striving to take eggs from a woman and fertilize them in a petri dish with sperm from a man, so that they could be reimplanted in the mother. Watson called this "routine test-tube conception of human eggs," a concept that became a reality some years later with the birth of Louise Brown on 25 July 1978. Thousands of women might need the procedure, Watson argued, but what about such ethical dilemmas as surrogate motherhood? And what about genetic manipulation of the eggs?

At a congressional briefing, Watson asked how one would balance the risk of "abhorrent abuse" with the "unhappiness" of thousands of couples unable to have children. After success with test-tube babies, there would be pressure to go further and tailor eggs genetically. But cloning would bring "despair" to most families. "The nature of the bond between parents and their children, not to mention everyone's values about the individual's uniqueness, could be changed beyond recognition." Social decisions about this, he said, were far too important for scientists and doctors to make by themselves. "If we do not think about it now, the possibility of our having a free choice will one day suddenly be gone."[9] He adapted the testimony for an article in *The Atlantic,* and editorial comment was favorable, with one exception. Philip Abelson, the crusty editor of *Science,* wrote, presciently: "Talk of dire social implications of laboratory-related genetic engineering is premature and unrealistic. It disturbs the public unnecessarily and could lead to harmful restrictions on *all* scientific research."[10] Watson and a host of colleagues would soon learn how right Abelson was.

"WAR" ON CANCER

Besides the risks of overfrightening people, biologists also faced the risks of overpromising. By the early 1970s, more and more Americans were living to the age at which getting cancer was increasingly likely. Thus, health researchers were focusing on cancer, even though cancer still was difficult to detect or cure. "Higher" cells, including human cells, were vital to this research, but biological scientists had a hard time making such complex cells as amenable to research as one-cell bacteria. Nonetheless, with the support of the cancer-researcher Sidney Farber, the medical philanthropist Mary Lasker, the lawyer Benno Schmidt, Senator Ralph Yarborough of Texas, and Senator Edward Kennedy of Massachusetts, President Nixon declared "war" on cancer.[11] So much knowledge had already been gathered, they reasoned, that a lot more money and the establishment of big centers around the country would mobilize existing technology and significantly reduce death rates from cancer. Watson, a member of the National Cancer Advisory board, which President Nixon set up to supervise the war, hoped that the effort might even bring in some basic science. Perhaps, he reasoned, dramatic early discoveries would repeat themselves, even though the tumor-virus work was expensive and "painfully" slow. He reflected later that he had been "far more optimistic than I should have been," given that biologists' tools "for directly analyzing DNA" were still primitive 20 years after DNA's structure had been revealed.[12]

Watson wanted to be in on it, and was eager to establish at Cold Spring Harbor Laboratory one of the research centers for the "war," but he quickly sensed danger. Sixteen of the new centers were clinical, amounting to little more than "more cheerful places to die." Cancer physicians faced "a lifetime of effectively impotent ward rounds," Watson suspected, and so it was natural to do clinical research. With so much money going to the clinical centers, the relatively inexpensive but long-term basic research that alone would bring fundamental change could get squeezed out. By 1972, he was writing about "the current American hysteria to conquer cancer," and worried whether the President and the public had "again been sold a bill of unattainable goods."[13]

A year later he was telling a House subcommittee that the "war" was mostly public relations and that by concentrating resources on what was called "direct" research in clinics, it was undermining basic research and the training of scientists.[14] At the end of the year, when he was still a member of the National Cancer Advisory Board, he wrote that it was "hard not to feel ineffectual." He warned against "prematurely placing emphasis on highly targeted research" and neglecting the 25-to-50-year task of mapping out "the behavior of corresponding normal cells." The cancer war could end up resembling "our Vietnamese debacle."[15]

Soon, he was calling the War on Cancer "a sham." At the dedication of a cancer-research building at MIT, he charged that "the American people are being sold a nasty bill of goods about cancer. While they are being told about cancer cures, the cure rate has improved only about 1 percent." The director of the National Cancer Institute rejoined that no "responsible scientist" had promised "a magic bullet" to Congress. In a letter to the *New York Times*, Benno Schmidt lamely denied that anyone had promised a "breakthrough."[16]

The basic problem, according to Watson, was "how to be realistic about what we [scientists] can do for society without seeming to reject the hands that may want to overfeed us."[17]

He had a hand in being "realistic." On the National Cancer Advisory Board one bone of contention was a special virus cancer program, controlled by National Cancer Institute insiders in Bethesda, whose funds were distributed by contracts to a small number of scientists, not by means of peer-reviewed grants. Many scientists disapproved of this way of parceling out money. Watson pushed for a review of the procedure (even though Cold Spring Harbor was receiving some of the monies), which took place and was led by Norton Zinder, who concluded that "it really was a gravy train" and recommended opening up the funds to many more scientists. The National Cancer Advisory Board approved and allowed the contract program to fade.[18]

By denouncing the very program that supported cancer research at Cold Spring Harbor, Watson risked offending legislative and executive patrons. Yet he persevered. It seems plausible that he regarded his campaign as a rear-guard action to protect long-term basic research.

ROBERT POLLACK HAS A FIT

Watson was not shy about considering and calling attention to possible risks attendant on the new work with tumor viruses. Not too much was known about them. The molecular biologists beginning to use them to probe the cell were not imbued with the culture of caution that had built over decades among microbiologists who worked with disease organisms. He exploded in wrath at the plan of Harvard colleagues to do work he considered hazardous, and even threatened to sue.[19]

The issue grew sharper in the summer of 1971 with accelerating advances at Stanford University, a hotbed of recombinant DNA studies. Janet Mertz, a researcher in Paul Berg's laboratory at Stanford, was helping to develop the technology. (Berg had left behind many years' work in bacteria to move into animal cells and tumor viruses.) She attended Robert Pollack's summer course on tumor viruses at Cold Spring Harbor, which included a lecture on safety. Like her fellow students, Mertz described an experiment she was planning, and, as Berg recalled it later, "sort of bragged about this great experiment." With Berg's encouragement, she hoped to use Mort Mandel's new technique of using a lot of calcium to boost the uptake of DNA by bacteria. With the help of the well-known phage called lambda, Mertz planned to transfer, into the bacteriophage called lambda, DNA from the tumor virus called SV40. Pollack also was working on SV40 with cells in culture. The DNA would be snipped open with the newly discovered "restriction" enzymes, and sewn back together with other enzymes called "ligases," discovered in 1967.

Mertz's plans frightened Pollack, who, according to Berg, "had a fit," fearing that cancer genes multiplying in the bacteria of the human gut would "unloose a plague." Mertz called home and communicated these concerns to Berg, who pooh-poohed them. Pollack called Berg up to berate him about the risks. "I thought it was bunk," Berg later recalled, and he tried to dismiss Pollack's repeated objections as "red herrings."

But after he talked to other scientists, he concluded that he couldn't prove the risk was zero, and so the experiment was put on the shelf.[20]

One consequence of the incident was the first, little-noticed conference on biohazards, attended by the tumor-virus and recombinant-DNA people in January 1973.[21] It was held at Asilomar, a state campground, on the California coast near Monterey. Previously a YMCA camp, it had a rugged, stone-walled chapel amid the pines and was not far from a foaming inlet of the sea. Pollack was an organizer of this conference on biohazards. At the conference, Berg recalled, "Jim was the most vehement in his concerns about work with tumor viruses."[22] Watson said that the National Cancer Institute was falling short, morally if not legally, when it declared all "the viruses we work with" as unlikely to cause long-term hazard, so that labs did not need high-security facilities. Watson was interested enough in the topic to have Cold Spring Harbor Laboratory quickly publish the conference deliberations.[23] Many years later, in conversation with Berg, Watson recalled that the conference would never have been held if Pollack hadn't raised the alarm, and Berg said the first conference was a "psychologically important" precedent for the second Asilomar conference in 1975.[24] Scientists had begun struggling in advance over possible risks from particular experiments, and they had begun holding meetings on the topic.

"ALMOST MAGICAL" TOOLS

The worries died down after Asilomar I, but the work at several universities in the United States and overseas, and particularly at Stanford and the University of California at San Francisco, kept pushing the revolutionary field of recombinant DNA forward. The flood of what Watson later called "the almost magical, sequence-specific DNA-cutting enzymes," swelled, thanks to such efforts as Richard Roberts's at CSHL. The enzymes recognized particular short sequences of DNA as the spot to cut; some enzymes cut the DNA every few thousand bases; others did their cutting a quarter of a million bases apart. These enzymes and ligases were used to cut and paste DNA. Tiny satellite rings of DNA in bacteria, called plasmids, became gene-transfer vehicles. To obtain ever-larger amounts of particular stretches of DNA, biologists could transplant genes from complex animal cells into simple, fast-growing bacteria. Scientists began hoping to "read" the sequences and envisaged many uses in medicine and agriculture.[25] They swiftly gained the ability to clone DNA fragments easily, so that Watson could report in 1981 that "work with tumor viruses now could move almost as fast as that with bacterial genes."[26] Molecular biologists began to taste the possibility that their frustrations could dissipate. They could see, in the words of François Gros in Paris, "revolutionary practical consequences [of recombinant-

DNA technology] in public health, animal husbandry, energy production, chemistry, and the environment."[27]

Important tools of what would become recombinant DNA were found and sharpened at Stanford, including several DNA-processing enzymes that were developed in the laboratories of Arthur Kornberg and I. R. Lehman. A key step occurred in 1969, when a graduate student, Peter Edward Lobban, who had been trained at MIT in both engineering and virus studies and was floundering about with his thesis topic, was encouraged to come up with a brand-new topic. According to Kornberg, Lobban came up with "the first technique for making recombinant DNA." Lobban, in Kornberg's eyes, also was one of the first, if not the first, to see the industrial applications.[28]

Some DNA-cutting enzymes cut the two strands straight across, creating "blunt-ended DNA" that seemed difficult to join. Lobban pushed for a way around this. He used a "polymerase" from Kornberg's lab to tack on artificial short stretches of DNA at one end of one strand and at the opposite end of another. To "anneal" DNA that had been cut straight across, Lobban had fabricated what were called "sticky ends."[29]

There was another approach: find a restriction enzyme that would make "sticky ends" naturally. Also at Stanford, Janet Mertz and Ronald Davis, just in from a dry spell in Watson's Harvard lab, found that a restriction enzyme called Eco RI would do it, cutting one strand a few bases away from the cut it made on the other.[30]

Meanwhile one of Paul Berg's graduate students, John Morrow, discovered that Eco RI cut the DNA of SV40 at one place along its 5,224 base-sequence. Berg and two postdocs, David Jackson and Robert Symons, used Eco RI so that foreign DNA could be inserted. [31]

At about this time, another Stanford scientist, Stanley Cohen, with Herbert Boyer of the University of California at San Francisco, decided to harness the plasmids Cohen had been using for studies of drug resistance to the task of moving genes into bacteria. They, too, used Mort Mandel's recipe for getting foreign DNA through the bacterial cell membrane. Along with colleagues, they published their method in 1973. The biologist Mark Ptashne was one of many who recalled Cohen and Boyer's achievement as "a huge event." Suddenly it was "so easy" to do things that had been difficult. Their method was patented, and the patent—a foundation stone of the biotechnology industry—was one of the most lucrative in history.[32]

LETTER TO COLLEAGUES: HOLD OFF

The achievements of Cohen and Boyer, on top of what Lobban and Mertz and Davis had done, set Watson and the other molecular biologists buzzing. The debate was entering a long political phase, during which Watson took contradictory positions and managed to exasperate most of his grave and reverend colleagues—and yet be able toward the end, to say, "I told you so." He may even have helped crystallize the fierce discontent almost all biologists came to feel with a scheme of cumbersome regulations of research practice that sprang up.

Those attending a Gordon Research Conference in New Hampshire in June 1973—less than six months after Asilomar I—heard an ebullient description of Cohen and Boyer's very simple way to create "sticky ends" that would allow foreign DNA, stitched into the tiny circles called plasmids, to be attached for export to bacteria. Genetic manipulation was beginning to look easy, and conference participants were nervous: Were they entering a new domain of hazard? On the last day of the meeting, after perhaps 40 participants had already headed home, 78 of 90 still present voted that a letter of concern about the need to examine possible risks to researchers and conceivably the public of particular recombinant experiments before they were performed should be sent to the National Academy of Sciences.[33] By a vote of 48 to 42, they also decided to send a letter to the widely distributed weekly journal *Science,* which published it in September 1973.[34] Because Berg had already faced the biohazards problem of recombinant DNA, he was drafted to look into the matter and form a committee. For this he recruited others, including David Baltimore, Richard Roblin, and Norton Zinder, to join the inquiry. Jim Watson also was invited because of his prior concerns about risk.[35]

Early in 1974, Cohen and Boyer published their method, by means of which plasmids, those tiny circles of DNA, could have very long stretches of foreign DNA stitched into them, whereas the lambda phage could "only take so much."[36] Not long afterward, John Morrow, who had been at Stanford, published with Cohen a paper on the success they had had using plasmids to transfer a gene which happened to be available in hundreds of copies in the toad *Xenopus,* from the toad into bacteria.[37] The bacteria were converted into factories to make many copies of the gene. The gene was "expressed" in bacteria. It was a powerful new tool, comparable to a new type of microscope. The jungle telegraph began throbbing. Cohen's phone began ringing off the hook as scientists called him to ask for some of his highly successful plasmid. Because the Cohen-Boyer method was very convenient, dozens of biologists were eager to use it. To Berg, "Things were moving extremely rapidly; the technology was burgeoning." Berg's little committee of inquiry needed to meet. And on 17 April 1974 it did, in David Baltimore's office at MIT. The result was an open letter to the biological community, asking for deferral of particular classes of recombinant DNA experiments until an international conference could be held at Asilomar in February 1975.

A strong force that day was Norton Zinder of Rockefeller University, Watson's old research rival and now a key scientist-trustee at CSHL. He recalls that he arrived "absolutely convinced that something had to be done. I didn't know what." The morning, he felt later, was largely wasted on "show and tell. It was a bunch of very clever guys sitting around a table and seeing . . . what they can do with this system." Zinder didn't see his colleagues focusing on action. "The clever guys were just sitting there and babbling away without getting down to the issue, because I think we were a little afraid of the issue." He recalled a year later, "I sometime or other threw my hands up and went stalking out of the room." He asked the others, "Can't we agree on anything?"[38] According to Berg, Zinder declared, "Well, if we had any guts at all, we'd tell people not to do these experiments." The experiments in question involved cancer-causing viruses and drug-resistant bacteria. This idea, Berg recalled, "sort of came as a shock."[39] The "group dynamics" began to change. "Brought back to order," participants got down to specifics. On the blackboard they wrote down a list of experiments that should be deferred. Zinder contributed "egregious scenarios . . . putting botulinus in something."[40]

Jim Watson, Zinder recalled, "was particularly quiet." Laconic as usual, "Jim participated by reading the newspaper" and made no objection.[41] Watson recalled having a stomachache that day, but added, "I don't believe anyone was very worried." According to Watson, two of those present, Daniel Nathans of Johns Hopkins, a later winner of the Nobel Prize in Medicine for his work on restriction enzymes, and Sherman Weissman of Yale, were "very lukewarm to the idea of alerting the public to the fact that we might be working against nature, if not God. Not wishing to appear irresponsible to the public good, they said they would go along."[42] It was relatively easy to ask for a deferral of certain experiments, Watson wrote later, because "recombinant DNA procedures were not yet a necessary ingredient of our day-to-day research."[43] But in hindsight, Watson reflected that "we presumably responsible molecular biologists . . . first gave the DNA scare its legitimacy."[44]

The need for a conference was so clear that Berg had already reserved the Asilomar campground for the following February. But that was a long time off. Something needed to be done soon. So many people were jumping in to the research that they could not be reached except by a public statement. The idea of the letter, which Zinder called "a statement of conscience, not of conviction," was born. There was much uncertainty about the risks—if any—but not much worry. To Zinder, the letter was "consciousness-raising." Because the technology was "too simple and too easy, and it's going to be ultimately too difficult to control," he thought, "moral suasion" was needed to get people to wait for the conference before embarking on experiments that might be risky.[45] Alerting the public of "a possible problem," those at MIT on 17 April 1974 intended "to protect the public, evaluate the hazard, and preserve the science."[46] Richard Roblin of the Massachusetts General Hospital was assigned to write the first draft.

It was then time to test the idea on the road. Zinder carried the conclusions to an international conference in late May in Ghent, in Belgium. The aim of the conference was to review progress on taking DNA apart and putting it together with enzymes, sequencing DNA, and making synthetic DNA sequences. An evening session was reserved to discuss the proposal of deferring experiments until the planned conference. Nathans was chairman. Zinder asked the 50 or 60 scientists present if they disagreed. Most seemed to agree, as Zinder remembered it. John Gurdon of Britain, famous for his manipulations of frogs' eggs, sounded a sour note: He thought the proposed moratorium was an American conspiracy to control recombinant-DNA science.[47]

Meanwhile Baltimore carried a nearly final draft to the executive committee of the Assembly of Life Sciences of the National Academy of Sciences and won unanimous approval. Early in June, he called a special session at the annual Cold Spring Harbor Symposium, that year on tumor viruses, which he had helped organize and which he would summarize in a closing talk. Reading the draft letter, Baltimore ran into the same objection from the Europeans: that the Americans were aiming at hegemony. Others worried "about the thing getting too formal" and giving rise to regulations. Renato Dulbecco and Walter Bodmer were skeptical. To them, it looked like "a tempest in a teapot." They didn't think that the problem was severe.[48] In the end, the Cold Spring Harbor session did not lead to any changes in the draft. With four additional signatures, all from California, the letter was readied for publication in both *Science* and *Nature*.[49]

Once the "dreadfully portentous" letter was published—and misunderstood—Watson began to regret it bitterly.[50] "It didn't occur to us that we were alarming the public unnecessarily. . . . We couldn't have been more naïve."[51] Revisiting Asilomar in February 2002, Watson exclaimed, "I wish we had never written the moratorium letter."[52]

GENETIC ENGINEERING AND THE BOMB

The signers of the letter were taking the risk that the press and public would react to their modest proposal in terms of fear of an Andromeda Strain as imagined in a popular novel. But they did not expect the new gene-splicing technologies, however many their potential uses, to be compared to the momentous applications of atomic physics in World War II and after.

That was what happened, however, in the immediate aftermath of the press conference that Paul Berg, David Baltimore, and Richard Roblin held in Washington, D.C., on the day the letter was published, 18 July 1974, in the fevered atmosphere leading up to President Nixon's resignation on 9 August to avoid impeachment. The three described an "extremely powerful technology" and the existence of widespread agreement with the idea of deferring two classes of experiments. Berg said, "We've taken this route because we feel that the scientific community should be given a chance to regulate itself in its movement into the future."

They said they thought this type of voluntary deferral of new experiments was unique in the history of science, but soon they were asked if there was any analogy to the effort in June 1945 by atomic scientists in Chicago to call off military use of the atomic explosives which were just weeks from being tested and used. Berg didn't think so. In 1945, the atomic scientists faced a moral issue about using something that already existed. Here, biologists were struggling "to assess probabilities without having the data to allow us to do it very accurately." Baltimore, however, said that "we all" had grown up with the great moral dilemma of using atomic bombs. Hence, he could see "a direct line of thinking" from the atomic bomb to recombinant DNA.[53]

Many biologists thought the subsequent coverage was "apocalyptic." Their request for voluntary deferral of particular experiments became in news stories a "ban," a "moratorium," a "halt," even though only narrow classes of experiments were involved and none had yet been done. Berg had months of explaining to do, in the United States and abroad. And Watson got off the train right then. He was not involved in organizing the forthcoming conference.

The press had used the analogy of recombinant-DNA technology to nuclear energy and a wide public took note.[54]

ASILOMAR II: "THE WORST WEEK OF MY LIFE"
Watson arrived at the old YMCA camp thoroughly disenchanted with recombinant-DNA regulation in general and the long-awaited conclave of "scientific bigwigs"[55] in particular, which began on 24 February 1975, amid the scent of the pines and the sound of the sea. Delegates could see an inscription, carved above the conference chamber walls, that began, "Sing, O Heavens and Be Joyful O Earth!" The organizers, led by Paul Berg and David Baltimore, did not think of the conference as a parliament to rule on the risks of recombinant DNA or to vote a set of regulations. They wanted a review of the scientific facts, by about 140 invited experts from the United States and abroad, to help them draft a report for the National Academy of Sciences. They had already been briefed by three subcommittees that had met before the conference and delivered sometimes lengthy reports.

After several "tense, often acrimonious" days in which it seemed likely that the chaotic conference would have no result, the delegates ended up voting nearly unanimously by a show of hands for the principle of defining experiments as more or less risky.[56] They rejected both Watson's opposition to any regulation and the opposite view, that all recombinant-DNA work be stopped.[57] The assembled scientists adopted Sydney Brenner's pet idea of "safe bugs," laboratory bacteria that had been crippled so that they would become a form of "biological containment" on top of physical barriers. The organizers, at first taken aback at the demand for a concluding vote, eventually realized that it reinforced their hope of biological scientists' showing themselves so responsible that legislation, with all its threat of permanence, would be avoided. Little did they realize that intense coverage of the conference in the press by the 16 reporters who had been invited would intensify the concern that had begun building with the issuing of the letter the previous July, and would open the way not only to strict federal rules that took years to soften, but also to years of noisy hearings and punitive bills at city, state, and federal levels—a legislative drive that was only narrowly blunted. A sharp battle began to keep the cops out of the lab.

Less than two years later, Watson confected a satirical account of the proceedings at Asilomar II, where "you eat cafeteria food at large round tables." He recalled that "the only aspect which cheered me up" was that the meeting "was to be held in California in February," which provided him "a formal excuse" to escape the wintry East. Unfortunately, he had brought a pair of tennis shoes half a size too small, and, after a hike in the hills near Palo Alto, had acquired a limp that made it uncomfortable to walk to and from his quarters in the Surf and Sand dormitory. Usually wearing a sweater without a jacket, and sporting an experimental mustache, Watson assumed the role of acerbic backseat driver.

In and out of the chapel, which was "grim, like any 20th-century church," Watson quickly realized that he was "the odd man out." He had decisively switched from his earlier calls for attention to risk to defense of the search for new knowledge. Almost at once he "blurted out that the moratorium should stop." Battling the wide agreement that regulations would be needed, he kept putting forward his message "that if you can't quantitate the danger, you still go ahead if you believe there may be long-term gain from the knowledge you may discover."

He found out that "the experiment I most wanted done soon," breaking up mouse DNA by a "shotgun" method and cloning the mouse genes for antibodies, was branded as potentially "medium" dangerous because of the danger that cancer genes might be spread. He was sure that cancer genes could be manipulated and studied more safely when they were embedded in bacteria. He derided the "ever intelligent" Brenner's idea of "disarmed 'safe' K-12." This strain of *E. coli* was already known to be "totally unsuitable to colonize the human gut." Scientists soon discovered that the safe bugs "are generally so miserable that they hardly can grow anywhere."[58]

The conference was beset by many conflicting forms of discomfort. Some of the younger scientists present silently feared that their careers, based on one of the most exciting technologies in the history of biology, would go down the tubes. Others sensed that "we really shouldn't be in this at all because we just don't know enough." Some, including Jim Watson, sensed a kind of logrolling in which many delegates were striving to define their own work as safe, while jettisoning other types of experiments to costly containment and even de facto prohibition. Ronald Davis thought, "No one wants their system put on the shelf because it's more dangerous."[59] The organizers feared that if they did not reach a "result," that society at large would condemn biologists as irresponsible people playing with fire. Others, including Watson, feared that a process designed to fend off regulation and statutes was actually stimulating interference. The Nobel Prize–winning biologist Joshua Lederberg, then of Stanford University, said that steps designed to keep "the feds" out were actually inviting them in. There was so much tension that a participant recalled, "The beach was 100 or 200 yards away and I never got down to the beach once."[60]

David Perlman, the science editor of the *San Francisco Chronicle,* soon afterward recalled the nonstop schedule. "We started meeting every morning at whatever it was . . . eight or eight-thirty, had a very brief break, went through till lunch, one hour for lunch, back in the meeting, ten minutes [break] in the middle of the afternoon, finish at six or later, and back in session at seven-thirty until ten or ten-thirty, and then talking in that room with the beer." He remembered Watson as one of those who "was always scolding people." During breaks, Watson and other leading scientists were available to talk with the reporters. Perlman remembered Watson as "certainly eager to talk and eager to make pronunciamentos, profane and otherwise." Watson recalled that the reporters were "officially silent but all too opinionated at the coffee breaks." All the reporters there, including Perlman, had agreed that no stories would be filed until the conference had concluded.[61]

It was a difficult conference to watch and report on. The days at Asilomar seemed dominated by putative unquantifiable risks for the health of laboratory workers and the general public. Less emphasis was given to how low the risks were from experiments using the favorite strain of *E. coli* bacteria called K-12. Also overshadowed were talks on the potential of recombinant DNA in tailoring leading food crops to need less fertilizer or smaller applications of pesticides. There seemed, as so often in public discussions of the new biology, far too much emphasis on remote risks and far too little on the need to use new techniques to fight disease and stave off food shortages. The discussions at Asilomar failed to find a good balance between risks that most agreed could not be quantified and needs that appeared urgent.

The four days at Asilomar left Watson feeling angry, put-upon, and futile. He thought of it as one of the low points of his life. But ultimately he wasn't deterred from reentering the fight.

Early in the conference, Watson began his attack on regulating risks that you couldn't measure. He continually annoyed the organizers. On the second day, Maxine Singer, then of the National Institutes of Health, demanded "more data" on what had happened since the previous July "to undo what we did then." Paul Berg was by now thoroughly convinced that biologists must regulate themselves or face having rules imposed on them by society. He said that containment was "common sense in the scientific community," and that scientists would be sued "if we're wrong."[62]

Between sessions, Berg went up to Watson to tease him. A couple of years earlier, Watson had threatened a Harvard colleague with a lawsuit if he "so much as brought an infected lymphocyte" into the Bio Labs. Berg said jocularly that he would sue Watson "for being irresponsible."[63] Watson countered with a question: How could one legislate against conjectural dangers?

Fearing rules that would spare the work of those at the conference while dumping other research, Watson spotted language in a subcommittee report that placed birds in category 3 of risk and whales in category 4. In a jab at the subcommittee's chairman, Donald Brown of the Carnegie Institution of Washington, who worked with the frog *Xenopus,* Watson asked why frog DNA was less dangerous than DNA from a calf. Didn't Brown know that frogs can cause warts?[64] He said acidly, "My intuition totally fails me." Paul Berg rejoined, "We can't come out of here with nothing."[65] But Watson pointed out that cancer viruses, which scientists already were handling at Cold Spring Harbor and Stanford, were closer to being a hazard than the exciting new genetic hybrids. He protested against the hue and cry: "You can't measure it. Don't put me out of business for something you can't measure."[66]

On the final morning of the conference, he perceived that the delegates wanted "both to be responsible and hopefully not to stop the research of anyone in the audience." As he left, Watson said later, his friends thought he "had freaked out." On the plane home, he reflected, "Boy, this has been the worst week of my life!" [67] He felt that everybody there, including himself, were "jackasses."[68] He was determined to keep a public silence about recombinant DNA from then on.[69]

HYSTERIA AFTER ASILOMAR

Discouragement did not prevent Watson from working behind the scenes. With an eye to his own concerns about possible risks from studying cancer viruses, he had quietly begun adding security facilities at Cold Spring Harbor, and invited scientists to come there to work from other institutions where they were hampered. At Harvard, for example, recombinant-DNA work was severely restricted by a departmental feud that led to a public fight in the Cambridge City Council. When U.S. regulations required facilities even more "secure" than those at CSHL, scientists like Joe Sambrook (leader of CSHL tumor-virus research) took off for England, where rules were strict but public hysteria was muted.[70]

Watson also became concerned about a related, potentially far larger source of public fear and political response. This was the speculation that profit-mad businesses were poisoning their own workers with a flood of new chemicals and driving up cancer rates faster than the aging of the population would have done by itself. The issue was a land mine for increasingly visible biologists. Unnecessary regulations would cost a lot of money, possibly aim at small risks when big ones were ignored, and conceivably offer the public little or no protection. On the other hand, in a climate of "carcinogen of the week," biologists were expected to provide the facts about risks and come up with remedies.

Amid the hysteria, the issue boiled down to questions of fact. What were the cancer rates? Did they vary widely from one country or environment to another? Were the rates in fact changing? What environmental factors actually influenced those rates? Which organs, which types of cancer were involved? Watson decided that the issue was too urgent to wait for the deliberate process of a National Academy of Sciences committee, struggling slowly toward a single verdict—which often was the product of hard-fought compromises. Instead, he would invent a new way to get at the issues. He would create a center for conferences about environmental cancer, bring the warring parties together, get them to present talks to each other for three days and submit papers based on those talks, tape-record the whole proceedings to enforce quick submission of manuscripts and capture the discussion—and publish the whole thing in six months. Thus a timely assembly of facts would appear alongside the often wide divergences among the scientists. To do all this, he decided to start off his Banbury Center with this program, and enlist John Cairns, who had become deeply interested in environmental cancer, as chief adviser.

To prepare the intellectual ground, he staged a massive conference at Cold Spring Harbor in the late summer of 1976, calling it "Origins of Human Cancer."[71] The conference laid out some parameters of the situation. Incidence of some major cancers, such as stomach or colon or breast or lung cancer, did vary—sometimes by a factor of 10—from one country to another. This was strong evidence of the influence of environment. The incidence of some cancers, notably lung cancer, was soaring, in direct proportion to the dosage of cigarette tar to which people exposed themselves. Other occupational cancers caused far, far fewer deaths than smoking. Problems were not limited to rich countries. The combined action of

viruses and bacterial toxins was tightly linked to liver cancer in such countries as Mozambique and Iran, and dietary nitrosamines were the principal suspect in esophageal cancer in China.

Just before the environmental-cancer conference, by which time
greater knowledge was greatly lessening concerns about risks from recombinant DNA, Watson was jolted by publication in the *New York
Times* of a blistering attack on recombinant DNA by a disgruntled scientist named Liebe Cavalieri, who had appointments at Cornell University
Medical School and the Memorial Sloan-Kettering Cancer Center. Yes,
biologists had opened an exciting research frontier, he argued, "but the
future will curse us for it." Cavalieri urged the Swedes never to grant a
Nobel Prize for recombinant DNA.[72] Although Zinder later character-
ized it as "a science-fiction story," [73] the article added fuel to the regulatory fire then raging in Cambridge, Massachusetts, and elsewhere.

Jim's alarm was increased by an invitation to speak at the Oyster Bay
Rotary club about "the dangerous bugs we might be letting escape from
our lab." The neighbors were worried. Shortly afterward he remarked,
"While you know that the *National Enquirer* publishes trash, the *Times* is
supposed to have some standards." He attacked Cavalieri: "He spent five
years in the early 1960s trying to convince the world that DNA had four
strands, long after everyone else had accepted the double helix. . . . He is
not one I would turn to for a perceptive feeling as to what science we
should do next." He derided Cavalieri's "amazing strikeout record."[74]
Watson was on the warpath.

"I FELT WE WERE IN REAL TROUBLE"

Despite his great fame, enormously increased by the success of *The Double Helix*, Watson's normal mode of operation was out of the public eye,
in the laboratory, in seminars, on boards and committees (particularly his
meticulously orchestrated meetings with the board at CSHL), and over
meals (spontaneous or carefully organized). With recombinant DNA,
however, he chose the route of colorfully, even intemperately, worded
speeches and articles. Detailed lobbying was left to others, notably Norton Zinder, who was convinced that the fear and shouting basically arose
from a conviction "that our hidden agenda was ultimately genetic engineering in man."[75] It is not clear how many minds Watson changed by
this campaigning, but it relieved his genuine anger and provided cover
for those who quietly felt the same way and who worked steadily to beat
back legislation and regulations.

In the fall of 1976, the legislative furor kept building and opponents
kept talking. At a hearing held by Senator Edward Kennedy, Robert Sinsheimer of Caltech made the dreaded analogy: "To my mind this [recom-
binant DNA] is an accomplishment as significant as the splitting of the

atom." In an interview afterward, he said that the atomic bomb was the first development that led to any question about the beneficence of knowledge.[76] The same hearing left Zinder depressed. Kennedy, seeking information from industry, complained that General Electric refused to send anyone to testify. But GE wasn't doing any recombinant-DNA research. Zinder said, "I felt we were in real trouble."[77]

New York was gearing up to pass a law regulating recombinant-DNA work. A month after the Kennedy hearing, opponents and supporters of recombinant-DNA research appeared at hearings before New York State Attorney General Louis Lefkowitz, who advocated a strict state bill. The supporters of the bill included Cavalieri, Jonathan King of MIT, George Wald of Harvard, and Watson's traditional bugbear, Erwin Chargaff of Columbia Physicians and Surgeons, all of whom advocated a total moratorium. Most of the scientists testifying, including David Baltimore, took another tack: They said the NIH guidelines promulgated in June provided adequate safeguards. One witness was Robert Pollack, who in a sense had started it all. The proposed New York law, he said, "would turn the county health commissioner into the sole arbitrator of our individual research efforts."[78]

Only one scientist, Jim Watson, testified "against any form of legislation."[79] He testified just after the perpetually embittered Chargaff. The organizers of the hearing, whom Jim referred to a month later as "two particularly twerpish types," were told that he had just come from the hospital where his five-year-old son Duncan had been taken because of a broken arm. He asserted, "You know we take in DNA every time we eat. Each time we bite on a raw carrot, healthy carrot DNA is going down our gullets."[80] Watson started off with a joke. He had told Sargent Shriver, brother-in-law of the late President Kennedy, that the whole subject was like "talking religion," the classic futile topic. Referring to a notable fiasco of the Kennedy administration, Watson told Shriver he thought "it was the most overblown thing in American science since his brother-in-law created the fallout shelter." He was referring to President Kennedy's hasty proposal to build shelters during the 1961 Berlin crisis. The shelter idea quickly was perceived as an excessive response.

In Watson's eyes, "The marginal danger of this thing is a joke compared to many other dangers." Referring to his sons, Rufus and Duncan, he said, "I'm afraid that by crying wolf about dangers we have no reason at all to worry about, we are becoming indistinguishable from my two small boys. They love to talk about monsters because they know they will never meet one."[81]

After describing the building of a P3-level high-security lab at Cold Spring Harbor, only one level below the P4 security of a germ-warfare laboratory, he was asked—in a prosecutorial manner, he thought—

whether anybody had gotten sick. He said he'd be glad to send in a color photograph of Tom Maniatis, who was using the facility to do recombinant-DNA experiments he couldn't do at Harvard, which hadn't completed its P3 facility. He invited the officials to judge whether Maniatis and his crew were abnormally pale.[82] He dismissed the furor: "You could say I should be careful never to have oysters when I have a pain in my stomach—when I buy Rolaids. There is a limit to where I will carry my own paranoia."[83]

At Cornell late in November Watson was, if anything, more dismissive of any danger connected to recombinant-DNA research. He asked rhetorically, "Is this the first time in eons that DNA has moved about between totally unrelated organisms?" His answer was, "Not in the slightest!" Molecular biologists had been "damned fools not to have blasted this evolutionary insight around the world before, not after, Asilomar." He said he "would eat grams of any K–12 strain carrying recombinant DNA rather than be licked by any neighbor's dog." He felt he was acting in "a real theater of the absurd in which the only professionals were a bizarre collection of kooks, sad incompetents, and down-right shits."[84] Some months later, that language was repeated in the press. Watson explained that the kooks opposed recombinant-DNA research on "doctrinaire religious, phony environmental" and other grounds. The incompetents he defined as "people who have been total failures themselves over the past 20 years of DNA research." The shits he said were "politically motivated groups" who had no fear of recombinant DNA but promoted it "as part of their general attack on the U.S. system and its science priorities."[85]

He continued in this vein in annual reports and magazine articles he wrote over the next several years. In 1978, he wrote that it was "time to bury Asilomar."[86] The following year he denounced a "DNA folly"[87] and "the DNA biohazard canard"[88] and wrote, "I believe we should quickly and resolutely abandon any form of DNA regulation."[89] Underlining a major theme, he asked, "Why should we be afraid of something that goes on in nature all the time?"[90] Soon he and John Tooze, an English biologist and writer based in Heidelberg, Germany, pulled together a collection of relevant documents about the controversy in what he referred to dismissively as "the DNA crapbook," which was published in 1981 as *The DNA Story.*

THE CRITICAL YEAR: 1977

Those who had raised the alarm in 1974 had become thoroughly fed up with the process they had unleashed. Berg said, "Our work is crawling." Baltimore complained, "At this point, 30 percent of my time is being taken up by talking on this issue." Baltimore said of Robert Sinsheimer, one of the critics of recombinant-DNA research, "I think [he] has allowed his imagination to run away with him."[91] Baltimore was one of 13 members of the National Academy of Sciences who drafted a resolution by the Academy expressing concern about forthcoming legislation.[92] Berg scoffed at the idea of waiting some years until a safer organism than *E. coli* could be identified. He said it was "unrealistic. We may never find another organism." Of Sinsheimer, he said, "For one guy, no matter how eminent he is, to go up and say 'I fear,' and then expect the whole world to stop. . . . It's not going to happen." Watson kept hammering at the insight that exchanges of information between bacteria and other species had been going on throughout evolution. "I don't think we're doing something which has never happened before." As for the time taken up by the controversy, Watson said he was totally dominated by it. "You have to be manic."[93] Stanley Cohen attacked "the vague fears of the unknown" that were hindering the use of recombinant DNA to attack real risks like disease, hunger, and environmental degradation.[94] Roy Curtiss III of the University of Alabama, a principal worker on "safe bugs," wrote the director of the National Institutes of Health that, painfully and reluctantly, he must acknowledge that the risks had been overstated.[95]

As 1977 opened, the Cambridge City Council received a report on recombinant-DNA research from a citizens' review committee and decreed that scientists in Cambridge, chiefly at MIT and Harvard, would be required to observe the NIH guidelines promulgated the previous June.[96] In both California and New York, legislators were pushing for statewide regulatory statutes providing for registration of all lab employees, reporting on details of research, regular inspections of labs, and fines and closure as penalties for violations.[97] Bills to accomplish similar aims, including regulation of industrial research, emerged in the U.S. House and Senate,[98] although they contained seeds of a future impasse, which represented an opening for lobbyists. On the one hand, Representative Paul Rogers of Florida wanted federal laws to preempt state and local regulations stiffer than the federal ones. On the other, the principal sponsor of the Senate bill, Edward Kennedy of Massachusetts, opposed preemption of state regulations.[99] Watson wrote soon afterward, "Happily, Congress got tangled up."[100]

Kennedy evidently accepted the analogy of recombinant DNA with the birth of atomic energy. On a visit to Stanford, he told Arthur Kornberg, "This time we want to be in on the takeoff as well as the landing." Kornberg was incensed. He wrote the NIH director, Donald Frederickson, in February 1977, "What we must now fear most is not the remote possibility of biological warfare but instead the war on biology." Looking back 20 years later, he said, "The worst thing you can do is curb the acquisition of new knowledge."[101]

Many leading scientists, in and out of the government, began to feel that some kind of legislation was inevitable and focused their attention on obtaining the mildest possible federal law, one which would preempt state and local action.[102] Environmental groups filed suits to stop all recombinant-DNA work.[103] A contentious public forum in the auditorium of the National Academy of Sciences in Washington in March seemed to demonstrate the ability of opponents of recombinant DNA to make noise and capture headlines, and made the excitable Zinder certain that "all was lost."[104] But in the spring, Robert Sinsheimer reflected, the discovery of split genes, showing the human genome to be organized in a far more complicated way than that of bacteria, helped make the idea of cancer epidemics even more remote. Sinsheimer had been a leading backer of a tough regulatory bill in California, but withdrew his support. After he did this the bill died.[105]

Scientific societies, most notably the American Society for Microbiology, voted to lobby hard in Congress for a mild bill, and collected endorsements from other societies. Philip Handler, president of the National Academy of Sciences, was also convinced that legislation was coming, even though the hazards "exist largely in the imaginations of a very small group of scientists."[106] Lobbying continued on other fronts. In June 1977, the annual Gordon conference on nucleic acids took a very different stance from the one taken at the 1973 conference. The participants drafted a vigorous letter saying that accumulating scientific data made it clear that the risks had been overblown. The Gordon conference attendees called for no new law and urged relaxation of the guidelines.[107] Stanley Cohen, one of those who had started it all, deprecated the risks in a letter that circulated widely on Capitol Hill.[108] But many still expected laws. Although the effort to pass a law in California fizzled after Watson advised a committee in Sacramento to "wait until someone gets sick at Stanford,"[109] the New York legislature passed a regulation law with penalties for noncompliance.

Now came the turning point. After behind-the-scenes lobbying by Zinder, who kept Watson informed about what he was doing, Hugh Carey, the Democratic governor of New York, vetoed the law on a matter of principle: academic freedom.[110] The scientists could not be told what work to do or not do. Air began slowly to hiss out of the regulation balloon. In the U.S. House, Representative Paul Rogers of Florida, principal sponsor of a recombinant-DNA bill, did not get on with his committee chairman, Harley Staggers of West Virginia. Letters went to Staggers.[111] Rogers's bill never emerged from committee. There was trouble also for Kennedy's harsher bill in the Senate. Senator Adlai Stevenson III of Illinois muscled in by organizing his own hearings in a separate committee. After a letter from 12 leading biologists at the University of Wisconsin,

Senator Gaylord Nelson of Kennedy's committee put forward what amounted to a substitute bill. Kennedy could sense support for his bill slipping away. Zinder observed, "I think Kennedy knows he's way out on a limb." It is said that he began to hear insistently from lobbyists for Harvard, who presumably explained that the bill would disrupt one of the most important industries in Massachusetts, medical research. In September, at a gathering of medical writers, Kennedy threw in the towel. He put off his bill for the year.[112] It turned out to be forever. Watson rejoiced at the "politically astute" action. It "almost certainly signifies that recombinant DNA no longer has any political value and that public hysteria cannot be maintained indefinitely in the absence of a credible villain."[113]

In the fall, biotechnology companies such as Genentech in South San Francisco began to show results. The manufacture of the hormone somatostatin in bacteria was soon followed by the manufacture of human insulin. Medical value could be seen. Money and jobs and continued competitiveness with Japan in biotechnology could be seen. The phrase "recombinant DNA" was giving way to the word "biotechnology."

The year closed with a meeting in Bethesda, Maryland, of the NIH Recombinant DNA Advisory Committee, called RAC. After hearing sometimes heated testimony from scientists who said that the farce had gone on long enough, the committee cut back on the regulations.[114] After a noisy and threatening few years, molecular biologists were on their way to regain full freedom of choice about their experiments. They could freely clone genes, and, thanks to new techniques pioneered by Walter Gilbert and Allan Maxam at Harvard, and Fred Sanger in Cambridge, England, sequence those genes 100 times faster than before.

WAIT UNTIL THE SUN BURNS OUT?

The crumbling of barriers to carrying out research with recombinant-DNA techniques did not, of course, suppress the undercurrent of unease with the new powers of modern biology. Opponents of recombinant-DNA research shifted their focus to trying to stop introduction of genetically modified crops into agriculture. As before, Watson stuck to the point he made constantly in the recombinant-DNA furor, that there are benefits "too great to forgo."

People should not shrink from the chemical and physical nature of life, he maintained. "It still bothers people that a disease could be caused by just a wrongly folded protein."[115] Genetic knowledge, he held, was the main hope of conquering diseases and it was not emerging fast enough. Inhibiting the work was immoral.[116] The new knowledge should be used quickly and widely. The effects were almost certain to bring far more benefits than risks. New things must be tried. If we waited to answer all the doubts, "we might as well wait for the sun to burn out."[117] Of course the genetically modified crops that had been discussed at Asilomar must be used to increase the world's ability to grow food.[118]

Watson was equally bold and forthright about an even more controversial issue in genetics. Failure to abort a fetus with a severe defect, he said, exposes the stricken child but also everyone caring for him or her to years, decades, of anguish—and hence is immoral.[119] Of course the incidence of severe defects should be reduced by abortion, well before people figure out—if they ever did—how to do "gene therapy" on tragically stricken children and adults. If germ-line gene therapy—the injection of genes into sex cells so that the change would become hereditary—ever became available, why shouldn't mothers try to endow their children with more intelligence or with a pigment that would make them less susceptible to skin cancer?

The response was expectable. Detractors shouted, "Eugenics!" They were not just referring to Hitler's infinitely hateful and devastating misuse of a pseudoscience to kill millions of Jews, gypsies, homosexuals, mental patients, and old people. As Watson frequently pointed out, there were campaigns for sterilizing the mentally ill in the United States and Sweden, and for preventing southern Europeans from immigrating to the United States. But while he frequently recited these misuses, he held that disuse was a bigger danger than misuse. Despite the shock, or perhaps because of it, he remained in demand as a speaker into the twenty-first century.

On 23 September 1981, at a taping of Phil Donohue's television talk show in Chicago, Jim Watson was asked what recombinant-DNA technology would be used for. He talked about improving prenatal diagnosis of severe deformities in a fetus, to offer the choice of the tested technique of abortion. Gene therapy lay in the indefinite future. A questioner disagreed: If she knew she had a deformed fetus, she would go ahead with the pregnancy. Watson snapped back, "I'd hate to be the child you were so eager to bring into the world!" Watson recalled two days later that the startled audience burst into applause.[120]

12

GENOME:
"IT IS SO OBVIOUS"

*When finally interpreted, the genetic messages encoded within
our DNA molecules will provide the ultimate answers to the
chemical underpinnings of human existence.*

JAMES D. WATSON, 2 AUGUST 1989

WHO?

Late in February 1988, the National Academy of Sciences published a re-
sounding scientific consensus on a proposed $3 billion project to map
the entire human genome.[1] Just afterward, James Wyngaarden, the director of the National Institutes of Health, convened an ad hoc group of 18
advisers on "complex genomes" at Reston, Virginia. They were to confront the dilemma of who should lead the now-inevitable enterprise. The
inevitable answer was Jim Watson.

Wyngaarden did not want the project to go by default to the big national laboratories of the Department of Energy. Other, more reluctant,
NIH leaders agreed, although they feared diversion of funds from the
rest of biomedical research. To chair the meeting, in what Watson
thought was a "brave" move, Wyngaarden chose a prominent skeptic,
David Baltimore of the Whitehead Institute at MIT in Cambridge,
Massachusetts. In 1987, Baltimore had attacked the "megaproject" as
"huge [and] low-priority," and "not crucial."[2] Shortly before the Reston
meeting he said in a speech in Boston, "The scientific rationale for sequencing the human genome is not particularly evident." Wyngaarden
considered it critical to get Baltimore involved, in the hope of muting his
public opposition.[3] His choice was successful; Baltimore and the other

committee members agreed in recommending a program to be based in Wyngaarden's own office. Watson would run it as an associate NIH director. Jim had a unique combination of the traits necessary to achieve success in this role: experience with building scientific programs and an inexhaustible appetite for gossip, intellectual influence and star power, guile and outspoken impatience, passion for basic knowledge, and political sensitivity. Only he could keep selling the project to a growing cast of supportive leaders in Congress, mobilize the necessary scientific talent, and ensure that the project's center of gravity would reside at NIH.

Scientists had mixed reactions to the choice of the highly visible Watson as leader of this unprecedented campaign in genetics. He was "really a celebrity, in a way that none of the rest of us will ever be." He was not afraid to push for a big goal. He was also not afraid to say that it would help people. The cancer virologist Harold Varmus, a winner of Nobel Prize and director of the National Institutes of Health from 1993 to 1999, recalled that publicity-hungry members of Congress uncharacteristically hung around a whole day at a hearing where Jim testified, and "everyone wanted their picture taken with Watson." Jim set "high standards and high goals" for the project, and made clear that "goal-oriented doesn't mean bad science." Yet his concern for medical benefit was grating to those for whom fundamental knowledge was the Holy Grail. "He is very sensitive to what really troubles people about the human condition, whether that's cancer or heart disease or mental illness. He sees the potential to diminish anguish through science. . . . Jim has always been unabashedly willing to say that what we do is good for something."[4]

Watson, by no means the project's first champion, had repeatedly intervened, mostly privately, in the debate about the urgency of building a dictionary of all the "instructions" to make a human being. Shortly before the Reston meeting he told a reporter, "It has got to go ahead; it is so obvious. The only question now is the rate and under whose auspices."[5] Marshaling arguments most likely to convince Congress, he told another reporter, "There are absolutely tremendous benefits in medicine and biology to be gained, as well as spin-offs for biotechnology."[6] Months earlier, however, he had publicly professed reluctance to take the job: "There are going to be bad years, the program is going to be under economic attack, and there's got to be someone there who sees all the components and fights for it. I couldn't think of a job I'd like less."[7]

But at Reston Jim defined the job in a way that seemed to exclude anyone else:

I strongly urged that the Associate Director position be filled by an active scientist, as opposed to an administrator, arguing that one person had to be visibly seen in charge and that only a prominent scientist was likely to reassure simultaneously Congress, the general public, and the scientific community that scientific reasoning, not the pork barrel, would be the dominant theme in allocating the soon-to-be-large genome monies.

Then I did not realize that I could be perceived as arguing for my own subsequent appointment. For many years, my most visible role had been that of an administrator dominated by the fund-raising activities needed to keep Cold Spring Harbor Laboratory at the forefront of DNA-based science. Whether I was still a real scientist was not at all clear.[8]

During the Reston meeting, Watson argued that the big projects in particle physics should be an example of how the Human Genome Project should be conceived. "They always have a scientist, not just a manager. They have a scientist who is a strong advocate of the project and a good manager, too." The idea of naming Watson is said to have crystallized during a lunch break. At a moment when Watson left the room, Wyngaarden and the others agreed that Watson was the only credible leader. His appointment would be seen as guaranteeing first-class science. It appeared that the job would have to be structured to attract him.[9] But Watson reflected a couple of years later: "If I [had] turned down the job, it was not clear that any prominent scientist active in the lab would take on the task."[10] In the weeks after the Reston meeting he began to reflect that "I would only once have the opportunity to let my scientific life encompass a path from the double helix to the three billion steps of the human genome."[11] He reflected, "My name was good."[12] In May, when Wyngaarden offered him the job, he said yes, and began figuring how in the world he could juggle jobs in Washington and Cold Spring Harbor.

A STARTLING IDEA

In 2000, attended by a host of controversies and hopes, the Human Genome Project and its commercial rival, Celera Genomics, achieved a "rough draft" of humankind's entire DNA endowment, with the goal of a final draft in 2003, the fiftieth anniversary of the double-helix discovery. Both efforts used new sequencing machines whose development was led by Michael Hunkapiller of the California-based company Applied Biosystems. By far the biggest undertaking in the history of biology, the project uncovered an astonishing degree of "conservation" of apparently indispensable genes and gene functions as species evolved over billions of years into more and more complex organisms.

Jim had not been an early convert to the idea of mapping the human genome, but eventually, his lively, even unique foresight and intellectual opportunism brought him once again, between 1988 and 1992, into the role of a detonator in biology. But the enterprise had been thrust into Jim's mind years earlier. This happened in 1985, when he said his official farewell to Joe Sambrook, who for so long had led the tumor-virus work at Cold Spring Harbor.

It was a "sizzling" hot September day, during a tumor-virus conference at Cold Spring Harbor. Appropriately, a new wing of James Laboratory, Joe's empire for many years, was dedicated as the Sambrook Laboratory. Renato Dulbecco of the Salk Institute, Jim's old lab mate at Indiana and a longtime pioneer in the laborious study of viruses that induce cancer, was the dedication speaker. Back at the start of Jim's adventure in leading Cold Spring Harbor, he had recruited Joe from Renato's lab.

The guests, including many alumni from Sambrook's lab, took their places on the metal chairs ranged for the ceremony on the newly laid asphalt of a parking lot. The tips of the thin chair legs sank into the tar, to everyone's mixed amusement and embarrassment. There was another surprise: Dulbecco did not give the usual dedication speech. Of course he was there to recognize Joe's achievements in science—but he quietly said he wanted to change the subject. He advocated a truly grandiose idea that, despite enormous obstacles, was becoming both possible and necessary. The intensifying hunt for genetic diseases dictated making a complete sequence of all the 3 billion "letters" in the genetic alphabet of human beings. The means—the knowledge of how to sequence DNA and the automated machinery—were there to do it.

Far more than before, biologists and physicians were thinking about human disease as a problem in genetics. Recently, discovery after discovery had shown that slightly altered genes, acting alone or in combination, not only lie behind rare inherited disorders but also have a role in causing diseases affecting millions, such as cancer. This was not just a matter of gross changes in the chromosomes. It was already known that important inherited diseases, like sickle-cell anemia, resulted from the change of a single "letter" in the DNA. Thousands of disease genes and dozens of cancer genes had been detected, but it was an exasperatingly slow process to locate them among the chromosomes and nail down their precise genetic makeup—the exact sequence of A and T and G and C in their DNA.

The DNA sequence was seen as a broad highway into the forest of worker proteins, enzymes, whose makeup was specified by the genes. Exploring proteins in their tens of thousands was a task much vaster than deciphering the genome: it was the principal preoccupation of biologists after 2000. The proteins have two salient roles. If the protein is missing or not abundant enough, as with insulin in diabetes, it could be supplied. If a mutated protein is harmful, perhaps a normal protein could be supplied or a drug could combat the defect. The hope was bright and immense. But because of the slow pace of finding and deciphering genes, much of the medical relevance of modern biology still seemed to lie discouragingly far in the future. Why not reach it sooner?

The concept was in the air, but Dulbecco did not know then that anyone else was thinking about the project. According to Heiner Westphal, an alumnus of Dulbecco's lab and Joe's, who was sitting in one of the metal chairs that hot afternoon, Dulbecco had not told "everybody openly about it until this event." Westphal thought it "quite likely that he got [the genome project] in Jim's brain on that day, that minute."[13]

GENETIC FOCUS ON DISEASE

The new focus on genetics intensified the ambitions of molecular biologists for relevance in the fight against disease. People began to imagine that the Human Genome Project not only could be done, but should be. It was perhaps the most dramatic instance yet of Brenner's slogan: "Think small, talk big."

Speed was always a concern and the lack of it a constant frustration, not just for Jim. As we have seen, the early tumor-virus work at the Salk Institute, at Cold Spring Harbor, and at many other cooperating and competing labs went terribly slowly, for want of an ability to get large supplies of particular stretches of DNA so that the genes could actually be "read." The recombinant-DNA techniques developed in the early 1970s had sped research almost miraculously, so that biologists could pick out and clone particular genes and move them at will into rapidly multiplying microbes. But the biologists' imaginations soon outran this great advance. Wasn't there a faster way to obtain large numbers of copies of genes of interest, so that biologists could work out the sequences of those genes, particularly disease genes?

If one could know the position and exact message of genes, a new era of personally tailored medicine could open up. Subgroups of patients could be detected who would have an adverse reaction to particular drugs, or who would need higher doses. It would be possible to identify newborns who should follow special diets and regimens to reduce their chances of contracting a disease to which they were predisposed. People at risk of transmitting a catastrophic mutation to their children could choose to be tested, not merely to identify fetuses for which an abortion was indicated, but also to confirm normal fetuses not carrying a particular inherited disease. No longer would couples from high-risk families have to consider never having children or aborting all pregnancies. One could avoid the tragedy of a young mother of a three-year-old boy who already showed early symptoms of muscular dystrophy bringing along to the doctor's office a second baby son who also carried the disease.[14]

MAPPING, RADIATION, CANCER, GENE THERAPY, AND MACHINES

Behind the promise of these great benefits lay several intellectual and technical developments.

Gene Mapping

There was a powerful new way to map the location of healthy or disease-carrying genes on chromosomes, such as the 23 pairs into which human genes are packed. The method emerged suddenly at a conference in 1978 in the ski resort of Alta, Utah, from the mind of David Botstein, then of MIT. The new tool took advantage of the fact that DNA-cutting enzymes would produce fragments of different lengths from different people. And so the technique acquired the tongue-twister name of "restriction fragment length polymorphisms" (RFLPs or "riflips"). In the next few years, with the help of RFLPs, the genes for such catastrophic and rare diseases as Huntington's chorea, cystic fibrosis, and muscular dystrophy were all located. But scientists also focused on cancer—and heart disease. A gene was found that boosted the cholesterol of certain families far above normal, and predisposed family members to early heart attacks. That finding earned Joseph Goldstein and Michael Brown a Nobel Prize in 1985, and crucially spurred the development of cholesterol-lowering "statin" drugs taken by millions in the United States alone by the end of the century. Even in the early 1980s, genetically based disease was being called "the next frontier of medicine."[15]

Radiation Risks?

A quite different impulse came from the nuclear industry and the field of radiation genetics—pioneered so long before by one of Jim's teachers at Indiana, Herman J. Muller. Key agencies like the U.S. Department of Energy wanted to know the risk of mutations to humans from nuclear radiation, whether from a cataclysmic event like a nuclear explosion or, at enormously lower levels, from the generation of nuclear power. For almost four decades, scientists from the United States, Japan, and other countries had sought answers by following some 12,000 children conceived after their parents had survived the atomic bombs of Hiroshima and Nagasaki. Did genetic mutations show up more often in the survivors' offspring than in the offspring of "control groups" of people who had not been exposed? The puzzling answer was that no extra effect could be detected with the relatively insensitive methods then available.

The question "Just how resilient are our genomes to various kinds of environmental damage?" was far from closed. An Energy Department official said later, "We realized that we didn't have any methodologies sensitive enough to detect the mutations if they were there"[16] The answers were vital to the nuclear industry, beset by fears of insidious or sudden leaks of radiation. Might there not be tiny but harmful alterations in the children's genes? To solve this, human geneticists began to realize that real answers must await a total human DNA sequence. Meeting on the subject in 1984, again at Alta, leaders in the field saw no immediate way to get there. Among those at Alta were David Botstein and such other future Human Genome Project leaders as Charles Cantor and Maynard Olson, but not Watson. They arrived discouraged at the lack of techniques to compare damaged and undamaged DNA, but left somehow excited. The brass hats of DOE research saw an opportunity looming to crack the nut.[17]

They assumed that the institutions that would solve the total DNA sequence were the huge laboratories in such places as Los Alamos, New Mexico, and Berkeley, California that the department had inherited from the Atomic Energy Commission, formed after World War II. Possessing some of the world's most intense experience with big computers and long familiarity with Big Science projects, the labs seemed natural environments for working out the human genome sequence. Besides, the Cold War missions of the laboratories were winding down.

Cancer Genes

A factor that increased the hunger for a complete sequence, the one Dulbecco stressed, was a torrent of discoveries since 1981 concerning two classes of genes associated with lab animal cancers. The first class, known as oncogenes, could mutate into slightly different forms to set cells growing uncontrollably. It was like pressing down on the accelerator. The second class, tumor-suppressor genes, or anti-oncogenes, inhibited formation of tumor cells when they were operating normally. But if they mutated, their suppressor functions were degraded. It was like taking the foot off the brake.[18] Scientists at Cold Spring Harbor, including Michael Wigler, Ed Harlow, and Earl Ruley, had figured importantly in discovering both types of genes.[19]

Gene Therapy?

Yet another factor was the much-publicized possibility of "gene therapy," which was pushed by some scientists but regarded skeptically by many—as a 1982 conference at Cold Spring Harbor's Banbury Center showed. The hope was to supply missing genes to particular individuals and thereby avoid many instances of therapeutic abortion, and reduce the severity and expense of special medicines and diets. At that stage the idea took the form of using benign viruses, if such could be found, to inject genes into people who lacked them, but not in a way that the change could be passed on to offspring. To some, at least, this looked like a blunderbuss approach, to be used only in extreme cases at great cost in money and time, and with great risk. Instead, why not sequence the entire genome, speed discovery of genes, and also open the way sooner to drugs aimed at particular proteins, inhibiting them or supplying those lacking?[20]

Sequencing Machines

Adding impetus was the accelerating development of automated machines capable of working out the sequences of A, T, G, and C in samples of DNA—a task that was still taking scientist-years for each gene in the typically small labs of biologists.

All of these developments, exploding over just a few years, had a powerful effect on fundamental biologists, who, like Jim, had argued for decades that their work would lead to ways to forestall or alleviate disease—but so far had little to show for it.[21]

SANTA CRUZ

Even before Dulbecco's talk at Cold Spring Harbor, others had begun thinking about the feasibility of a human genome project. The virologist Robert Sinsheimer, chancellor of the University of California at Santa Cruz, got to thinking about sequencing the entire human genome, a project larger than biologists had dreamed of. Biology, he hoped, could move faster toward "a rational approach," not only for human genetic diseases but also for all human growth and development. He thought this bold concept might recapture a huge endowment for astronomy, which, at the end of many twists and turns, he was losing. What about using the money for something equally dramatic? To be sure, many biomedical labs sequenced bits and pieces of DNA, but now Sinsheimer wanted an assessment by experts of doing the whole thing. Without this, he could not shoot for the money.

In March 1985, in the Babbling Brook Inn near the Santa Cruz campus, Sinsheimer and Robert Edgar brought together "a handful of dreamers and a few skeptics."[22] Among them were several who later became leaders of the Human Genome Project. As far as winning back the endowment, the meeting did not convince the president of the University of California system. Sinsheimer had told him the genome project would produce "a permanent and priceless addition to our knowledge," but the president wasn't interested. The scientists at the Babbling Brook Inn, however, found their interest mounting—much to their surprise. They cast aside their initial deep pessimism and concluded that the vast project was not only worth doing but doable. As Sinsheimer recalled: "Feasibility was no longer the issue."

Influential scientists had begun wondering whether a dose of Big Science could complete the human DNA sequence in 15 years. Among those at Santa Cruz were the ubiquitous David Botstein, the intellectual father of the RFLPs for mapping; Leroy Hood of Caltech, leader in developing sequencing machines; Walter Gilbert of Harvard, who shared a 1980 Nobel Prize with Fred Sanger for developing ways of sequencing DNA; and John Sulston of Cambridge, England, later head of the world's biggest sequencing lab. They could envision a biological equivalent of a huge telescope for finding disease genes instead of stars and galaxies.

But despite the enthusiasm, major pitfalls were visible. Botstein mused that concentrating uniquely on the human DNA sequence would produce a complete set of Egyptian hieroglyphics but no Rosetta Stone to decipher them. Without being able to compare the human data with sequences from other organisms, such as yeast, the human sequence could be unreadable. The meeting concluded that such parallel sequencing was essential, but that the technology for an immediate mass attack on the human sequence had not yet arrived. There were reasons why only short DNA sequences had been achieved so far. Nonetheless, Leroy Hood came away fired up with developing the sequencing machines further. And Gilbert wrote Sinsheimer suggesting a $3 billion, 30-year project at a single institute. He declared, "The total human sequence is the Grail of human genetics. It would be an incomparable tool for the investigation of every aspect of human function."[23]

The conversation was beginning to shift from "Why?" and "Whether" to "How?"

SANTA FE

In April 1986, Dulbecco published in *Science* the thoughts he had spelled out at Cold Spring Harbor the previous year. In March 1986, another player joined him and Sinsheimer. The Department of Energy held its own meeting of 50 people, in Santa Fe, New Mexico. Again Gilbert was there, but not Sinsheimer or Watson. The project looked like "a natural offshoot" to the modest library of genetic sequence data that Los Alamos had accumulated.

At Santa Fe, a major topic was organizing the project. Should it be done in one central facility, in the Energy Department style, or the many decentralized labs biologists were used to? Gilbert pushed for a Human Genome Institute that would at first focus solely on the routine and boring work of grinding out sequences. Sherman Weissman of Yale placed more emphasis on distributing the work in labs across the world, which could lead to more creativity. Overall, however, the consensus was relatively sunny: "The objective was meritorious, obtainable, and would be an outstanding achievement in modern biology." The Santa Fe meeting participants also began spelling out a timetable. The first step would be an intensive mapping of genes while the sequencing machinery was streamlined and the necessary computer technology was advanced. The major sequencing campaign would start later.

Encouraged by "near unanimous enthusiasm," the Energy Department lost no time in contacting White House budget officials and influential members of Congress. One of these was Senator Pete Domenici of New Mexico, whose interest in the continued robustness of Los Alamos was intense.[24]

STORM CLOUDS

Despite the growing enthusiasm, long shadows lay over a genetic approach to disease. Some people, moved by religious fundamentalism, opposed any interference in God's plan. Others, with Aldous Huxley's *Brave New World* in mind, feared that new genetic knowledge would lead inexorably to new forms of oppression of the many by the few. Further, biologists and physicians alike were still sore from the "War on Cancer" of the early 1970s. That massive push for cancer treatment not only had spread false optimism among politicians and the public but had also threatened to engulf the basic research that was a far likelier source of more effective, more targeted, and gentler diagnoses and treatments.

Biologists were also gun-shy after the recombinant-DNA fight of the 1970s, with its accompanying irrationalism, threats of legislation, and years of struggling to beat back regulations. The events of the 1970s carried the lesson that biologists must tread as carefully as physicists in neither overpromising nor overthreatening. Some scientists, including Jim, were determined to avoid repeating what they saw as mistakes.

EMOTION AT COLD SPRING HARBOR

The next step after Santa Fe was a special session during the annual Symposium, organized as always by Watson, at Cold Spring Harbor in June 1986. There, biologists became embroiled in often-contradictory emotions. The question of whether to do the project was increasingly difficult to divorce from questions of how it would be done, and who would lead it. Many who gathered in the new 350-seat Grace Auditorium hated the idea of Big Science in biology. They not only questioned the urgency of obtaining the full sequence but also claimed that the spending for it would inevitably siphon off money for grants for Small Biology. But if there must be a genome project, the Department of Energy mustn't be allowed to lead it, because it lacked experience with crucial techniques and didn't award grants on the basis of peer review, as did the National Institutes of Health.

It was obvious that the Department of Energy, despite its aggressive push on the genome, was not the principal supporter of biological research in the United States—the National Institutes of Health were. Thousands of biomedical scientists were supported by NIH and participated in the peer-review study sections that ranked competing proposals according to scientific merit. Biologists began to fear that a project of central importance to biology would be captured by a top-down agency run by administrators, not scientists.

Jim had come to see the validity of many of these fears. He was convinced that the project was necessary, but equally convinced that the Department of Energy must not control it. "It is not possible to do the project without a lead agency. There is only one genome."[25] Jim "wasn't arguing that it shouldn't be done, but that it was too important for the NIH not to be involved." He told Energy Department representatives at the Symposium that they wouldn't lead the program if he had anything to do with it. The program must be run as "an A effort, not a B effort." Jim was convinced that "it must be done by scientists. You can't build a bridge without calling in the engineers."[26] To reach the right result, consensus must be built among biologists—and among legislators. Jim was getting deeper into the genome project.

Part of the problem was NIH, which was uncomfortable with such a Big Science project. Further, NIH tended to think proteins were more important than the genes in DNA that specified the makeup of the proteins. Watson liked to joke that the "protein" people who led NIH really didn't like DNA.[27] Only one leading NIH official, Vincent DeVita, the head of the National Cancer Institute, responded to an invitation to Santa Fe, but he didn't show up.

Debate on hot issues in biology was a specialty at Cold Spring Harbor, and 1986 was a particularly appropriate time for a discussion there. Mindful of the explosion of human genetic knowledge, Jim had organized the annual Cold Spring Harbor Symposium around "The Molecular Biology of *Homo sapiens*." The catchy title—Jim called it "sweeping"—drew more than the usual number of science journalists to join more than 300 scientists (including both Botstein and Gilbert, and a galaxy of disease-gene discoverers) at the traditional weeklong, exhaustingly detailed meeting. That year's Symposium, Jim wrote, celebrated "a turning point in modern biology, . . . our newfound ability to study ourselves at the molecular level." He reflected, "DNA science had at last come of age."[28]

The Symposium opened with forceful endorsements of a human genome project from Walter Bodmer, a leader of cancer research in England, who stressed that this "most exciting human endeavor" would start producing medical benefits long before completion,"[29] and Victor McKusick, a leading human geneticist, who said, "Complete mapping of the human genome and complete sequencing are one and the same thing."[30]

Lee Hood also spoke enthusiastically. Little more than a week later, he and colleagues at Caltech, including hardworking graduate students and postdocs, publicly announced a breakthrough in the sequencing machines they had pioneered in California over the previous five years. They had found a way to label each of the bases with a different fluorescent dye and sort the bases with the help of lasers and computers. Sequencing with the help of radioisotopes could be left behind. Hunkapiller, the principal pioneer in designing successive generations of sequencing robots, led the work of designing and building the $90,000 machine for Applied Biosystems in Foster City, California. It could sequence 1,000 bases an hour, that is, do in one day what had been taking a scientist a whole year. Applied Biosystems was already popular on Wall Street.[31]

Adding to the excitement at the Cold Spring Harbor Symposium was the colorful, even bizarre, Kary Mullis of Cetus Corporation in California, who shared the 1993 Nobel Prize for Chemistry. On Saturday evening, 31 May, just after the traditional picnic on the lawn below Airslie, Mullis exploded a methodological bombshell: polymerase chain reaction, or PCR. Many of those listening were hearing for the first time about this enormous speedup of DNA research, what Mullis called "a process that provided all the DNA one could want." In the next few years, biologists gradually taught themselves to use it. With PCR, Watson wrote later, one could pick out genes, those "rare targets in a complex genome."[32]

The following Wednesday afternoon, in Grace Auditorium, the storm broke. Paul Berg and Walter Gilbert, co-Nobelists of 1980, were chairmen of the session on the total-sequence program—the need for it, and its organization, pace, and scope. Unaware of either the Santa Cruz or Santa Fe meetings, Berg had suggested the public discussion, and Watson recruited Gilbert with a telephone call.[33] Jim clearly hoped that the assembled biologists would not only be excited by the idea in general but would insist that NIH take a dominant role.

Instead, doubts and even anger spread through the auditorium. Mapping the location of genes looked more urgent than mass-scale automated sequencing, which looked too expensive. David Botstein, who was there, reflected that small-scale research would be starved, just as the Space Shuttle had sapped the vitality of planetary astronomy for 20 years.[34] As Watson put it sardonically later, there was fear that the project would be "so dull it would be done by people you wouldn't want to have dinner with." The hope of "knowing everything there is to know about the human genetic blueprint," a reporter for *Science* magazine wrote, had aroused "a significant level of doubt."[35]

What would it cost? Gilbert went up to the blackboard and wrote a cost figure, "$3 billion." In an instant, money became the central issue for a stunned audience. Sensing the considerable momentum behind the project, Botstein pushed forward to the stage and said, "It endangers all of us, especially the young researchers." He added, "What we really want is a physical map of the genome."[36] Berg, expressing "qualified strong support," tried to turn the discussion back from money to desirability. "Is it worth the cost, not in terms of dollars but in terms of its impact on the rest of science?"[37] Looking back sadly a year later, Berg thought he had been shouted down. "We could hardly get to the science because of the ominous views people had about the project."[38] Soon after the meeting, David Baltimore, who had not been present, weighed in: "I think we should only have mega-projects if there is a crying need for them."[39] The believers knew they had their work cut out for them.

Backers of the project, like Watson, were worried by the controversy dividing molecular biologists, "the very scientists whose expertise in manipulating and analyzing DNA would lie at the heart of its successful completion."[40] Although senior scientists such as Watson, Gilbert, and Berg "voiced the opinion that now was the time to start the project, there was much less enthusiasm, if not downright hostility, by many younger scientists."[41] Though most over 50 were backers, "almost everyone under 50 was against it who wasn't a human geneticist," Jim recalled. The young scientists could see that money, their money, would certainly be diverted from high-quality biological and medical research. On another plane, however, they favored NIH over the Department of Energy.

But many asked: Why do the whole DNA sequence anyway? Didn't the work since the split-gene discovery of the late 1970s show that at least 95 percent of human DNA doesn't code for proteins?[42] Jim didn't take these objections lying down: "Though many young hotshots argued that the time for the project had not yet arrived, those of us a generation older were seeing at too close hand our parents and spouses falling victim to diseases of genetic predisposition. And virtually all of us knew couples rearing children whose future was clouded by a bad throw of the genetic dice."[43]

These were not abstract reflections for Jim. He was acutely aware of a tendency to develop cancer in his own family—his father and sister both died of it—and, even more painfully, of the undiagnosed psychiatric problems of his older teen-age son, Rufus. That very week, Rufus, who had recently been sent home from secondary school, had run away, and Jim and Liz had issued public pleas for his return. In conversations during the conference, he spoke freely about his anxiety over his son.

HAMMERING OUT THE DEAL:
BRUCE ALBERTS' COMMITTEE

The grapevine began to vibrate. The genome project swiftly went onto the agenda for a meeting of the Board of Basic Biology of the National Academy of Sciences, which met on 5 August at the Academy's summer house in Woods Hole. On hand were Jim and other heavy hitters, including Gilbert, Hood, Cantor, Ray White, and Frank Ruddle, who had been chairman of the Energy Department's Santa Fe meeting. They concluded that a blue-ribbon committee was needed right away. A surprised Bruce Alberts, a biologist at the University of California at San Francisco, was told that he had been tapped as chairman.[44] The money would come from the James McDonnell Foundation, whose president, Michael Witunsky, just happened to have attended the Cold Spring Harbor meeting in June.[45]

As often happens in the scientific community when scientific consensus and political backing are needed, the Alberts committee became the central forum for winning the necessary majority support among biologists and making it clear to NIH that it would have to step up to the plate. In the committee's hands, the project became both more sophisticated and less grandiose, and one that called for "new money." Under Alberts, the committee became one of the most significant in the history of biology, and it helped prepare him for his future task as president of the U.S. National Academy of Sciences. He later recalled the experience as "the most fun of any committee I have worked on."[46]

The membership was chosen carefully to pull in skeptics as well as backers, and to represent not only major centers of research and major technologies that could figure in the project, but also opinion abroad. Most of them knew Jim, also a member, very well, and had taken part in conferences at Cold Spring Harbor. Alberts himself was an agnostic on the genome project and, as Jim put it, "a known foe of big biology labs." According to Watson, "he could talk to other biologists and be on their side. People could trust him. That was important. He also knew the science very well." [47]

From abroad came Sydney Brenner of Cambridge, a member of the inner club of molecular biology, the father of studies of the roundworm *C. elegans* and a veteran of the recombinant-DNA wars of the 1970s, and John Tooze, a scientist and writer from the European Molecular Biology Organization in Heidelberg who had coauthored with Jim a book on the recombinant-DNA battle.[48] Among the skeptical members were the yeast geneticist David Botstein and the mouse geneticist Shirley Tilghman (later president of Princeton). Besides Jim there were two other Nobel Prize winners, Gilbert and Daniel Nathans of Johns Hopkins, one of the pioneers in the discovery of the "restriction" enzymes that cut up DNA. Robert Cook-Deegan, a staff member of an evaluation of genome sequencing by the Congressional Office of Technology Assessment, sat in.[49]

The committee included Victor McKusick, who had long been the sparkplug of a global effort to identify human disease genes. Two members most identified with genome project technologies were Cantor of Columbia and Hood of Caltech. Others were Russell Doolittle of the University of California at San Diego, a student of the evolution of proteins; Stuart Orkin of Harvard Medical School, a human geneticist; Leon Rosenberg, dean of Yale Medical School; and the human geneticist Ruddle, also of Yale. Olson of Washington University, who worked on the structure and function of yeast genes, joined the committee later, when Gilbert resigned to try starting a company he called Genome Corp.[50]

The task of the Alberts committee, which met in Washington, D.C., was to decide whether some kind of project should be done, and, if so, to define the strategy, pace, and scope. To Alberts, the first task was self-education. The committee needed to bone up on the facts and, with the help of invited scientists testifying at a series of hearings, let a consensus evolve and put off the possibly divisive votes until afterward.[51] What were the technologies of genetic mapping, physical mapping, and sequencing? Would a project focused on the genome take a bite out of biological and medical research monies just when its funding seemed to have hit a ceiling? Was a special project needed?

Jim strongly influenced the committee's membership and procedure, according to more than one witness. Zinder remembered Watson "standing astride it all, like a colossus."[52] Botstein observed, "Jim absolutely grasped immediately that what was required was to get a consensus of the academic folks." Thus, Botstein had been named a member not only because of his role in genetic mapping but also because he represented "the radical fringe" instead of the outspoken Baltimore. With Baltimore on the committee, Botstein reflected later, "there would have been a lot more serious sparks flying." The architects of the committee probably thought "that if they could convince me, they could convince anybody."

Gilbert led the charge for industrializing the sequencing effort "right now, right now, someplace," Botstein recalled. Some on the committee feared the idea, remembering what happened when the U.S. space agency bet most of its money on the shuttle in the 1970s and 1980s. "It sucked the entire vitality for a couple of generations out of planetary astronomy. . . . The concern was that the big Wally Gilbert factory would . . . take up the whole box."

Watson and Brenner posed the issue more suavely. Of course, there would be a complete sequence someday, but they asked whether "an extraordinary effort" made sense now. Watson spoke rarely. "He was not big in the technical issues," Botstein recalled, "but he would make impassioned speeches about how he wanted to see the sequence before he died."

Alberts also did not speak much, according to Botstein, but his contribution was crucial— "enormous." In Botstein's view, Alberts had the incalculable advantage of not having taken a position—and yet very well known as being on the side of students. "He would not act against the interest of the new young people."

The committee ended up strongly favoring significant emphasis on sequencing "model" organisms like yeast, the roundworm *C. elegans*, and the fruit fly *Drosophila*. The big concern was what use could be made of the completed human sequence. Nathans "made the argument that only by understanding the model organisms would we ever be able to interpret the [human] sequence."

Botstein fervently agreed. The committee knew that the smaller organisms could be sequenced. "Not only could we do it, we would learn a whole hell of a lot" by doing so. Interpretation, relating the genetic code in DNA to the amino acid sequence in proteins, was the worry Botstein had expressed at Santa Cruz back in 1985. "I used to say in lectures [that] it's like having the secret of life in Urdu, and then being told I could translate it into Uzbek."[53]

During the period that the Alberts committee was meeting, Botstein told a reporter, "We don't have to know all of it in order to interpret it." Looking back later, Shirley Tilghman approved highly of the Alberts committee's decision to push for nonhuman as well as human sequences: "Model organisms were an extraordinary investment. We learned how to sequence on these simpler organisms. And more important, we got a preview of the human genome."[54] In retrospect, Jim occasionally denied that the committee's decision was so influential. He told Botstein it was obvious. Botstein rejoined "that it may have been obvious to him but it wasn't obvious to anyone else."[55]

The committee went a long way in the direction of Botstein's own interest in genetic mapping, urging that mapping should be the main emphasis during the project's early years, while sequencing technology was pushed to a much lower cost per base pair. Indeed, as the committee heard scientists' testimony and discussed what it meant, the entire project was redefined away from an all-or-nothing bet on sequencing to something less grandiose, something that would yield results along the way—like a railroad being used to building a railroad.

The genome project was often called biology's moon shot, but few who made this comparison actually understood how apt the analogy was. In the Apollo moon project, one big rocket for "direct ascent" to the moon gave way to a smaller rocket bearing a flotilla of craft with limited jobs, one to orbit around the moon, another to land on the surface, and yet another to rise back to lunar orbit. Each was discarded when it had accomplished its purpose. It was a system that provided at least two ways to get home if there were a failure, and at the same time added to knowledge at each step.[56]

As the meetings went into 1987, scientists' enthusiasm still lagged behind that of the public and politicians, but it was swinging in favor of the genome project. As Watson recalled, this shift had four principal causes: more awareness than before of potential medical benefits; the speedup of disease-gene finding from the mapping drive; the low cost compared to the total spending for biology and medicine; and the prospect of speeding up research on the small organisms.[57]

The committee got down to the question of money—new money. Botstein reported a subcommittee's view of how much staff and research centers would cost each year, and projected how long, at a given spending level, it would take to reach the goal line. At $50 million a year, it would take until 2025. Botstein's group mentioned $100 million a year or a "crash" program of $200 million. Watson immediately objected. He feared the committee would want to seem reasonable by picking the middle number. He proposed that the top number be $500 million a year, a "crash crash." As Jim probably intended, the committee settled for $200 million in fresh funds above existing health-research spending. The committee had concluded that "it could be done for a reasonable cost and in a reasonable period of time." Watson recalled, "At the end of it, arguing against it was like arguing against motherhood."[58]

It was time to start selling the idea intensively outside the scientific community, particularly to Congress. In May 1987, Watson and David Baltimore, shepherded by Bradie Metheny, a Washington lobbyist for biological research, made a round of calls on leaders of the House and Senate Appropriations Committees, including the "adulatory" William Natcher, chairman of House Appropriations.[59] Baltimore asked for hundreds of millions more for research against AIDS. Watson spoke of what the genome project could do for millions of sufferers from Alzheimer's disease or breast cancer, and said that $30 million in new money would let NIH start "a serious genome effort."[60] To understand "the genes behind breast cancer" and "the genes which were going wrong and leading, say, to senility," biologists needed to "get" the genes.[61] Long before the full human DNA sequence was deciphered, Watson said, biologists would have cloned genes that predisposed families to cancers or senile dementia. Congress got the message faster than some of Jim's colleagues.[62]

COLLIDING WITH DAVID BALTIMORE

Getting and keeping political support for the grand enterprise was a source of anxiety for Watson. This may have influenced him in a quarrel with David Baltimore that went on for over a decade. The issue was Baltimore's handling of the long conflict over a paper on immunology that he had cowritten for the journal *Cell* in 1986. In the face of a challenge by a postdoctoral fellow, Margot O'Toole, regarding the results of the experiments that the paper reported, Baltimore defended his collaborator on the paper, Tereza Imanishi-Kari, who had done the experiments. O'Toole, who worked in Imanishi-Kari's lab, said she, O'Toole, couldn't reproduce Imanishi-Kari's experimental results. Baltimore was much censured for facing O'Toole down and pushing the paper through rather than pulling it back until criticisms could be addressed. Watson and Harvard colleagues such as Paul Doty, John Edsall, Walter Gilbert, and Mark Ptashne thought that Baltimore had pulled rank.

The intensity of the ill-feeling between Watson and Baltimore continues to excite speculation among scientists. Some put it down to rivalry between the Harvard and MIT biology departments. Others attribute it to Jim's irritation, even jealousy, at Salva Luria's forceful patronage of Baltimore at MIT. Still others see the skeleton of an old grudge from the recombinant-DNA fight. Watson had passed up the notion of locating the Whitehead Institute near Cold Spring Harbor, most likely because the institute might overshadow his own lab. It was eventually located at MIT, where it flourished. And Baltimore's skepticism about the genome project coincided with his very public troubles over the *Cell* paper.

Just as the genome project began growing, the "Baltimore affair" was taken up in Congress. Congressman John Dingell of Michigan held hearings about whether scientists who were spending so much taxpayers' money could police themselves. The U.S. district attorney of Maryland considered prosecuting Imanishi-Kari. Secret Service agents were set to analyzing her data-tapes to see whether they had been intentionally misdated.[63] Imanishi-Kari was exonerated in 1996.

At a congressional hearing in the spring of 1988, Baltimore was pilloried in his absence.[64] By January 1989, there was so much hostility between Dingell's committee staff and many scientists that Watson, having just taken over the genome project, staged a conference on the issue of fraud at the Banbury Center. The conference, attended by associates of Baltimore, was a sign that Watson sensed a major danger to congressional support just as the genome project was taking off.[65] President George Bush was proposing a big jump in the genome budget to $130 million for the year beginning in October 1989, $100 million of it for NIH. In the end NIH got only $60 million.[66]

Baltimore raised the stakes when, in May 1989, he finally appeared at a Dingell hearing. He furiously and directly challenged Dingell, who had sought to gavel him into silence. Baltimore denied that there should be laws to police the enforcement of professional norms in science. He maintained that any scientific finding of importance would be tested by other scientists, and hence fraud would be uncovered promptly.[67] To Watson, it looked as though Baltimore was tickling the dragon's tail.

In 1989 Baltimore was named president of Rockefeller University,[68] but by the end of 1991, the furor over the *Cell* paper reached such a peak that Baltimore resigned.[69] He returned to MIT for a few years before being named president of Caltech—and being restored to his role of leading scientific adviser to the U.S. government in the struggle against AIDS.[70]

Jim remained anxious about the possible impact of the matter. In 1991 he wrote, "Drawing the wagons around us to protect our sinners from punishment is not the way to ensure that the voices of scientists are heard more, not less, in the Washington corridors of power." Two former Harvard colleagues, Paul Doty and John Edsall, were writing publicly about disapproval of Baltimore's refusal to withdraw the *Cell* paper. Doty decried a "gradual departure from traditional scientific standards." Jim told the Nobel Prize–winning biologist Paul Berg that his late mother was Irish and that he believed O'Toole because she was "a good Irish girl." His anger against Baltimore went so far as to lead him to accuse Baltimore in private of stealing the work for which he shared the Nobel Prize in 1975—although Jim retracted this notion when Horace Judson, a historian of biology, told him he was wrong.[71]

In the fall of 1999, there was an official gesture of reconciliation. When the new Watson graduate school was dedicated at Cold Spring Harbor, Baltimore received its first honorary degree. But the old story resurfaced in 2001, when an NBC television science reporter, Robert Bazell, in reviewing a biography of Baltimore, asserted, "James Watson, the codiscoverer of the structure of DNA and the embodiment of biology for the U.S. public, actually campaigned to have Baltimore's Nobel Prize rescinded and to have him expelled from the National Academy of Sciences."[72]

Watson replied stiffly a month later, denouncing as "false" the idea that he led "a campaign to destroy Baltimore," but stating that he shared other scientists' dismay at what he called Baltimore's "intransigence." He referred to a "long-discredited rumor that David knew of Howard Temin's discovery of avian reverse transcriptase before he isolated its murine equivalent." Jim admitted, "I had inappropriately repeated what turned out to be an unfounded story."[73]

START-UP

As head of the genome project, Watson, like most executives of a start-up operation, had to oscillate, at times unpredictably, between the role of "Mr. Outside" and "Mr. Inside." It is difficult to assess which of these roles was his greater contribution to the success of the project.

A good ambassador writes his own instructions. Serving on the Alberts committee, which he helped create and energize, Jim was no exception. With his participation, the scientific community had given him a mandate, a set of specifications, and he used it. He set up small staffs both at the genome project's headquarters at the NIH and in Cold Spring Harbor to allow him to shuttle back and forth. The staff leaders in Washington were Elke Jordan and Mark Guyer, wise in the ways of NIH.[74] At Cold Spring Harbor, Rich Roberts, Bruce Stillman, and Terri Grodzicker kept things running scientifically.

Jim started off with a bang at his first press conference as chief of the Human Genome Project in October 1988. The $3 billion project would eventually involve 2,000 scientists, he estimated, and his job was to get the best people on board.[75] The money would be spent in universities around the country, in many political jurisdictions.[76] But it would not just be a series of grants. "This is a project with a real goal."[77] From the beginning, he was determined that the project would also be international.[78] Mindful of patrons in Congress, he said, "I would hope that all the major industrial nations of the world would contribute to the cost."[79] His involvement was personal. "We are the genetic disease hunters. . . . I'd like to get the answer while I am still alert enough to appreciate it."[80]

Congress wanted a five-year plan, and the task of hammering out its details was turned over to a genome advisory committee, headed by Norton Zinder and including Alberts, Botstein, Hood, and McKusick. The group met first in January and again in June of 1989, and then held a crucial "retreat" at Cold Spring Harbor in August. Zinder recalled the retreat as "total war between those who want to study genetic diseases and those who wanted to do the infrastructure of the genome." Watson favored the latter approach.

Genetic diseases had such powerful advocates as Thomas Caskey of Baylor University, Francis Collins of the University of Michigan (he was a leading hunter of the cystic fibrosis gene and was Jim's eventual successor in running the program), Uta Francke of Yale University Medical School, and Ray White of the University of Utah, a leader in gene mapping. On the "infrastructure" side were such scientists as Botstein of Stanford, Eric Lander of the Whitehead Institute at MIT, Maynard Olson of Washington University in St. Louis, and Shirley Tilghman.

After heated debate, Watson signaled Zinder to call a vote, and the infrastructure advocates won. The assembled scientists then approved the idea of "comprehensive" genome centers, not just administrative ones, and picked five model organisms: the colon bacillus, the roundworm, the fruit fly, yeast, and the mouse. It was a watershed meeting, controlled almost silently by Watson.[81]

PREEMPTIVE STRIKE:
"SOME VERY REAL DILEMMAS"

In blunt recognition of the potential misuses of new genetic information, Watson announced, without consulting with the NIH director, James Wyngaarden, that at least 3 percent of the project's money would go toward studying such issues.[82] The basic principle, he said, was that each person owns her or his DNA, and must be protected against use of the information without consent by employers and insurers. "You do not want a group of people labeled as genetically damned."[83] He named Nancy Wexler, a principal sleuth of the gene for Huntington's chorea, to run the program on ethical, legal, and social issues. His very visible commitment on ethics was a politically fruitful preemptive strike.

Most observers agree that this was one of Jim's shrewdest moves. After all, as the project unfolded people would ask more and more insistently what would be done with the gigantic treasure of information Watson proposed to accumulate. He admitted that "some very real dilemmas" exist already about the privacy of DNA. "The problems are with us now, independent of the genome program," he said, "but they will be associated with it. We should devote real money to discussing these issues. People are afraid of genetic knowledge instead of seeing it as an opportunity."[84] In 1989 he told a gathering of 700 geneticists in New Haven that there must be a "rigorous dialogue" on ethical issues or there would be "a nasty backlash" from people demanding that the whole genome enterprise be shut down.[85] He said that people should become more literate about genetic knowledge. On the uses of this knowledge he said, "We cannot make policy . . . all we can do is educate."[86] The genome program faced the burden of the "pseudoscience" of eugenics, and the fears about genetic privacy.[87] Of the Nazi horrors, Watson said in London, "We need no more vivid reminders that science in the wrong hands can do incalculable harm."[88] Nonetheless, "I want to go ahead with the Human Genome Project. I don't feel [I am] a Hitler-like individual. I know there will be ethical dilemmas and will have to be prepared for them."[89] Jim told another interviewer in London, "The potential benefits of unraveling disease are too great to forgo."[90]

His unblushing advocacy of getting and using genetic knowledge was not universally admired. In *The New Republic* Robert Wright asked rhetorically in 1990 whether Watson was a "mad scientist." He concluded that he wasn't, but described him this way:

> Watson's eyes are intense and bugged out, yet unfixed; they wander cryptically as he talks, floating from one point in the visual landscape to another and another. His head floats slightly, too, so that altogether he conveys the sort of ethereal detachment that typically signifies either profound inner peace or utter disorientation. And then there's the inappropriate laughter. At times his laughs—quiet, sudden nasal or guttural exhalations—come almost as regularly as the ends of his sentences, but most of them don't follow observations that would strike the average person as funny, and many are entirely inscrutable.

Wright was repelled when Watson told him, "I think knowledge is going to come out and people are going to use it, but I don't know what sort of knowledge [laugh] will come out." Talking of the future possibility of selecting children's qualities, Watson said, "I doubt we'll be doing that for so long I'm not going to think about it [double laugh]. I think there'll be a great diversity of opinions between people on . . . what is an unacceptable genetic disease [laugh]."[91]

A SPLASH OF JINGOISM

Jim quickly showed that he could make political hay out of such issues as genetic disease, competitiveness in biotechnology, and plain jingoism. He told the Germans to get going instead of being "so horrified of what it did in the past that it's afraid to move into the future."[92] His principal target was Japan, redoubtable competitor in making both automobiles and electronics, which was threatening to start its own industrialized 10-year push for the total sequence.[93] Despite Japanese foot dragging a decade earlier, which had scuttled the establishment of an Asian molecular biology organization with Watson as its head, Watson was interested in the possibility of dividing up the human chromosomes among different labs, including those in Japan.[94] But the Japanese were slow in committing funds to what he and others such as Victor McKusick regarded as a global effort.

Despite scrambling by colleagues to soften the tone, Watson wrote threatening letters to Kenichi Matsubara, who was coordinating the Japanese work—and struggling to get money for it. Unless the Japanese government paid up, Jim implied, it could find itself shut off from biological data of commercial importance. He wrote, "I have little enthusiasm for being a supplicant before a nation that should much more spontaneously begin to return the generosity with which America has treated Japan ever since the war ended." He added, "They keep saying that we should remember their sensitivities, but they should remember that we remember Pearl Harbor. I'm old enough to have lived through it." At one point in the controversy, Matsubara asked acidly whether big science projects had ever produced anything. Watson shot back: "The Manhattan Project." This reference to Hiroshima and Nagasaki struck Tom Maniatis as "probably the most outrageous and insensitive thing he said."[95]

The letters were leaked in Washington, where political leaders smiled. A professor in Japan called Watson's threat "blackmail," and Matsubara said, "We don't work according to his suggestions." But Japan, the second largest economy in the world, was spending less than $7 million a year on the project when the United States was spending $87 million.[96] The attention-getting language offended people in Japan deeply. Norton Zinder, as chairman of Watson's genome advisers, found himself in the position of trying to soothe feelings. In his mollifying letters to the Japanese, Zinder recalled, he was mindful that "you take all the onus of the blame because the Japanese can never be guilty of anything." The team of good cop–bad cop appears to have boosted Matsubara's clout with the Japanese government, and smoothed the way to bigger Japanese cooperation. An exasperated Zinder told a reporter that Watson "administers by petulance." Long afterward he ruminated over "how much trouble [Watson's behavior] has caused in my lifetime."[97]

OPPONENTS TRY AGAIN:
"I'M SITTING HERE STARVING"

In 1990, while genome centers were created at several American universities[98] and work began on sequencing nonhuman genomes, Watson had to beat down a serious attempt to destroy the scientific consensus on which the project was going forward. Martin Rechsteiner of the University of Utah and Michael Syvanen of the University of California at Davis launched letter-writing campaigns against the Human Genome Project, and Bernard Davis and colleagues at the Harvard Medical School published an attack on the program in *Science*. The *New York Times* reported that the project was "arousing alarm, derision, and outright fury," and also published Watson's uncompromising reply: "Our project is something we can do now, and it's something we should do now. It's essentially immoral not to get it done as fast as possible." The opponents said that even the first stage, a drive on gene mapping, was years behind schedule and that if scientists ever completed the sequence, it would be uninterpretable. Rechsteiner said a young chemist had told him, "The fat cats are getting all the cream while I'm sitting here starving." Syvanen said the gene for sickle-cell anemia had been known for 20 years but "it has yet to lead to a cure for the disease." Davis said the Human Genome Project was growing when the number of younger researchers receiving grants was shrinking. "You can't prove it, but I think it is widely felt that the two are in competition."[99] By the time Zinder and his fellow advisers met in June 1990, he recalled, "the atmosphere was very unhealthy." Watson, negotiating for the first "real" genome budget, was afraid that Congress would not give it to him. Characteristically, he threatened to resign, and, if he announced this at the meeting, Zinder planned a diversion of pushing over a pitcher of water onto the meeting table, so that a colleague could hustle Jim out of the room. But Jim didn't resign, Congress approved of the five-year plan, and the program got its money.[100]

Davis promptly got roasted at a U.S. Senate hearing in July, where he was the only witness opposing the Genome Project. After hearing his testimony, Senator Pete Domenici of New Mexico burst out, "I am thoroughly amazed, Doctor, at how the biomedical community could oppose this project. I am absolutely amazed. . . . It is beyond my comprehension." Davis rejoined that he wasn't just speaking for himself. "I would urge you not to underestimate the extent of the unhappiness in the biomedical community over the issue."[101] But earlier skeptics such as Baltimore and Botstein had changed their minds. Botstein said, "If it were not Jim—if it were headed by an ordinary person—then I believe the project would not have acquired the degree of acceptance that it has in the scientific community."[102]

The danger from the opposition was not theoretical. The Senate placed a far higher value on a fast pace for the project than did the House of Representatives. Over several years, the eventual compromise appropriations were far lower than the White House requests. In 1990, the administration asked $108 million and the Senate approved this figure. The House cut it to $71 million, and Senate-House conferees agreed on $87.6 million.[103] Zinder and Watson wrote a letter to the *New York Times* defending the project as essential to identifying "molecular defects that result in disease," and a way to speed "a new era in drug development."[104] But the project was growing more slowly than anticipated, and, in the eyes of Ronald Davis of Stanford, whom Watson pressed to sequence baker's yeast, too little money went for development of sequencing techniques that were more advanced than those eventually used.[105]

In the hope of changing his opponents' minds, Jim invited Bernard Davis and another leading opponent, Donald Brown of the Carnegie Institution of Washington, to speak at the 1990 Genome Project meeting in San Diego. Davis objected to justifying the science of the project just because "it's there—like Mount Everest." He wanted a ban on full-scale sequencing in favor of mapping the location of genes in the genome. Watson shot back that this was like saying, "Let's get married, but no sex." Brown objected that despite Watson's advocacy of peer review, a large proportion of HGP money was being spent on "top-down" contracts and grants that were not reviewed by peers. Watson rejected objections to the size of the program: "The human genome is so big, so how can we keep the genome project small? We want to be big enough to get it done in a reasonable period of time." As to targeting, Jim said it wasn't wrong per se. The important question was "whether you have the wrong target." He told Brown, much of whose funding at the Carnegie Institution came from an endowment, that he was "the product of rich-man's science. . . . You should be honest with the way you exist." The genome effort was analogous to physics: "You can't make a discovery until you give investigators a big machine. You have to get more genes." Most of those who were listening were impatient and wanted to get on with the science, including the issue of how accurate a sequence needed to be to be useful.[106]

RECRUITING

From the beginning, Watson knew that his most important task was to sign up the best people. He was "the head of a new branch of science, but without any scientists." To rope them in, he employed all the techniques he had worked out over many years at Harvard and Cold Spring Harbor. One of the most effective was to say, "Show me." Among the many challenges was the sequencing of nonhuman organisms to pave the way for the main event, the human sequence. But where would the first big effort be made? The devotees of a tiny worm called *C. elegans* stepped forward. An old enthusiasm of Sydney Brenner's, the creature had a small number of cells, 959 to be exact, each of whose fates could be followed right through the development of the transparent worm. A physical map of the DNA of the worm, about 97 million base pairs long, had been made. Could sequencing *C. elegans* be the model for doing the human sequence? John Sulston of the Sanger Centre in England and Robert Waterston of Washington University in St. Louis, both of whom had worked with the effervescent Brenner, thought so, and talked to Watson about it. As so often, he was skeptical. He thought the fruit fly *Drosophila* would be more suitable. The worm community was furious and determined to prove Watson wrong. Waterston and Sulston asked for a second chance and decided "to blow him away." They brought their map to an annual meeting on *C. elegans* at Cold Spring Harbor in May 1989. The map took up an entire wall. Jim was impressed and told Waterston and Sulston, "The fly can wait. You're in." By making Sulston and Waterston think he might shut them out, Watson had gotten them "involved, excited. In other words, he tricked us," Waterston later recalled. After getting approval, Sulston and Waterston went to the Syosset railroad station to go back to New York. On the platform, Sulston told Waterston they'd been given "just enough rope to hang ourselves." Sulston could hear "the prison door" clanging behind them.[107]

"I WAS FIRED"

Soon after he became head of the NIH project, Watson began signaling that he wouldn't stay long. In 1989 he rejoiced that telephone and electrical lines now ran underneath the "sylvan path" of Bungtown Road. He wrote that his days on the Pan Am shuttle would not result in a permanent move to Washington. "A Bungtown Road restored and the new pond beside it make my every return here a delight. No one need further imagine that the corridors of real power will tempt me."[108] In 1990 he remarked to a reporter, "After a year or two, the job will become just one of micromanagement, and I'm not interested in that, so I'll leave."[109] Although "his ego got involved in the genome project," and he "gave the program real visibility for a while," it was clear to a Cold Spring Harbor alumnus who chatted with Jim on one of his innumerable visits to Bethesda "that he wasn't going to be here very long. . . . He wasn't meant for that position. . . . He just didn't fit. His weight was needed elsewhere." In this view, Watson functioned better as a backstairs counselor than in "being the front man."[110]

In the spring of 1991 the project continued growing. The Bush administration was asking $110 million for the NIH project, and another

$59 million for the Department of Energy's genome work.[111] A seventh U.S. genome center, in Philadelphia, was approved.[112] In May, at a meeting of microbiologists in Dallas, Jim "disarmed" a hostile audience when he asserted that the Human Genome Project was not a blind, brute-force effort that took no account of cost. Afterward, according to David Botstein, attendees crowded around Jim "almost as if he were a rock star."[113] At about this time Paul Berg replaced Norton Zinder as chairman of Jim's advisory committee; Zinder described his job as "interpreting Jim Watson to the rest of the world."[114]

At a Senate hearing in July, Watson reported the finding by Thomas Caskey of Baylor College of Medicine and his colleagues in the United States and abroad of the gene responsible for a genetic disorder called fragile X. Afflicting about 2,000 boys born in the United States each year, the disorder keeps the boys from maturing to a mental age beyond five years. The annual bill for the care of each fragile X boy was $100,000. The total for each annual group of boys was $200 million, the same as the funding goal for the Genome Project. "This project," Watson said, "will pay for itself if we can go beyond the discovery of this one gene to doing something about it."[115] The project cost about 1 percent of the NIH budget. "Is this one percent going to make the other 99 percent better spent?" Watson rhetorically asked an English reporter. "The answer is yes."[116]

But trouble was on its way. At the center of controversy was the patenting of genes, a highly charged subject, not only among the general public but also among scientists. Some argued that without patenting drug companies would never invest in new medicines, and others argued that patenting would not only stifle cooperation among scientists on the genome, but also trigger international disputes. To its later regret, the Alberts committee had neglected to focus on such potential problems. The committee "lacked expertise on intellectual property issues and was misguided in its belief that DNA sequences without a thoroughly explored utility could not be patented."[117]

In April of 1991, Craig Venter, a scientist in NIH's brain science institute despairingly withdrew his twice-rejected proposal for a multi-million-dollar sequencing effort, which had languished for two years at the Genome Project.[118] He was following a different tack. He would go after the active parts of brain genes by "a high-tech shortcut," using "expressed sequence tags (ESTs)" to speed up the work; he would leave out the vast stretches of silent, unexpressed, "junk" DNA and patent the fragmentary sequences he found. In a paper in *Science*[119] he announced that he had identified more than 300 genes, and he decided to apply for patents on them. The newly installed NIH director, Bernadine Healy, asked her lawyers whether Venter's gene sequences were patentable, and they said yes.[120] Venter said that this process of finding genes was "a bargain in comparison to the genome project."

Watson regarded Venter's proposed shortcut as a mortal threat to the genome project. He dismissed the EST approach as "cream-skimming." He said Venter knew nothing about the genes' function, and hence the genes he had found had none of the "utility" required by patent law.[121] Random sequences did not deserve a patent.[122] Venter didn't even know how many genes gave rise to his ESTs. His approach was "strongly biased against rarely expressed genes," and would never make sense for the "model organisms" that were to throw light on the human genome.[123] This type of patenting would stifle the painstaking work of finding what a particular gene did, and cut off the effort to find out the functions of so-called "junk DNA."[124] "A genetic gold rush" that would stifle international cooperation must be prevented.[125] Watson urged the European approach of not patenting a gene itself but its application.[126] He told a Senate hearing that the DNA-sequencing machines Venter was using "could be run by monkeys."[127] The work was "brainless," and the proposed patenting was "sheer lunacy."[128]

The dispute became far more public in October. Scientists in the United States and abroad expressed disgust at the move to patent EST's,[129] and the biotechnology industry protested.[130] Britain's Medical Research Council swiftly made a countermove—applying to patent sequences analogous to Venter's.[131] Many were unmoved by the arguments of Healy and her technology transfer officer that they were applying for patents to preserve options. Watson and Healy sparred directly about patenting at a House hearing. Speaking as an individual, Jim repeated his point that "these small pieces" of DNA should not be patented because they did not constitute "a useful product."[132] Healy sought, in public at least, to play down the brouhaha: "One of Jim's strengths is that he is deliciously blunt and has strong opinions on things. My belief, at the present time, is that this is a bit of a tempest in a teapot. I think the discussion has become a little silly because it has been elevated to the level of some moral, science policy, ethical issue."[133]

Venter and Healy soon raised the stakes, by applying for an additional 2,375 gene patents.[134] And Healy seized on a new angle, accusations that Watson faced a conflict of interest. A businessman from Norwalk, Connecticut, Frederick Bourke, sought to start a new company, in alliance with Microsoft and Leroy Hood in Seattle, to carry out Waterston and Sulston's *C. elegans* sequencing project, whose full funding was in question. Bourke made offers to both scientists. Watson intervened furiously to kill the idea, asking the Medical Research Council in Britain to increase Sulston's funding, which it did. Bourke riposted that Watson had not only been "excessively profane and vulgar," but held shares in the biotechnology companies Amgen, Glaxo, Eli Lilly, Merck, and Oncogen, a fact that had been disclosed before.[135] Asked to sign a waiver that would allow Watson to continue owning the shares, Healy refused. In a public statement, she said she didn't "have the luxury of ignoring ethical questions, not even for a Nobel Prize winner."[136] After Healy talked over the situation with Daniel Nathans of Johns Hopkins, where Healy had worked, Nathans talked with Watson, who concluded that he was being forced out.[137]

Watson was shown Bourke's letter, another sign of Healy's hostility. "After I saw the letter, I decided I was out. . . . I have a fine reputation and they are trying to soil it."[138] Not long afterward, Jim resigned without an interview with Healy, by means of a brief fax from Cold Spring Harbor.[139] Healy responded with a statement that implied he was a troublesome has-been. Watson, she said, was a "historic figure in the annals of molecular biology. We have been fortunate to have had his expertise and scientific judgment, which have been invaluable."[140]

Years later, Jim said that the basic reason he was fired was because "I had no respect for her." He recalled, "She's bright, but she didn't *know* anything. She didn't really want advisors. . . . It is very dangerous to have power and exercise it in the absence of knowledge."[141] Early the following year, when the Clinton administration took office, Healy, a Republican, yielded her place to the Nobel Prize winner Harold Varmus.

By then, just a few months after Jim's departure, Craig Venter had stormed off to the private sector, first joining a nonprofit research institute subsidized by Human Genome Sciences, a company formed to patent disease-related genes. HGS was headed by William Haseltine, a veteran of the Harvard Bio Labs. Venter became the major competitor to the global nonprofit genome project, garnering tons of publicity from the sequencing of a dazzling array of micro-organisms. After a number of years at the Institute for Genomic Research (whose initials were TIGR), Venter went on to head the highly publicized Celera, whose aggressive push on sequencing stunned the public project and probably speeded the achievement of the human sequence by years. Watson later reflected, "That it occurred faster is due to the fact that industry came in. Wall Street decided that genes really were useful. You could make money if you patented them."[142] Wall Street funded private, industrial competition to the worldwide nonprofit effort. Industry's "strong participation," speeding progress, was "a surprise." Further, the much-maligned ESTs proved "enormously valuable, enabling biologists to quickly identify families of related genes, and to shuttle back and forth between organisms to study gene function." Bruce Alberts and his colleagues on the National Academy committee had failed to appreciate this possibility.[143]

Haseltine, looking back in 2001, not only disapproved of the policy of going slow on sequences, which he saw as delaying medical payoffs for years, but also strongly disapproved of Watson's stands on patents. He saw Venter's DNA tags as "unique identifiers of functional genes that make the critical components of our human body [that] can lead immediately to medical application." Venter's approach promised "a gusher of medical value." But Watson, according to Haseltine, feared that if Venter's short-cut was followed, "nobody will pay us to do the genome." Further, Watson feared "an intellectual-property problem that will blow apart my consortium." And so Watson "deliberately" rejected patents as "interfering with scientific progress," Haseltine suspected, in part because he placed basic knowledge ahead of conquering human diseases. Haseltine professed a different attitude: "I love science, but I love curing disease more." Jim had forgotten, in his view, the iron law of pharmaceutics: "No patent, no drug." By placing a "stigma" on patenting genes, Jim had done "a deep disservice to medicine . . . a colossal mistake."[144]

Jim's departure from the Human Genome Project in April 1992 was a farce, but an affordable farce because the project was safely launched. Moreover, one could regard the psychic gain for Jim as large. He had done the start-up and didn't want to hang around micromanaging. The constant shuttling to and from Washington was frazzling, and the project clearly needed a full-time boss. In fact, he was getting pretty tired of management altogether (and soon would give up the directorship at Cold Spring Harbor). He may not have liked being maneuvered out by the time-honored Washington maneuver of accusing an opponent of "conflict of interest," but he was able ever after to delightedly speak of being fired because of the "ignorance" of an NIH director determined to press forward with the "lunacy" of patenting strips of DNA. He could think of himself as having resigned on a matter of principle.

"THANK YOU, SIR"

Watson did not drop out of the genome effort entirely. Behind the scenes he lobbied for more money and a faster pace, not always successfully. Ironically, it was Venter's formation in 1998 of Celera, the hell-for-leather sequencing company, that spurred the public project, headed since 1993 by Francis Collins, to hurry up dramatically. There was a moment when U.S. leaders thought of acceding to Venter's suggestion that they drop the human and go for the mouse genome. But the nonprofit project's other major backer, the Wellcome Trust in England, threatened to go ahead alone. The issue came to a head at the 1998 annual genome science meeting at Cold Spring Harbor at which Venter threw down his challenge. To counter it the Wellcome Trust hurried a top official across the Atlantic. And Jim exploded. He exclaimed that Venter was "Hitler."

It was like Munich in 1938, he angrily told Francis Collins, adding, "Are you going to be Churchill or Chamberlain?"[145]

In a television interview in June 2000, Jim was asked what the completed human genome sequence showed. "You can say it's the script for life. It's the information for the play of life. And if you have the script . . . you know what the actors are going to do."[146]

Shortly after 10 o'clock on the morning of 26 June 2000, a few days after this interview, the completion of a rough draft of the total human DNA sequence was celebrated in hyperbolic language by President Clinton at a White House ceremony. The occasion was embellished with a closed-circuit television link to Prime Minister Tony Blair in the state dining room of 10 Downing Street in London. Blair spoke of a development that was "almost too awesome fully to comprehend . . . a breakthrough that takes humankind across a frontier and into a new era," an example of the science that is "fast-forwarding us into the future." Clinton called the sequence "a map of even greater significance" than the one brought back from the American West two hundred years earlier by Meriwether Lewis. "Without a doubt, this is the most important, most wondrous map ever produced by mankind." To Clinton, it made it "conceivable that our children's children will know the term cancer only as a constellation of stars." It was a moment of theater, including a staged reconciliation between the publicly funded and private-enterprise genome projects. Feature roles in Washington went to Francis Collins, the U.S. leader in the global project since 1993, and Craig Venter of Celera Genomics, the maverick who probably speeded the effort up by years. Less prominent at the White House was Eric Lander of the Whitehead Institute, kingpin of the American non-profit push. John Sulston (Nobel Prize, 2002), a leading figure in the British effort that was so heavily supported by the Wellcome Trust, Fred Sanger, pioneer of sequencing and Jim's colleague long ago at the Cavendish, and Max Perutz, who died in 2002, were honored in London.[147]

Michael Hunkapiller, the chief technological hero of the occasion, was not present. He was kept home in California by the chicken pox, and sent his daughter, a biologist in Venter's Celera, to represent him.[148] The sequencing machines made by Hunkapiller's company, Applied Biosystems, were allowing both the public and private genome drives to record, as Collins said, "one thousand letters of the DNA code per second, seven days a week, twenty-four hours a day." The latest robots had produced most of the sequence in the previous nine months.

An arcane industrial-scientific achievement was being treated with almost as much hoopla as the landing on the moon or the signing of a major law or treaty, and perhaps more than the usual presidential greetings to winning sports teams. The event symbolized a "tie" between public and private enterprises racing to be first to read what Collins called "the human book of life." The excitement could have been dismissed as premature, given the decades lying ahead that it would take to use the completed sequence to find genes, to elucidate how they are controlled, to spell out their roles in disease and normal development, and to develop new drugs. But it was a milestone in biological research all the same, even if Sydney Brenner impishly said it really was "an entrepreneurial accomplishment, a great managerial achievement" involving no "new science."[149]

Jim sat silently at the ceremony in the East Room, out of range of the television cameras. During the ceremony he heard President Clinton jovially congratulate him for having found "the elegant structure" of DNA, and praise him for his understatement about its "considerable biological interest." To applause the President added, "Thank you, sir."[150]

EPILOGUE:
"I'M AN OPTIMIST"

Jim's explosive resignation from the Human Genome Project was not his only renunciation of power. The handover to Bruce Stillman at Cold Spring Harbor Laboratory occurred soon afterward. Becoming president of the lab, Watson made a mutually generous arrangement with the laboratory's trustees, who thereby kept their champion fund-raiser on hand. The lab bought an adjacent property, the house on it was torn down, and Jim and Liz Watson built a Regency-style President's House with a large living room for entertaining. They donated the house to the lab.[1] Their elder son, Rufus, lived with them there. Over the succeeding years, Stillman would gradually assume more and more the principal role at CSHL, although Watson never lost his taste for micromanaging the plantings on the grounds or his big-game hunter's instincts to build a small university.

Having given up the agonies and excitements of power—having been, like Pauling, "dethroned"[2]—Watson turned to a life of giving talks all over the world, dedicating research buildings (including one in London named for Rosalind Franklin and Maurice Wilkins)[3], and giving interviews in the media. He lectured at Oxford as a visiting professor, and in Kyoto. Audiences and journalists, inundated with comments about potential dangers of the new biology, were curious to hear what this very famous man, an aging icon, would say on the subject. His irreverence continued to make him particularly popular with youthful audiences. Addressing students at Wally Gilbert's old school, Sidwell Friends', in 1993, he advised, "Don't learn too much—you'll just deceive yourself." To a student's question about animal rights, he rejoined: "The logical conclusion is we don't do any research and will spend all our resources making monkeys happy. I don't like monkeys."[4] The same year he predicted that "our children will more be seen, not as the expression of God's will, but as results of the uncontrollable throw of genetic dice."[5]

He told a UNESCO conference in Paris of "such great possibilities for improving human life through genetics that we must not lose our purpose because of noisy minorities."[6]

He pulled together a collection of essays called *A Passion for DNA*, wrote a memoir of his young manhood called *Genes, Girls, and Gamow*, and started an account of his years at Harvard. As the fiftieth anniversary of the double-helix discovery approached, he tackled a book to accompany a five-part television series on the DNA revolution which he helped outline and in which he frequently appeared. He referred to his life as going around talking.

Watson's fame, which he had so carefully cultivated, kept growing. As usual, reporters felt compelled to give descriptions. One writer visited him in his "sunny office" with "a panoramic view" of the harbor: "Watson's tie is askew beneath a rumpled sweater and a rumpled sport coat, the pocket of which bulges with pens and pencils. He speaks very softly and his words are punctuated by long pauses. His wispy hair appears electrified."[7] A Malaysian reporter wrote of "tentative smiles and uncertain glances" as Watson waited to start speaking.[8] A reporter in Singapore noted, "wisps of white hair and a paunch from too much good food." On that occasion Watson said, "Being famous is better than not being famous."[9] But he admitted the attention was "much, much too much. You feel as though you're dead. I don't think any of this has changed my character at all. I'm still as flawed as I was at thirteen."[10]

Watson's fame showed itself in many ways. In the 1980s, the billionaire software entrepreneur Bill Gates read *Molecular Biology of the Gene*. Reportedly his reaction was to make sure he invested in biotechnology companies.[11] In 1997, President Clinton awarded Watson a National Medal of Science, remarking, "Five years ago, the mystery of the human genetic system was only partly known. Today, government-funded scientists have discovered genes linked to breast cancer and ovarian cancer, and our human genome project is revolutionizing how we understand, treat and prevent some of our most devastating diseases." Watson was praised for "five decades of intellectual leadership," from the double helix to the Human Genome Project.[12] Glitzy occasions multiplied. He attended the seventy-fifth anniversary celebration of *Time* magazine at Radio City Music Hall in 1998, an occasion that also drew Sophia Loren, Lauren Bacall, Raquel Welch, Toni Morrison, Diane Sawyer, Nancy Kerrigan, Dorothy Hamill, Carol Channing, and Anita Hill—to say nothing of President Clinton and the former Soviet president Mikhail Gorbachev and a host of others.[13] In the summer of 2001, he and Liz were among a group of celebrities (including the television homemaker Martha Stewart) whom Gates's billionaire partner, Paul Allen, flew to Helsinki. From there they sailed over to St. Petersburg in a steamship Allen had rented, so the guests could visit galleries and attend a party in the Catherine Palace. While aboard, Watson met Jeff Goldblum, the actor who played him in the television movie *Life Story,* back in the 1980s.[14]

There were times when Watson's fame got him into hot water. The most famous instance occurred in 1998, when he hyperbolically commented on the agonizing subject of drugs to fight cancer, thereby causing a media firestorm and a one-day, billion-dollar Wall Street frenzy. At a dinner party in March 1998 he had spoken enthusiastically about the progress being made by Dr. Judah Folkman of Children's Hospital in Boston in developing drugs that cut off the blood supply to tumors in mice, and made the tumors disappear. This progress had been reported six months earlier in several newspapers, including twice in the *New York Times.* At the party was the energetic and aggressive *New York Times* science reporter Gina Kolata, who heard Watson say, "Judah is going to cure cancer in two years." Normally, such remarks were discounted as typical unguarded Watsonisms, but this time they were taken seriously.

Perhaps understandably, Kolata did not call Watson back to give him a chance to offer qualifying comments. Watson's statement became the centerpiece of a front-page story by Kolata in May, in which she correctly reported growing excitement in the cancer community about the Folkman findings. The enthusiasm certainly was there: Folkman had been the keynote speaker in February at the annual meeting of Cold Spring Harbor Laboratory's local donors.

At the *Times,* the science editor, Cornelia Dean, knew the story "would generate a lot of shock waves" and held it for a week so that Kolata could add qualifications, or caveats. Unquestionably, Watson's comment at a social occasion helped get the story on the front page, where his categorical assertion was placed. Appearing inside were most of the cautions, such as the painful fact that cancer drugs that did well in mice often failed in humans, and that human trials had not yet begun.

The salience of cancer in so many lives, the *Times'* extraordinary influence over the news judgment of all media, and perhaps the general quiet of a Sunday without much other news combined to give the story maximum force. It was the lead item on at least three evening network television news programs, and went on the cover of the major U.S. national news weeklies. The day following the *Times* story, the stock of a company making the drugs quadrupled, amid the exchange of about a billion dollars on the market. The phones of cancer physicians across the country and of an agonized Folkman himself were jammed with thousands of desperate calls seeking the unobtainable new drugs.

Criticisms rained down on Watson. The comment of Dr. Timothy Johnson of ABC News was typical: "Everybody knows he is a kind of verbal loose cannon; a genius, but one of those people who make grandiose statements." Watson at once wrote a letter to the *Times* about the quotation in which he noted the frequent failures of cancer drugs in human trials. But he did not quite deny saying what Kolata reported, and reiterated his belief that Folkman's work was "the most exciting cancer research of my lifetime." He told one journalist that he wasn't aware his comments would be repeated and was "horrified" when they were. The *Times* itself, while defending the story's accuracy and prominence, not only ran minor corrections but also another reporter's much more skeptical story. Kolata's agent told her he could get her $2 million for a book, which stunned her fellow science writers, but after discussions with her employers she turned it down, which stunned her colleagues even more.[15]

Nonetheless, Watson was not deflected from taking his typically light-hearted, blunt, "politically incorrect" stands. At a conference in Madras in 1998, Watson said genetic engineering was "a very safe technology." Of its opponents he said, "Let the Greens walk into the sea. . . The only person who has been harmed by DNA is Bill Clinton."[16] Invited to a congress in Berlin to celebrate the opening of the Max Delbrück Center and Germany's belated entry into the Human Genome Project, he berated the Germans for appropriating too little money for it. For some 20 years, he said, "the totally unjustified German public perception that recombinant DNA per se is evil," rooted in horror at Nazi genetics to be sure, had held back both medicine and industry. In effect, he was telling the Germans "that the world wants them to finally shape up about their Nazi genetic past." Some 1,500 scientists, instead of gasping, applauded vigorously.[17]

Over and over, he spoke and wrote on the basic question of what people should do with the new genetic knowledge in the age of the double helix. He constantly insisted that the exploration of how genes work must be pressed in spite of the dangers of misuse. Refusing to reject germ-line gene therapy, Watson wrote, "You should never put off doing something useful for fear of evil that may never arrive."[18] While uttering words that shocked many, Watson gave expression to what thousands of people thought and did. There were signs, indeed, that his audiences were a good deal less repelled by what he had to say than a swarm of ethicists enjoined them to be.

But not everybody liked the message or the manner. One reviewer of Watson's *A Passion for DNA* observed in 2000, "Watson is at his best when he writes about science. The reader gains the impression that sociologic issues are of little interest to him."[19] A more caustic reviewer, denouncing Watson's "consumer eugenics," commented, "He seems to have no time for the precautionary principle."[20]

Watson also got into hot water with a talk in October 2000 in a jammed Berkeley, California, auditorium, when he touched the third rail of modern genetics: the idea that genes specify behavior.[21] But then he added race to the subject. Discussing derivatives of a little known protein called POM-C, he linked hormones having to do with skin color (melanin), plumpness (leptin), and a sense of happiness (endorphins) with sexual arousal. When he had tried the idea out in England a few months before, he mostly received the kind of lighthearted treatment given to loopy remarks by someone famous,[22] but at least one person present complained, "At this point in the lecture, most of the audience began to prickle at the tone of the argument."[23]

At Berkeley, many squirmed at Watson's implication that plump black people have better sex lives than thin and ambitious people such as the sad-faced English model, Kate Moss, whose picture he displayed. Michael Botchan, a Cold Spring Harbor alumnus and now Berkeley's Biology Department chairman, had invited Watson, at Watson's suggestion, to fill in as speaker at the department's regular colloquium. Botchan was amused at some of the talk but felt that "many of the things were clearly over the line." He even interrupted from the front row, "This is not politically correct." Watson's reply was, "Ha!" Over a beer afterward, Botchan told Watson he was in danger of falling somewhere between the vitamin C cultism of Linus Pauling and the racism of William Shockley. In his reply Watson implied that Botchan was "sort of a wimp" because he was in academia. He said that was why he had left Harvard.[24]

One Biology Department faculty member who heard the talk complained to a newspaper, which put a disapproving story on page 1. Gossip about the episode raced through the scientific community. Sydney Brenner remarked archly, "If you discover the structure of DNA, you're allowed to say just about anything."[25] Watson apparently thought so. On occasion over the years, individuals and groups heard him express national, racial, religious, and sexual prejudices.

As the DNA revolution became wider and louder, Watson was, in effect, doing battle with a continuing strain of dark futurism about biology, as imagined by Aldous Huxley in *Brave New World*, in which a ruling Alpha class clones subhuman, subject laborers—a sort of reverse "designer babies," in today's terms—in a modern but more systematic version of slavery.

Getting old seemed not to bother Watson too much. When he turned 70 and colleagues at Harvard threw him a birthday party, he said his birthday "sort of scared me . . . but it's passed and I feel the same." He thought being 70 "would be very depressing, but it's not really, because there's so much exciting still to hear."[26]

To be sure, age brought the death of coworkers, like Jack Richards, who for 30 years helped him rebuild Cold Spring Harbor Laboratory. Until almost the end of four years of battling prostate cancer, Richards drove himself into New York for radiation or chemotherapy treatments, which, Watson wrote, "largely kept him out of pain." As Richards weakened, Watson found that "always the best way to bring back Jack's spirit was to talk about buildings in the course of construction, or designs that soon must be made."

Watson's tribute to Richards was heartfelt:

> Though Jack was finally in need of a cane to move about, his face until the end was never that of an old or sick person. His brain remained functioning at high capacity, and his advice was as sensible and to the point as ever. Luckily, his end came swiftly and painlessly in Florida, where he went to enjoy the sun and its warmth after still one more round of treatments. For a week, he could enjoy looking out upon the green surrounding him. A bacterial infection then swiftly filled his lungs and his eyes and face sparkled no more.[27]

As the twentieth century ended, the Watson family sent their friends Christmas cards with pictures of themselves. On one were sketches by an Australian artist of Jim and Liz, and Rufus and Duncan. The next year, the four posed in front of a flamboyant sun image by the glass artist Dale Chihouly. Liz wore a radiant smile. Rufus, tall and husky, looked detached. He continued to suffer from what his father called severe learning disabilities, which were "of great anguish to him and all those who love him."[28] Duncan, who worked for a small software startup in San Francisco, had on a baseball cap worn backward. With a grin, Watson did, too. For the Christmas 2000 card the family posed on a brilliant green lawn not far from the President's House. To one side was a 15-foot-tall image, *Spirals Time—Time Spirals,* created by the postmodern artist Charles Jencks, an American living in Scotland, and fashioned by a local blacksmith near Jencks's home in Dumfriesshire. Watson had met Jencks a year earlier at the home of Matt Ridley, the chairman of the International Centre for Life in Newcastle, England. The creation of a large sculpture of a double helix was a condition of Watson's and Crick's patronage of the Centre. When Jim saw the drawings, he decided to order a second one to donate to CSHL.[29] Rufus stood a bit to one side and forward of the rest. In all the cards, the wordless message was the same: despite everything, they were together. The tone was of defiance and stability, of courage and love.

In January 2002, Watson received a high honor from the England he loved. He was designated an honorary knight in the New Year list of honors in the United Kingdom (another American on the list was New York's Mayor Rudolph Giuliani). As an American Jim couldn't be addressed as "Sir," but as a Knight of the Order of the British Empire he could write KBE after his name. The decision to grant the award may have been influenced by his leading part in one of the greatest scientific discoveries ever made in Britain and for his encouragement of Britain's big role in deciphering the genome. Watson was delighted. He wasn't invited to Buckingham Palace to receive his knighthood from the Queen (as Giuliani was), but on 20 June 2002 the British Embassy threw a lavish party for him, his family, and some 20 associates, 8 of them British, whom he flew there for the occasion. Those present said it was a magical evening.

In 2000, Watson said he had never broken away from his father's dismissal of miracles. "Miracles," he told the television interviewer Charlie Rose, "are worked by people like Jonas Salk," who deserved sainthood for making polio disappear. "I think science can improve human life, and I'm an optimist and want to use it." He was convinced that his type of science stretched hundreds of years into the future, and he remained uncompromisingly reductionist. Although he said his greatest hope was the conquest of the many forms of cancer, the biggest question in pure science was, "How is a telephone number stored in my brain?"[30]

JAMES DEWEY WATSON:
A BRIEF CHRONOLOGY

1928   Born 6 April in Chicago, the son of James Dewey Watson and Jean Mitchell Watson.

1934   Begins attendance at Horace Mann elementary school in Chicago.

1935   Christmas, receives bird book.

1940   Is confirmed, leaves Catholic Church, devoting Sundays thereafter to bird-watching expeditions with his father. Appears on *Quiz Kids* radio program but is eliminated on his third broadcast.

1941   Enters ninth grade at University of Chicago High School. Of his adolescence, Watson recalls, "I didn't fit in. I didn't want to fit in. I basically passed from being a child to an adult."

1943   Enters the University of Chicago under the early-admissions policy of President Robert Hutchins. Learns to go directly to original sources, to respect the importance of theory in science, and to think instead of to memorize. Attends course given by leading geneticist Sewall Wright.

1945   In his third year of college, Watson begins to feel at ease. Reads Erwin Schrödinger's *What Is Life?* and takes away the message: Genes are the key components of living cells, so "we must know how genes act."

1947   Begins graduate study in genetics at Indiana University in Bloomington. His supervisor is Salvador Luria, an Italian refugee who, with Max Delbrück, a German refugee, in 1941 founded the study of bacteria and their viruses, called bacteriophages.

1948   Spends the summer at the biological laboratories in Cold Spring Harbor, N.Y., studying phages with Delbrück. To Watson, the only focus in this "paradise" was the nature of the gene.

1950   Completes his thesis on treatment of phages with X-rays. He moves to the Copenhagen laboratory of Herman Kalckar. He spends most of his time across town in the laboratory of Ole Maaloe at the State Serum Institute.

1951   Receives permission to travel to an international conference in Naples, during which Maurice Wilkins of King's College, London, shows his clear X-ray pictures of DNA. Determined to work on DNA structure, Watson moves to Sir Lawrence Bragg's biophysics unit of the Cavendish Laboratory at Cambridge, England, where he meets the brash biophysicist Francis Crick. Although Crick is doing a thesis on hemoglobin, the principal protein of blood, he and Watson jump into DNA structure right away. Misinterpreting findings of Rosalind Franklin, also of King's College, Watson and Crick devise a model of DNA that is at once shown to be wrong, and they are banned from competing with King's on DNA structure. Earlier in the year, Linus Pauling of Caltech has humiliated the Cambridge group by discovering that much of the chain of proteins is in the form of a helix.

1952   At a conference at Oxford, Watson describes Alfred Hershey's discovery that the genetic material of viruses is DNA, comparing the DNA in virus heads to "a hat in a hatbox." Soon after, still thinking continually about DNA, he and Crick have a disastrous meeting with a scornful Erwin Chargaff of Columbia University, who has discovered the ratios of the amount of the DNA bases: adenine always equals thymine and guanine always equals cytosine. Meanwhile, Rosalind Franklin takes the clearest picture yet of the biologically important, "wet," B form of DNA.

1953   Pauling, using outdated X-ray pictures of DNA, devises a mistaken triple-helix model. He sends a report of it to be published, and to his son and Lawrence Bragg in Cambridge. Shown the manuscript, Watson and Crick resolve to tackle DNA structure again, and right away, before Pauling discovers his error. When Watson visits London, Wilkins shows him Franklin's B form picture with clear evidence of a helix. Soon after, Max Perutz of the Cambridge group shows Crick and Watson a report with more of Franklin's data. Getting close to a double-helix model, Watson receives a crucial correction from Jerry Donohue, recently of Caltech, and goes on to find pairing of the DNA bases that explains Chargaff's ratios. Papers on DNA from both Cambridge and London are swiftly published in *Nature,* and Watson makes the first public talk on the double helix in the new Vannevar Bush Auditorium at CSHL. In the autumn, he moves to Pauling's empire at Caltech, where Delbrück expects him to continue work on bacteriophages.

1954   Declining to study bacteriophages, Watson tries unsuccessfully for an X-ray-derived structure for RNA, and his association with Delbrück dissolves. George Beadle, leader of Caltech's biology division, arranges a fellowship for Watson. That summer he teaches the physiology course offered annually at the Marine Biological Laboratory in Woods Hole, and hears plans for the Matthew Meselson–Franklin Stahl experiment to prove that DNA is duplicated as he and Crick have proposed.

1955 After a mumbling trial lecture, but strong recommendations from Delbrück and others, Watson is appointed assistant professor of biology at Harvard. He immediately takes a sabbatical to work again with Crick at Cambridge on the structure of small viruses, while making a final stab at RNA structure.

1956 Watson and Crick attend a conference organized by Pauling at Baltimore, the first to take the double helix fully into account. Watson opens his laboratory on the third floor of the Biology Labs at Harvard, and plunges into studies of ribosomes, the site of protein manufacture in cells.

1957 Jean Mitchell Watson dies.

1958 The Meselson-Stahl experiment succeeds, and opens the way to the Nobel Prize for Watson and Crick. Watson and Alfred Tissières demonstrate that ribosomes are made of two units, one larger than the other. But studies by Pardee, Jacob, and Monod demonstrate that ribosomes do not carry the gene's information to the location of protein assembly.

1959 In a Massachusetts General Hospital lecture, Watson urges that attention be directed to viruses as a key to understanding cancer, but is greeted with indifference.

1960 Watson joins the race to find the short-lived "messenger" RNA, attracting the Harvard physicist Walter Gilbert to do his first experiments in biology. Crick and Watson receive Lasker Awards, often a prelude to a Nobel Prize.

1961 Watson receives an honorary degree from the University of Chicago; soon afterward he attends a biochemistry congress in Moscow, where Marshall Nirenberg of the National Institutes of Health announces the first crack in the genetic code.

1962 Elected to the U.S. National Academy of Sciences. Travels with his father and sister to Stockholm to receive the Nobel Prize along with Crick and Wilkins.

1963 Receives honorary degree from Indiana University. Gilbert, returning to Watson's lab, demonstrates that ribosomes line up along the "messenger" RNA copy of the genes and move along it, allowing multiple copies of proteins to be made in a short time.

1964 Addresses fiftieth-anniversary meeting of the French Biochemical Society in Paris.

1965 Publishes *Molecular Biology of the Gene,* the textbook defining the rapidly expanding field of molecular biology.

1966   With Crick, attends NATO science course on the Greek island of Spetsai. Colleagues Walter Gilbert and Mark Ptashne are the first to isolate "repressors" that shut genes down.

1967   To moderate quarrels between "old" and "new" biology Harvard creates a new molecular biology and biochemistry department. Soon after, the more traditional biologists also form a department of their own.

1968   Publishes *The Double Helix* to a storm of comment, much of it negative. Maximum attention, and big sales. Is appointed director of Cold Spring Harbor Laboratory (CSHL) and announces it will concentrate on cancer viruses. He marries Elizabeth Lewis, a Radcliffe undergraduate, in La Jolla, California. Recruits Joe Sambrook to lead the cancer virus work. As Watson spends more and more time at CSHL, leadership of his Harvard laboratory shifts to Walter Gilbert.

1969   Grant to Watson totaling $1.6 million from National Cancer Institute provides funds to launch the CSHL cancer virus project. Watson obtains American Cancer Society professorship for former CSHL director John Cairns. Delbrück, Luria, and Hershey win Nobel Prizes.

1972   Prompted by CSHL trustee Edward Pulling, a neighbor of the Lab, Charles Robertson, begins discussing $8 million endowment for research. Richard Roberts arrives from Harvard and begins building a large collection of DNA-cutting restriction enzymes.

1973   With the help of trustee Walter Page, the decision to establish the Robertson Research Fund is confirmed. CSHL trustees agree to buy harbor marina to prevent further development, after Watson threatens to resign over the issue. Concerned about possible risks from new recombinant DNA techniques, scientists at a New Hampshire conference in June ask for a National Academy of Sciences study.

1974   April, Watson attends meeting of National Academy of Sciences committee at MIT where a decision is taken to write a public letter urging colleagues to defer certain recombinant DNA experiments. Letter is published 18 July. CSHL scientists develop a new technique, soon widely adopted, of using fluorescent antibodies to locate cell proteins.

1975   In February, Watson attends the conference in Asilomar, California, called for by the July 1974 letter to consider risks of recombinant DNA and begin crafting rules to minimize the risks. Watson and a few others oppose such rules. Charles Robertson agrees to donate his 50-acre estate to the laboratory, along with an endowment of $1.5 million for maintenance.

1976   Watson leaves Harvard to work full-time at CSHL. Conference at CSHL on environmental causes of cancer.

1977   Researchers at Cold Spring Harbor and MIT, including later Nobel Prize winners Richard Roberts of CSHL and Phillip Sharp of MIT, independently find that the coding sequences of genes are not continuous but interrupted by long stretches of "unexpressed" DNA. Crick says this totally surprising discovery gives "our ideas a good shake." Banbury Center for small conferences on Robertson estate is dedicated. Scientists lobby successfully to kill state and federal legislation regulating recombinant-DNA research. Watson is awarded Presidential Medal of Freedom by President Jimmy Carter.

1978 June, receives honorary degree from Harvard.

1979 Installs group of scientists at an estate near CSHL, called Fort Hill, to begin drafting a new textbook, *Molecular Biology of the Cell* (its fourth edition appears in 2001). One member of writing team, the biologist Bruce Alberts, describes the experience as "the summer from hell."

1979–80 Invited to Japan to help establish an Asian equivalent of the European Molecular Biology Laboratory in Heidelberg, itself modeled in part on CSHL. Plans fail because of rivalries among Japanese scientists.

1980 Ahmad Bukhari, who studies viruses, and James Hicks, leader of CSHL's yeast group, organize the annual Symposium on the topic "Movable Genetic Elements." The meeting reflects an explosion of studies backing up long-ignored findings and concepts of Barbara McClintock.

1981 The CSHL scientist Michael Wigler is one of the first three scientists to clone cancer genes.

1982 Exxon signs $7.5 million agreement with CSHL that includes building two wings of Demerec Laboratory.

1983 Barbara McClintock receives Nobel Prize. Her work, Watson says, is "prophetic and individual . . . a victory of radical thought over scientific orthodoxy."

1983–84 Sabbatical in England. Considers assuming new role of president of CSHL. Joe Sambrook serves as Acting Director.

1985 CSHL signs a research contract with Pioneer HiBred, one of the world's leading seed companies, to undertake plant science; experiments will take place on a portion of the nearby Uplands Farm estate and in the new Arthur and Walter Page Laboratory, on the CSHL grounds.

1986 The new 350-seat Grace Auditorium is dedicated at CSHL on 1 June. Days later, discussion in the auditorium of the Human Genome Project, vigorously advocated by Walter Gilbert, is contentious. Many biologists foresee dangers to funding of the small research groups of "little science." In wake of discontent, Watson helps arrange and serves on Bruce Alberts' National Academy of Sciences committee to consider whether and how to set up a Human Genome Project.

1987 Alberts' committee concludes that the project must be funded with "new money" not taken from other biology projects; genes will be mapped before full-scale DNA sequencing is done; and other plant and animal genomes will be sequenced to make sense of the human data. Watson visits congressional leaders to sell idea for funding the Human Genome Project.

1988   Publication of the Alberts committee's report. The NIH, based in Bethesda, Maryland, decides to take the lead in the Human Genome Project; it names Watson first director of the effort. Watson continues as director of CSHL, and shuttles weekly between Cold Spring Harbor and Bethesda. Begins frequent travel to recruit scientists for the program, and to respond to criticism. Celebrates sixtieth birthday at Cold Spring Harbor and Harvard. Fourth and final edition of *Molecular Biology of the Gene* is published.

1989   Separate human genome center created at NIH, with granting authority. Watson loudly criticizes Japan and Germany for lagging support.

1990   Beats back a new campaign by opponents of the Human Genome Project.

1991   Angrily and publicly deplores plans of scientist Craig Venter and the NIH director, Bernadine Healy, to patent gene fragments. Scientists on both sides of the Atlantic also protest vehemently. At CSHL, the Neuroscience Center is dedicated, culminating more than 20 years' effort to build neurobiology work at CSHL.

1992   NIH director Healy forces Watson out as director of Human Genome Project on pretext of conflicts of interest. Watson says, "I was fired."

1993   Turns over directorship of CSHL to Bruce Stillman, who has worked there since 1978, and assumes presidency. Builds house on adjacent land purchased by the lab and donates the house to the lab. After many years of agonizing by participants, institutions, and the Nobel Committee over credit for the 1977 split-genes discovery, Richard Roberts (formerly of CSHL) and Phillip Sharp of MIT share the Nobel Prize.

1994   On sabbatical in England, lectures at Oxford.

1997   16 December, receives National Medal of Science from President Bill Clinton. Is cited "for five decades of scientific and intellectual leadership in molecular biology ranging from his co-discovery of the double helical structure of DNA to the launching of the Human Genome Project."

1999   Lecturing to an overflow crowd at Harvard at the opening of its genome center, Watson is applauded when he demands that women students be tested for "fragile X" gene, which they could pass on to sons.

2000   Publishes collection of essays, *A Passion for DNA*. Attends 26 June White House celebration of near completion of the human genome sequence; President Clinton says, "Thank you, sir." Arouses furor in a talk about hormones at Berkeley.

2002   Over objections of many colleagues and a long struggle to find a publisher, his memoir of the mid-1950s is published as *Genes, Girls, and Gamow: After the Double Helix*. Receives honorary British knighthood.

INTERVIEWS BY THE AUTHOR

Except where noted, citations from these interviews are made from the author's transcripts of tape recordings, and page numbers in the notes refer to these transcripts.

Bruce Alberts, National Academy of Sciences. Washington, D.C., 28 March 2001 (author's notes).

Paul Berg, Stanford University. Palo Alto, Calif., 13 February 2001.

Steven Blose, Bio-Rad Corporation. San Francisco, 18 February 2001.

Michael Botchan, University of California, Berkeley. Berkeley, Calif., 15 February 2001.

David Botstein, Stanford University. Palo Alto, Calif., 12 February 2001.

Sydney Brenner, Salk Institute. La Jolla, Calif., 14 February 2001.

Richard Burgess, University of Wisconsin. Telephone interview, December 2000 (author's notes).

Keith Burridge, University of North Carolina, Chapel Hill. La Jolla, Calif., 4 February 2001.

Lan Bo Chen, Dana Farber Cancer Center. Boston, 14 December 2000.

Robert J. Crouch, National Institutes of Health. Bethesda, Md., 26 March 2001.

James Darnell, Rockefeller University. New York City, 15 June 2000 (author's notes).

Ronald Davis, Stanford University. Telephone interview, 17 January 2001 (author's notes).

Paul Doty, Harvard University. Cambridge, Mass., 9 September 1999 (author's notes), and 29 August 2001 (transcript).

Gerald Fink, Whitehead Institute for Biomedical Research, MIT. Cambridge, Mass., 25 January 2001 (author's notes).

Julian Fleischman, Washington University, St. Louis. Cambridge, Mass., 22 December 2000.

James Garrels, Proteome Corp. Beverly, Mass., 16 January 2001.

Walter Gilbert, Biogen Corporation. Cambridge, Mass., 26 July 1982 (author's notes), and of Harvard University, Cambridge, Mass., 22 December 1999 (author's notes).

Alfred Goldberg, Harvard Medical School. Boston, 4 January 2001.

Susan Gottesman, National Institutes of Health. Bethesda, Md., 26 March 2001.

François Gros, Pasteur Institute. Paris, 17 April 1964 (author's notes).

Douglas Hanahan, University of California, San Francisco. Cambridge, Mass., 3 May 2001.

William Haseltine, Human Genome Sciences. Washington, D.C., 25 March 2001.

Nancy Hopkins, Massachusetts Institute of Technology. Cambridge, Mass., 22 September 1999 (author's notes), and 28 June 2000 (transcript).

Robert Horvitz, Massachusetts Institute of Technology. Cambridge, Mass., 21 December 2001.

François Jacob, Pasteur Institute. Paris, 13 April 1964 (author's notes).

Joel Kirschbaum, University of California, San Francisco. San Francisco, 17 February 2001.

Michael Konrad. San Francisco, 17 February 2001.

Charles Kurland, Uppsala University. Cambridge, Mass., 14 July 2001.

Tom Maniatis, Harvard University. Cambridge, Mass, 18 December 2000.

Ron McKay, National Institutes of Health. Washington, D.C., 29 March 2001.

Matthew Meselson, Harvard University. Cambridge, Mass., 22 March 2001.

Carel Mulder, University of Massachusetts, Worcester. Telephone interview, 29 January 2001 (author's notes).

Masayasu Nomura, University of California, Irvine. Irvine, Calif., February 2001.

Arthur Pardee, Dana Farber Cancer Center. Boston, 14 December 2000.

Robert Pollack, Columbia University. New York City, 16 June 2000.

Mark Ptashne, Memorial Sloan-Kettering Cancer Center. New York City, 16 June 2000.

Richard Roberts, New England BioLabs. Beverly, Mass., 15 December 2000.

Phillip Sharp, Massachusetts Institute of Technology. Cambridge, Mass., 12 December 2000.

Charles Stevens, Salk Institute. La Jolla, Calif., 1 February 2001.

Robert Tjian, University of California, Berkeley. Berkeley, Calif., 16 February 2001.

Harold Varmus, Memorial Sloan-Kettering Cancer Center. New York City, 19 June 2000.

Heiner Westphal, National Institutes of Health. Bethesda, Md., 28 March 2001.

Daniel Wulff, State University of New York, Albany. Cambridge, Mass., 14 January 2001.

Norton D. Zinder, Rockefeller University. New York City, 24 February 2000.

David Zipser, University of California at San Diego. La Jolla, 31 January 2001.

NOTES

For more detailed information on interviews conducted by the author, please see the list in the preceding section.

PROLOGUE: 19 OCTOBER 1962

1. Sydney Brenner, "Think Small, Talk Big," *Times Literary Supplement*, 19 October 2001, 6–7.
2. Victor K. McElheny, "James Dewey Watson and the Double Helix," *Boston Sunday Globe Magazine*, 19 May 1968, 28.
3. Nicholas Wade, "Scientist at Work: James D. Watson, Impresario of the Genome, Looks Back with Candor," *New York Times*, 7 April 1998, F1.

CHAPTER 1 BOOKS AND BIRDS: "GROWING UP" IN CHICAGO

1. Nicholas Weinstock, "How to Raise a Genius," *New York Times Magazine*, 8 April 2001, 42–43.
2. JDW, *Talk of the Nation/Science Friday*, National Public Radio, 2 June 2000 (transcript).
3. JDW, CSHL Annual Report, 1998, 2.
4. André Lwoff, "Truth, Truth, What Is Truth (About How the Structure of DNA was Discovered)?" *Scientific American*, July 1968, 136.
5. JDW, *Genes, Girls, and Gamow: After the Double Helix* (New York: Knopf, 2002), 237. Watson could trace his paternal ancestry back five generations to William Weldon Watson, who was born in 1794 in New Jersey and later settled in Nashville. Succeeding generations, including his grandfather Thomas, all lived in Illinois or Wisconsin.

6. Ibid., 252.

7. Carolyn Hong, "How Beautiful It Was, This Thing Called DNA," *New Straits Times* (Malaysia), Focus Newsmakers, 1 December 1996, 15.

8. Will Bradbury, "Genius on the Prowl: Nobel Prize Winner James Watson of 'Double Helix' Fame Tangles with Another Enigma—Cancer," *Life*, 30 October 1970, 57–66; JDW, "Values from a Chicago Upbringing," *Annals of the New York Academy of Sciences*, 758 (30 June 1995): 194–197, reprinted in JDW, *A Passion for DNA: Genes, Genomes, and Society* (Cold Spring Harbor, N.Y.: Cold Spring Harbor Laboratory Press, 2000), 3–5.

9. JDW, CSHL Annual Report, 1998.

10. JDW, *Talk of the Nation/Science Friday.*

11. JDW, *The Double Helix: A Personal Account of the Discovery of the Structure of DNA* (New York: Atheneum, 1968), 221–222.

12. Lwoff, "Truth, Truth."

13. JDW, *Genes, Girls, and Gamow*, 40.

14. Lee Edson, "Says Nobelist James (Double Helix) Watson, 'To Hell with Being Discovered When You're Dead,'" *New York Times Magazine*, 18 August 1968, 31, 34.

15. Larry Thompson, "The Man Behind the Double Helix: Gene-Buster James Watson Moves on to Biology's Biggest Challenge, Mapping Heredity," *Washington Post*, 12 September 1989, Z12.

16. JDW, "Values from a Chicago Upbringing."

17. JDW, *Genes, Girls, and Gamow*, 39–40, 102, 124.

18. Victor K. McElheny, "James Dewey Watson and the Double Helix, *Boston Sunday Globe Magazine*, 19 May 1968, 34.

19. JDW, sixtieth-birthday remarks, Grace Auditorium, Cold Spring Harbor Laboratory, 16 April 1988.

20. JDW, "The Dissemination of Unpublished Information," lecture, Museum of History and Technology, Washington, January 1974, reprinted in JDW, *A Passion for DNA* Cold Spring Harbor, N.Y.: Cold Spring Harbor Laboratory Press, 2000, 91–103.

21. Sydney Brenner, interview by author.

22. JDW, "Values from a Chicago Upbringing."

23. Thompson, "Man Behind the Double Helix."

24. JDW, CSHL Annual Report, 1998.

25. JDW, CSHL Annual Report, 1976.

26. Thompson, "Man Behind the Double Helix."

27. JDW, sixtieth-birthday remarks, Cold Spring Harbor.

28. Felicity Barringer, "Nobel Laureate Warns, Don't Learn Too Much," *New York Times*, 10 March 1993, B8.

29. JDW, sixtieth-birthday remarks as n.27.

30. Keith Burridge, interview by author.

31. Horace Freeland Judson, *The Eighth Day of Creation: Makers of the Revolution in Biology,* rev. and expanded ed. (Cold Spring Harbor, N.Y.: Cold Spring Harbor Laboratory Press, 1996), 28–29.

32. JDW, sixtieth-birthday remarks as n.27.

33. JDW, *A Passion for DNA*, 91–103.

34. Ibid., 105–108.

35. JDW, *Talk of the Nation/Science Friday.*

36. Edson, "Says Nobelist James (Double Helix) Watson"; JDW, sixtieth-birthday remarks, Cold Spring Harbor; Barringer, "Nobel Laureate Warns."

37. JDW, "Values from a Chicago Upbringing."

38. JDW, interview by Anthony Liversidge, *Omni*, May 1984, 118.

39. Ibid.

40. John Dunning, *On the Air: The Encyclopedia of Old Time Radio* (New York: Oxford University Press, 1998).

41. Thompson, "Man Behind the Double Helix."

42. JDW, *Talk of the Nation/Science Friday.*

43. JDW, remarks, Century Club, New York, 15 February 1968 (author's notes).

44. Paul de Kruif, *Microbe Hunters* (New York: Harcourt, Brace, 1926), 122.

45. Ron Chernow, *Titan: The Life of John D. Rockefeller, Sr.* (New York: Random House, 1998), 478.

46. Sinclair Lewis, *Arrowsmith* (New York: Harcourt, Brace & World, 1952).

47. JDW, CSHL Annual Report, 1998, 2.

48. Gunther Stent, ed., *The Double Helix* (New York: Norton Critical Edition, 1980), xiii.

49. Oswald T. Avery, M.D., Colin M. MacLeod, M.D., and Maclyn McCarty, M.D., "Studies on the Chemical Nature of the Substance Inducing Transformation of Pneumococcal Types: Induction of Transformation by a Deoxyribonucleic Acid Fraction Isolated from Pneumococcus Type III," *Journal of Experimental Medicine* 79 (1 February 1944): 137–158, reprinted in J. Herbert Taylor, ed., *Selected Papers in Molecular Genetics* (New York: Academic Press, 1965), 157–178.

50. JDW, Time Inc. Workshop on DNA, Banbury Center, Cold Spring Harbor Laboratory, 3 May 1981, session 3 (transcript), 10.

51. Erwin Chargaff, "Preface to a Grammar of Biology." *Science* 172 (1971): 639–640.

52. JDW, *Double Helix,* 13–14.

53. JDW, Time Inc. DNA workshop, 10–11; JDW, CSHL Annual Report, 1986, 1.

54. Stent, *Double Helix,* xv.

55. Richard A. Knox, "The Men Who Shed the First Light on DNA," *Boston Globe*, 3 October 1994, 25.

56. Franklin W. Stahl, "Hershey," in *We Can Sleep Later: Alfred D. Hershey and the Origins of Molecular Biology* (Cold Spring Harbor, N.Y.: Cold Spring Harbor Laboratory Press, 2000), 3–4.

57. JDW, CSHL Annual Report, 1998, 2.

58. Herbert Black, "Shy, Sensitive Dr. Watson Popular with Students and Associates," *Boston Sunday Globe,* 21 October 1962.

59. JDW, "Succeeding in Science: Some Rules of Thumb," *Science* 261 (1993): 1812–1813.

60. Francis Crick, interview 28 January 1971 by Horace Judson, in Judson, *Eighth Day of Creation*, 87–88.

61. JDW, *Double Helix*, 13–14.

62. JDW, "Discovering the Double Helix," lecture, Grace Auditorium, Cold Spring Harbor Laboratory, 2 November 1999; posted on www.cshl.org.

63. JDW, *Talk of the Nation/Science Friday.*

64. JDW, CSHL Annual Report, 1998.

65. Sydney Brenner, Warren Lecture, Massachusetts General Hospital, Boston, 1968 (author's notes).

66. JDW, *Talk of the Nation/Science Friday.*

67. Paul Weiss to Robert Olby, 25 April 1973, quoted in Robert Olby, *The Path to the Double Helix* (Seattle: University of Washington Press, 1974), 297.

68. JDW, CSHL Annual Report, 1998.

69. JDW, *Talk of the Nation/Science Friday.*

70. Ibid.

71. JDW, speech at "Winding Your Way through DNA Symposium, University of California at San Francisco, 25 September 1992; posted at www.accessexcellence.com/AB/CC /watsonpres.html.

72. Ibid.

73. JDW, "Values from a Chicago Upbringing."

74. Ibid.

75. JDW, CSHL Annual Report, 1998.

76. JDW, "Values from a Chicago Upbringing."

77. JDW, CSHL Annual Report, 1989, 1; JDW, "Values from a Chicago Upbringing."

78. JDW, "Winding Your Way Through DNA" speech.

Chapter 2 Target, The Gene: Bloomington and "Paradise"

1. Horace Freeland Judson, *The Eighth Day of Creation: Makers of the Revolution in Biology*, rev. and expanded ed. (Cold Spring Harbor, N.Y.: Cold Spring Harbor Laboratory Press, 1996), 29–29.

2. Ibid., 30–31; JDW, "Values from a Chicago Upbringing," *Annals of the New York Academy of Sciences*, 758 (30 June 1995): 194–197.

3. Ibid; JDW, "S. E. Luria," *Nature* 350 (14 March 1991): 113.

4. JDW, "Growing Up in the Phage Group," in JDW, *Phage and the Origins of Molecular Biology*, rev. and expanded ed. (Cold Spring Harbor, N.Y.: Cold Spring Harbor Laboratory Press, 1992), 239; Robert Olby, *The Path to the Double Helix* (Seattle: University of Washington Press, 1974), 297–298.

5. JDW, interview by author (telephone).

6. JDW, "Values from a Chicago Upbringing."

7. Olby, *Path to the Double Helix*, 297–298.

8. JDW, *Talk of the Nation/Science Friday,* National Public Radio, 2 June 2000.

9. JDW, "The Dissemination of Unpublished Information," lecture, Museum of History and Technology, Washington, D.C., January 1974, reprinted in JDW, *A Passion for DNA* (Cold Spring Harbor, N.Y.: Cold Spring Harbor Laboratory Press, 2000), 91–103.

10. Olby, *Path to the Double Helix,* 300–301.

11. JDW, *Talk of the Nation/Science Friday.*

12. JDW, Time 100 Scientist and Thinker, Time Room, on-line session, 24 March 1999; posted at www.time.com/time/community/transcripts/1999/032499watsoncricktime100.html.

13. JDW, CSHL Annual Report, 1989, 1–9.

14. Judson, *Eighth Day of Creation,* 45.

15. JDW, *Talk of the Nation/Science Friday.*

16. Judson, *Eighth Day of Creation,* 48.

17. "Two Minds After a Double Helix," *The Economist,* 27 December 1980, 59.

18. Ibid.

19. JDW, "S. E. Luria."

20. S. E. Luria, 4 October 1973, cited in Judson, *Eighth Day of Creation,* 45.

21. JDW, Time 100 Scientist and Thinker, Time Room, on-line session, 24 March 1999.

22. JDW, speech, "Winding Your Way Through DNA," symposium, University of California at San Francisco, 25 September 1992; posted at www.accessexcellence.com/AB/CC/watsonpres.html.

23. JDW, CSHL Annual Report, 1997, 3.

24. Luria, 3 October 1973, quoted in Judson, *Eighth Day of Creation,* 45.

25. JDW, "Growing Up in the Phage Group," 240.

26. Ibid.

27. Ibid.

28. Abraham Pais, *Niels Bohr's Times, in Physics, Philosophy, and Polity* (Oxford: Clarendon Press, 1991), 441–442, 450.

29. Neville Symonds, "Geneticist Against Hegel: Obituary of Salvador Luria," *Guardian,* 11 February 1991.

30. Gunther Stent, "The DNA Double Helix and the Rise of Molecular Biology," in Gunther Stent, ed., *The Double Helix* (New York: Norton Critical Edition, 1980), xiii.

31. Robert L. Sinsheimer, *The Strands of a Life: The Science of DNA and the Art of Education* (Berkeley: University of California Press, 1994), 83.

32. JDW, "Growing Up in the Phage Group," 240.

33. Judson, *Eighth Day of Creation,* 45.

34. Sinsheimer, *Strands of a Life,* 82–84.

35. François Jacob, *The Statue Within: An Autobiography,* trans. Franklin Phillip (Cold Spring Harbor, N.Y.: Cold Spring Harbor Laboratory Press, 1995), 265–266.

36. Joseph Palca, "Genome Projects Are Growing like Weeds," *Science* 245: 131.

37. Judson, *Eighth Day of Creation,* 39–40.

38. JDW, CSHL Annual Report, 1990, 3–4; Norton Zinder, personal communication.

39. JDW, CSHL Annual Report, 1998, 1.

40. JDW, CSHL Annual Report, 1997, 4.

41. Joseph Frazier Wall, Andrew Carnegie (New York: Oxford University Press, 1970), 860–863; JDW, lecture, Science Center B, Harvard University, 30 September 1999.

42. Sydney Brenner, interview by author, 19.

43. Jacob, *Statue Within,* 269.

44. Elizabeth L. Watson, *Houses for Science: A Pictorial History of Cold Spring Harbor Laboratory* (Cold Spring Harbor, N.Y.: Cold Spring Harbor Laboratory Press, 1991), 143.

45. Carol Strickland, "Watson Relinquishes Major Role at Lab," *New York Times,* section 13, LI, 1.

46. JDW, CSHL Annual Report, 1997, 1.

47. JDW, *Talk of the Nation/Science Friday.*

48. JDW, CSHL Annual Report, 1997, 2.

49. JDW, "The Lives They Lived: Alfred D. Hershey; Hershey Heaven," *New York Times Magazine,* 4 January 1998, 16.

50. JDW, "Growing Up in the Phage Group," 242.

51. Salvador Luria to Max Delbrück, 20 January 1949, Max Delbrück Papers, California Institute of Technology Archives (hereafter cited as Delbrück Papers), box 14, folder 29.

52. JDW, CSHL Annual Report, 1997, 3.

53. Ibid.

54. Gunther S. Stent, "Reminiscences," in Franklin W. Stahl, ed., *We Can Sleep Later: Alfred D. Hershey and the Origins of Molecular Biology* (Cold Spring Harbor, N.Y.: Cold Spring Harbor Laboratory Press, 2000), 89–91; Phillip Sharp, personal communication.

55. JDW, CSHL Annual Report, 1997, 3–4.

56. Judson, *Eighth Day of Creation,* 48.

57. Errol C. Friedberg, *Correcting the Blueprint of Life: An Historical Account of the Discovery of DNA Repair Mechanisms* (Cold Spring Harbor, N.Y.: Cold Spring Harbor Laboratory Press 1997), 23.

58. Olby, *Path to the Double Helix,* 306; Judson, *Eighth Day of Creation,* 48; Max Delbrück to H. M. Weaver (National Foundation for Infantile Paralysis, New York), Delbrück Papers, box 23, folder 19; JDW, CSHL Annual Report, 1998, 3; Salvador Luria, *A Slot Machine, a Broken Test Tube: An Autobiography* (New York: Harper & Row, 1984), 88; JDW, "Discovering the Double Helix," lecture, Grace Auditorium, Cold Spring Harbor Laboratory, 2 November 1999, posted on www.cshl.org; "Two Minds After a Double Helix," *The Economist,* December 27, 1980, 59; Michel Morange, *A History of Molecular Biology,* trans. Matthew Cobb (Cambridge, Mass.: Harvard University Press, 2000 [paperback]), 107; JDW, speaking at Time Inc. Workshop

on DNA, Banbury Center, Cold Spring Harbor Laboratory, 3–4 May 1981, session 1 (transcript), 10.

59. JDW, CSHL Annual Report, 1998, 3.

60. JDW, "Growing Up in the Phage Group," 244; Judson, *Eighth Day of Creation*, 26–27.

61. JDW, CSHL Annual Report, 1998, 3.

62. JDW, "Growing Up in the Phage Group," 244.

63. JDW, "Hershey Heaven."

64. JDW, Time Inc. Workshop on DNA, session 1 (transcript), 13.

65. JDW, "Discovering the Double Helix."

66. JDW to Max and Manny Delbrück, 22 March 1951, Delbrück Papers, box 23, folder 20.

67. Ibid.

68. Ibid.

69. JDW, "Discovering the Double Helix."

70. JDW, "Growing Up in the Phage Group," 245.

71. Olby, *Path to the Double Helix*, 307.

72. JDW, speech at "Winding Your Way Through DNA" symposium.

73. JDW, "The Race for the Double Helix," *Nova*, WGBH Boston, 22 February 1976, written and produced by Graham Chedd (transcript of videotape) 7–8.

74. JDW, speech at "Winding Your Way Through DNA" symposium.

75. Olby, *Path to the Double Helix*, 353; Maurice Wilkins, interview, 1974, quoted in Franklin H. Portugal and Jack S. Cohen, *A Century of DNA: A History of the Discovery of the Structure and Function of the Genetic Substance* (Cambridge, Mass.: MIT Press, 1977), 249.

76. JDW, speech at "Winding Your Way Through DNA" symposium; Jeremy Bernstein, "Confessions of a Biochemist," *The New Yorker*, 13 April 1968, 172; JDW, "Discovering the Double Helix."

77. Olby, *Path to the Double Helix*, 308–309; Watson, "Growing Up in the Phage Group."

CHAPTER 3 STUMBLING ON GOLD: TWO SMART ALECKS IN CAMBRIDGE

1. Crick, *What Mad Pursuit: A Personal View of Scientific Discovery* (New York: Basic Books, 1988), 64–65.

2. Sheryl Stolberg, "Chasing the Mysteries of Life," *Los Angeles Times*, 28 February 1994, 1.

3. Carol Strickland, "Watson Relinquishes Major Role at Lab," *New York Times*, 21 March 1993, section 13, Long Island ed., 1.

4. JDW, *Talk of the Nation/Science Friday*, "NPR, 2 June 2000.

5. JDW, "Involvement of RNA in the Synthesis of Proteins," *Science* 140 (5 April 1963): 17–26 (adapted from Nobel lecture, December 1962).

6. Stolberg, "Chasing the Mysteries of Life."

7. R. V. Jones, *Most Secret War* (London: Hodder & Stoughton/Coronet Books 1979; reprint 1990), 639, 657.

8. Francis Crick to Robert Olby, 5 June 1969, cited in Robert Olby, *The Path to the Double Helix* (Seattle: University of Washington Press, 1974), 310.

9. Ibid.

10. Olby, *Path to the Double Helix,* 316; Franklin H. Portugal and Jack S. Cohen, *A Century of DNA: A History of the Discovery of the Structure and Function of the Genetic Substance* (Cambridge, Mass.: MIT Press, 1977), 250.

11. JDW, in "The Race for the Double Helix." *Nova,* WGBH Boston, 22 February 1976, written and produced by Graham Chedd (transcript of videotape).

12. JDW, "Involvement of RNA in the Synthesis of Proteins."

13. JDW, letter to Max Delbrück, 20 May 1952, Max Delbrück Papers, California Institute of Technology Archives (hereafter cited as Delbrück Papers), box 23, folder 21.

14. Robert K. Merton, "Making It Scientifically," *New York Times Book Review,* 25 February 1968, 1, 41–43; reprinted in Gunther S. Stent, ed., *The Double Helix* (New York: Norton Critical Edition, 1980), 213–218.

15. "Two Minds After a Double Helix," *The Economist,* 27 December 1980, 59.

16. Dennis L. Breo, "At Large: The Double Helix—Watson and Crick's Freak Find of How Like Begets Like," *Journal of the American Medical Association* 269 (24 February 1993).

17. Francis Crick, in "Race for the Double Helix," *Nova.*

18. Francis Crick, quoted in Stephen S. Hall, "James Watson and the Search for Biology's 'Holy Grail,' " *Smithsonian* 20: 40–49, 154 (bibliography).

19. Breo, "At Large."

20. JDW, letter to Max and Manny Delbrück, 22 March 1951, Delbrück Papers, box 23, folder 20.

21. Portugal and Cohen, *A Century of DNA,* 268.

22. P. B. Medawar, "Lucky Jim," review of *The Double Helix,* by James Watson, *New York Review of Books,* 28 March 1968, 3–5, reprinted in Stent, *Double Helix,* 218–224.

23. JDW, *Talk of the Nation/Science Friday.*

24. JDW, "Involvement of RNA in the Synthesis of Proteins."

25. JDW, "Minds That Live for Science," *New Scientist,* 21 May 1987, 63–66, reprinted in JDW, *A Passion for DNA* (Cold Spring Harbor, N.Y.: Cold Spring Harbor Laboratory Press, 2000), 17–22.

26. "Two Minds After a Double Helix," 59.

27. JDW, *A Passion for DNA,* 17.

28. Ibid., 22.

29. Ibid., 18.

30. Horace Freeland Judson. *The Eighth Day of Creation: Makers of the Revolution in Biology,* rev. and expanded ed. (Cold Spring Harbor, N.Y.: Cold Spring Harbor Press, 1996), 4.

31. Stephen S. Hall, "Old School Ties: Watson, Crick, and 40 Years of DNA," *Science* 259 (12 March 1993): 1533.

32. JDW, letter to Max Delbrück, 9 December 1951, Delbrück Papers, box 23, folder 20.

33. Francis Crick, letter to JDW, 13 April 1967.

34. JDW, *A Passion for DNA*, 17–22.

35. JDW, "Early Speculations and Facts about RNA Templates," in Raymond F. Gesteland, Thomas R. Cech, and John F. Atkins, eds., *The RNA World: The Nature of Modern RNA Suggests a Prebiotic RNA World*, 2nd ed. (Cold Spring Harbor, N.Y.: Cold Spring Harbor Laboratory Press, 1999), xviii.

36. Harrison Echols, *Operators and Promoters: The Story of Molecular Biology and Its Creators* (Berkeley: University of California Press, 2001), 4–5.

37. JDW, "Discovering the Double Helix," lecture, Grace Auditorium, Cold Spring Harbor Laboratory, 2 November 1999, posted on www.cshl.org.

38. Crick, *What Mad Pursuit*, 59.

39. Ibid., 60.

40. Echols, *Operators and Promoters*, 4, 6.

41. Crick, *What Mad Pursuit*, 74.

42. Ibid., 59–60.

43. Ibid., 69.

44. Ibid., 70.

45. Francis Crick, *Morning Edition*, National Public Radio, 15 March 1993 (transcript).

46. "Two Minds After a Double Helix."

47. Crick, *What Mad Pursuit*, 74.

48. JDW to Delbrück, 9 December 1951.

49. Larry Thompson, "The Man Behind the Double Helix: Gene-Buster James Watson Moves on to Biology's Biggest Challenge—Mapping Heredity," *Washington Post*, 12 September 1989, Z12.

50. JDW to Delbrück, 9 December 1951.

51. Farooq Hussain, "Did Rosalind Franklin Deserve DNA Nobel Prize?" *New Scientist* 68 (1975): 470; see also Portugal and Cohen, *A Century of DNA*, 268.

52. Brenda Maddox, "The Dark Lady of DNA?" *The Observer*, 5 March 2000, Observer Review Pages, 1.

53. "Two Minds After a Double Helix"; Brenda Maddox, *Rosalind Franklin* (New York: HarperCollins), 2002.

54. Olby, *Path to the Double Helix*, 331–333.

55. Aaron Klug (19 January 1976), cited in Judson, *Eighth Day of Creation*, 97–100.

56. Melvyn Bragg, "The Dark Lady of DNA," *The Independent* (London), 15 March 1998, 1, 2; JDW, sixtieth-birthday remarks, Science Center C, Harvard University, 13 May 1988.

57. Olby, *Path to the Double Helix*, 346; Melvin Bragg, "The Dark Lady of DNA"; Maddox, "The Dark Lady of DNA?"

58. Judson, *Eighth Day of Creation*, 624.

59. Olby, *Path to the Double Helix*, 344, 347.

60. Maddox, "The Dark Lady of DNA?"

61. Judson, *Eighth Day of Creation*, 625–626.

62. Maurice Wilkins, in "Race for the Double Helix," *Nova*.

63. JDW, in "Race for the Double Helix," *Nova*.

64. Olby, *Path to the Double Helix*, 347–348, 357; "Race for the Double Helix," *Nova*; JDW, "Discovering the Double Helix."

65. JDW, "Discovering the Double Helix."

66. Olby, *Path to the Double Helix*, 312.

67. JDW, remarks, Science Center B, Harvard University, 30 September 1999 (transcript).

68. JDW, "Discovering the Double Helix."

69. Crick, in "Race for the Double Helix," *Nova*, 22.

70. Olby, *Path to the Double Helix*, 351.

71. JDW, *The Double Helix: A Personal Account of the Discovery of the Structure of DNA* (New York: Atheneum, 1968), 69.

72. JDW, "Discovering the Double Helix"; JDW, in "Race for the Double Helix," *Nova*, 22.

73. JDW, "Discovering the Double Helix."

74. Olby, *Path to the Double Helix*, 336–337.

75. JDW, in "Race for the Double Helix," *Nova*, 22–23.

76. JDW, 30 September 1999 Science Center B remarks.

77. JDW, in "Race for the Double Helix," *Nova*, 25.

78. Francis Crick, in "Race for the Double Helix," *Nova*, 27.

79. Bragg, "The Dark Lady of DNA."

80. Maurice Wilkins, in "Race for the Double Helix," *Nova*, 26.

81. JDW, "Discovering the Double Helix."

82. JDW, *Double Helix*, 97; JDW, "Discovering the Double Helix."

83. JDW, address, American Neurological Association, Marriott Copley Place, Boston, 17 October 2000 (transcript).

84. Erwin Chargaff, "Chemical Specificity of Nucleic Acids and Mechanism of Their Enzymatic Degradation," *Experientia* 6 (1950): 201–209, reprinted in J. Herbert Taylor, *Selected Papers on Molecular Genetics* (New York: Academic Press, 1965), 245–253.

85. JDW, 30 September 1999 Science Center B remarks.

86. JDW to Delbrück, 9 December 1951.

87. Ibid.

88. JDW in "Race for the Double Helix," *Nova*, 27.

89. JDW, "Discovering the Double Helix."

90. JDW, *Double Helix*, 125.

91. A. D. Hershey and M. Chase, "Independent Functions of Viral Protein and Nucleic Acids in Growth of Bacteriophage," *Journal of General Physiology* 36: 39–56; Francis Bello, "Great American Scientists: The Biologists," *Fortune*, June 1960, 230, 235.

92. JDW, "Hershey Heaven," *New York Times Magazine*, 4 January 1998, 16.

93. JDW, speaking at Time Inc. Workshop on DNA, Banbury Center, Cold Spring Harbor Laboratory, 3–4 May 1981, session 1 (transcript), 15; Walter Gilbert, personal communication.

94. Stent, *Double Helix*, xv.

95. JDW, "Early Speculations and Facts about RNA Templates," xvii.

96. Olby, *Path to the Double Helix*, 319.

97. Judson, *Eighth Day of Creation*, 109.

98. JDW, *Double Helix*, 119.

99. François Jacob, *The Statue Within: An Autobiography*, trans. Franklin Phillip (Cold Spring Harbor, N.Y.: Cold Spring Harbor Laboratory Press, 1995), 264.

100. JDW, *Double Helix*, 119.

101. Jacob, *Statue Within*, 262–263.

102. Ibid., 264.

103. Crick, in "Race for the Double Helix," *Nova*, 29.

104. JDW, "Discovering the Double Helix."

105. Francis Crick, in "Race for the Double Helix," *Nova*, 29.

106. JDW, *Double Helix*, 117.

107. Olby, *Path to the Double Helix*, 369.

108. Maurice Wilkins, letter, *Science* 164 (27 June 1969): 1539.

109. Judson, *Eighth Day of Creation*, 114.

110. JDW, "Discovering the Double Helix."

111. Georgina Ferry, *Dorothy Hodgkin: A Life* (Cold Spring Harbor, N.Y.: Cold Spring Harbor Laboratory Press, 2000).

112. JDW, "Discovering the Double Helix."

113. JDW, letter to Max Delbrück, 20 May 1952, Delbrück Papers, box 23, folder 21.

114. Ibid.; JDW, "Discovering the Double Helix."

115. JDW, *Genes, Girls, and Gamow: After The Double Helix* (New York: Knopf, 2002), 172.

116. Judson, *Eighth Day of Creation*, 121.

117. Francis Crick, in "Race for the Double Helix," *Nova*, 32–33.

118. JDW, *Double Helix*, 126.

119. Francis Crick, interviews by Robert Olby, 8 March 1968 and 7 August 1972, in Olby, *Path to the Double Helix*, 387.

120. JDW, *Double Helix*, 126.

121. Olby, *Path to the Double Helix*, 387.

122. JDW, 30 September 1999 Science Center B remarks.

123. JDW, *Double Helix*, 129–131.

124. Olby, *Path to the Double Helix*, 388–390; Judson, *Eighth Day of Creation*, 119–120; "Race for the Double Helix," *Nova*, 35.

125. JDW, "Discovering the Double Helix"; JDW, 30 September 1999 Science Center B remarks.

126. Francis Crick, cited in Olby, *Path to the Double Helix*, 388.

127. JDW, "Discovering the Double Helix."

128. André Lwoff, "Truth, Truth, What Is Truth (About How the Structure of DNA was Discovered)?" *Scientific American*, July 1968, 134; see also Judson, *Eighth Day of Creation*, 121–122.

129. JDW, "Discovering the Double Helix."

130. JDW, letter to Max Delbrück, 22 September 1952, Delbrück Papers, box 23, folder 21.

131. John Cairns, "Through a Magic Casement," review of *What Mad Pursuit*, by Francis Crick, *Nature* 336 (17 November 1988): 268–269; JDW, "Early Speculations and Facts about RNA Templates," xviii.

132. M. F. Perutz, "DNA Helix," letter, *Science* 164 (27 June 1969): 1537–1538; M. F. Perutz, letter, *Scientific American*, July 1969, 8; Wilkins, letter, *Science* 164 (27 June 1969): 1539; JDW, letter, *Science* 164 (27 June 1969): 1539; JDW, letter, *Scientific American*, July 1969, 8; Judson, *Eighth Day of Creation*, 131, 625.

133. JDW, letter to Max Delbrück, 15 January 1953, Delbrück Papers, box 23, folder 22.

134. JDW, *Double Helix*, 156; Peter Pauling, "DNA—The Race That Never Was?" *New Scientist*, 31 May 1973, 558–600; Francis Crick, cited in Olby, *Path to the Double Helix*, 393–394; Judson, *Eighth Day of Creation*, 131–133; JDW, "Discovering the Double Helix."

135. "Race for the Double Helix," *Nova*.

136. JDW, "Discovering the Double Helix."

137. Jeremy Bernstein, "Confessions of a Biochemist," *The New Yorker*, 13 April 1968.

138. "Race for the Double Helix," *Nova*, 40.

139. Olby, *Path to the Double Helix*, 394–395.

140. JDW, "Discovering the Double Helix."

141. JDW, Time 100 Scientist and Thinker, 24 March 1999, Time Room, on-line question-and-answer session available at www.time.com/time/community/transcripts/1999/032499watsoncrick.html.

142. JDW, remarks, Sanders Theatre, Harvard University, 11 March 2002.

143. Medawar, "Lucky Jim," 3–5.

144. Robert K. Merton, "Behavior Patterns of Scientists. " *American Scholar* 38 (Spring 1969): 197–225.

145. JDW, "The Dissemination of Unpublished Information," in JDW, *A Passion for DNA*, 91–103.

146. Francis Crick, quoted in Portugal and Cohen, *A Century of DNA*, 269, 271 (citation from *The Listener* [BBC], 11 July 1974, 43).

147. Olby, *Path to the Double Helix*, 395–397.

148. Judson, *Eighth Day of Creation*, 147–148

149. Hall, "Old School Ties," 1533.

150. JDW, 11 March 2002 Sanders Theatre remarks.

151. JDW, "Discovering the Double Helix."; JDW, *Genes, Girls, and Gamow*, 10.

152. Ibid.

153. JDW, 11 March 2002 Sanders Theatre remarks.

154. Maddox, "The Dark Lady of DNA?"

155. Ibid.; JDW, *Genes, Girls, and Gamow*, 10; JDW, 11 March 2002 Sanders Theatre remarks.

156. Judson, *Eighth Day of Creation*, 138; JDW, "Discovering the Double Helix."

157. JDW, 11 March 2002 Sanders Theatre remarks.

158. Olby, *Path to the Double Helix*, 398–399.

159. Ibid.

160. JDW, in "Race for the Double Helix," *Nova*, 45.

161. JDW, interview by Anthony Liversidge, *Omni*, May 1984, 118.

162. "Race for the Double Helix," *Nova*, 46; Judson, *Eighth Day of Creation*, 138–139.

163. Olby, *Path to the Double Helix*, 399–400; "Two Minds After a Double Helix"; Crick, interview by Robert Olby, 10 July 1968, cited in Judson, *Eighth Day of Creation*, 139; JDW, *Double Helix*, 177–178.

164. Pauling, "DNA—The Race That Never Was?"

165. Olby, *Path to the Double Helix*, 400–401.

166. Judson, *Eighth Day of Creation*, 141.

167. M. F. Perutz, "DNA Helix," letter, *Science* 164 (27 June 1969); 1537–1538; Perutz, letter, *Scientific American*, July 1969, 8.

168. JDW, *Science*, 27 June 1969, 1539.

169. Judson, *Eighth Day of Creation*, 628.

170. Francis Crick, letter to Olby, 13 April 1973, cited in Olby, *Path to the Double Helix*, 402.

171. Francis Crick, interviews on 8 March 1968 and 7 August 1972, cited in Olby, *Path to the Double Helix*, 404.

172. JDW, "Discovering the Double Helix."

173. JDW, letter to Max Delbrück, 20 February 1953, Delbrück Papers, box 23, folder 22.

174. JDW, *Double Helix*, 190.

175. Ibid., 209.

176. Francis Crick, "The Double Helix: A Personal View," *Nature* 248 (26 April 1974): 766–769; reprinted in Stent, *Double Helix*, 137–145.

177. Judson, *Eighth Day of Creation*, 147.

178. Ibid., 148.

179. JDW, *Double Helix*, 195–196; Olby, *Path to the Double Helix*, 412.

180. JDW, CSHL Annual Report, 1986, 1.

181. JDW, *A Passion for DNA*, 20.

182. JDW, 30 September 1999 Science Center B remarks.

183. Crick, *What Mad Pursuit*, 65.

184. Jerry Donohue, "Old Twists to an Old Tale," *Nature* 290 (23 April 1981): 648–649.

185. Echols, *Operators and Promoters*, 8.

186. JDW, Time 100 Scientist and Thinker.

187. JDW, Time Inc. DNA Workshop, 18–19.

188. Crick, *What Mad Pursuit*, 66.

189. Portugal and Cohen, *A Century of DNA*, 264, 268.

190. Liversidge interview, 119; JDW, *Talk of the Nation/Science Friday*.

CHAPTER 4 A BEAUTIFUL MOLECULE: BEING BELIEVED

1. JDW, *The Double Helix: A Personal Account of the Discovery of the Structure of DNA* (New York: Atheneum, 1968), 200.

2. JDW, interview by Anthony Liversidge, *Omni*, May 1984; Jan Witkowski, "Mad Hatters at the DNA Tea Party," *Nature* 415 (31 January 2002): 474.

3. "DNA Discoverers Celebrate 40th Anniversary of Finding," *Morning Edition*, NPR, 15 March 1993; JDW, remarks, Science Center B, Harvard University, 30 September 1999; Amy Barrett, "Weird Science: Questions for James D. Watson," *The New York Times Magazine*, 3 February 2002, 9.

4. JDW, *Double Helix*, 200; Nicholas Wade, "Erwin Chargaff, Pioneer in Research of DNA, Dies at 96," *New York Times*, 30 June 2002, 23, citing interview with JDW.

5. JDW, *Double Helix*, 206.

6. JDW, "Early Speculations and Facts about RNA Templates," in Raymond F. Gesteland, Thomas R. Cech, and John F. Atkins, eds., *The RNA World, The Nature of Modern RNA Suggests a Prebiotic RNA World*, 2nd ed. (Cold Spring Harbor, N.Y.: Cold Spring Harbor Laboratory Press, 1999, xvii–xxv.

7. Larry Thompson, "The Man Behind the Double Helix," *Washington Post*, 12 September 1989, Z12.

8. JDW, CSHL Annual Report, 1982, 5.

9. Wade, "Erwin Chargaff."

10. Stephen S. Hall, "Old School Ties: Watson, Crick, and 40 Years of DNA," *Science* 259 (12 March 1993): 1533.

11. JDW, sixtieth-birthday remarks, Grace Auditorium, Cold Spring Harbor, 16 April 1988.

12. Sydney Brenner, "Think Small, Talk Big: The Heroic Days of Molecular Biology," *Times Literary Supplement* (London), 19 October 2001, 6–7.

13. Gunther S. Stent, "That Was the Molecular Biology That Was," *Science* 160 (26 April 1968): 390–395, reprinted in John Cairns, Gunther S. Stent, and James D. Watson, *Phage and the Origins of Molecular Biology*, rev. ed. (Cold Spring Harbor, N.Y.: Cold Spring Harbor Laboratory Press, 1992), 345.

14. JDW, address, American Neurological Association, Marriott Copley Place, Boston, 17 October 2000; Horace Freeland Judson, *The Eighth Day of Creation: Makers of the Revolution in Biology*, rev. ed. (Cold Spring Harbor, N.Y.: Cold Spring Harbor Laboratory Press, expanded edition 1996), 151; JDW, *Double Helix*, 206, 208–210; Brenda Maddox, "The Dark Lady of DNA?" *The Observer* (London), 5 March 2000, Observer Review, 1; Brenda Maddox, *Rosalind Franklin* (New York: HarperCollins, 2002), 208.

15. Maddox, *Rosalind Franklin,* 208.

16. JDW, *Double Helix,* 210; Judson, *Eighth Day of Creation,* 627–629; JDW, "Discovering the Double Helix," lecture, Grace Auditorium, Cold Spring Harbor Laboratory, 2 November 1999, posted at www.cshl.edu; Maddox, *Rosalind Franklin,* 205–206.

17. JDW, "Discovering the Double Helix."

18. Maddox, *Rosalind Franklin,* 210.

19. JDW, remarks, Sanders Theatre, Harvard University, 11 March 2002 (transcript), 12.

20. JDW, Liversidge interview, 118.

21. Hall, "Old School Ties."

22. JDW, *Double Helix,* 204–205; Witkowski, "Mad Hatters," 474.

23. Judson, *Eighth Day of Creation,* 151.

24. Ibid.

25. JDW, *Double Helix,* 214.

26. Judson, *Eighth Day of Creation,* 151.

27. JDW to Delbrück, 12 March 1953, Delbrück Papers, box 23, folder 22, California Institute of Technology Archives (hereafter cited as Delbrück Papers); also reproduced in Watson, *Double Helix.*

28. Judson, *Eighth Day of Creation,* 152.

29. JDW, *Double Helix,* 222–223; Judson, *Eighth Day of Creation,* 154; JDW, "Discovering the Double Helix."

30. Olby, *Path to the Double Helix,* 422.

31. JDW to Delbrück, 12 March 1953.

32. Ibid.

33. Ibid.

34. Judson, *Eighth Day of Creation,* 153–154.

35. JDW to Delbrück, 22 March 1953.

36. JDW, *Double Helix,* 223.

37. Judson, *Eighth Day of Creation,* 152.

38. Delbrück to JDW, 14 April 53, Delbrück Papers, box 23, folder 22.

39. Judson, *Eighth Day of Creation,* 238.

40. Sydney Brenner, interview by author.

41. J. D. Watson and F. H. C. Crick, "Molecular Structure of Nucleic Acids: A Structure for Deoxyribose Nucleic Acid," *Nature* 171 (25 April 1953): 737–738.

42. White House, Office of the Press Secretary, "Remarks. . . on the Completion of the First Survey of the Entire Human Genome Project."

43. JDW, 11 March 2002 Sanders Theatre remarks.

44. Hall, "Old School Ties."

45. JDW, 30 September 1999 Science Center B remarks.

46. Hall, "Old School Ties."

47. Judson, *Eighth Day of Creation,* 155.

48. JDW to Delbrück, 5 May 1953, Delbrück Papers, box 23, folder 22.

49. Judson, *Eighth Day of Creation,* 154, 169.

50. JDW, *Double Helix,* 221–222.

51. Hall, "Old School Ties."

52. Gunther S. Stent, ed., *The Double Helix* (New York: Norton Critical Edition, 1980), xvi.

53. Hall, "Old School Ties."

54. P. B. Medawar, "Lucky Jim," *New York Review of Books*, 28 March 1968, 3.

55. J. D. Watson and F. H. C. Crick, "The Structure of DNA," *Cold Spring Harbor Symposia on Quantitative Biology* 18 (1953): 123–131.

56. F. H. C. Crick and J. D. Watson, "The Complementary Structure of Deoxyribonucleic Acid," *Proceedings of the Royal Society* 223, series A (1954): 80–96.

57. Abraham Pais, *Einstein Lived Here* (New York: Oxford University Press, 1994), citing stories in *The Times* (London), 7 November 1919, and *New York Times,* 9 and 11 November 1919.

58. "Form of 'Life Unit' in Cell is Scanned," *New York Times,* 16 May 1953.

59. "Clue to Chemistry of Heredity Found," *New York Times,* 13 June 1953, 17.

60. JDW to Delbrück, 21 May 53, Delbrück Papers, box 23, folder 22.

61. Olby, *Path to the Double Helix,* 421; Judson, *Eighth Day of Creation,* 260; JDW to Crick, 9 October 1953, in Crick's possession.

62. F. H. C. Crick, "The Structure of the Hereditary Material," *Scientific American,* October 1954.

63. "Ten Top Young Scientists in U.S. Universities," *Fortune,* June 1954.

64. John Engstead, *Vogue,* August 1954, 126–127 (photograph).

65. Judson, *Eighth Day of Creation,* 279.

66. JDW, *Genes, Girls, and Gamow: After The Double Helix* (New York: Knopf, 2001), 106.

67. JDW, CSHL Annual Report, 1984, 13–14; JDW, CSHL Annual Report, 1993, 1–10; Elizabeth L. Watson, *Houses for Science: A Pictorial History of Cold Spring Harbor Laboratory* (Cold Spring Harbor, N.Y.: Cold Spring Harbor Laboratory Press, 1991), 145–155, 157–158; JDW, "Amyas Ames Obituary," CSHL Annual Report, 1999.

68. Francis Crick, interview by Carolina Biological Supply Company, 1989, posted at www.AccessExcellence.org/AE/AEC/CC/ Crick.html.

69. Delbrück to JDW, 1 May 1953, Delbrück Papers, box 23, folder 22.

70. Ibid; Max Delbrück, introduction, *Cold Spring Harbor Symposia on Quantitative Biology* 18 (1953); Judson, *Eighth Day of Creation,* 239–240; JDW, *Genes, Girls, and Gamow,* 18–19, 25.

71. Holmes, *Meselson, Stahl, and the Replication of DNA: A History of "The Most Beautiful Experiment in Biology"* (New Haven, Conn.: Yale University Press, 2001), 24–29.

72. JDW, *Genes, Girls, and Gamow,* 19.

73. Holmes, *Meselson, Stahl, and the Replication of DNA,* 25.

74. Matt Ridley, "Genome," *Daily Telegraph* (London), 23 August 1999, 13; JDW, *Genes, Girls and Gamow,* 19.

75. JDW, quoted in Jonathan Weiner, *Time, Love, Memory: A Great Biologist and His Quest for the Origins of Behavior* (New York: Knopf, 1999), 47.

76. François Jacob, *The Statue Within: An Autobiography,* trans. Franklin Phillip (Cold Spring Harbor, N.Y.: Cold Spring Harbor Laboratory Press), 1995), vii, 269.

77. Ibid., 270–271.

78. Alfred D. Hershey, introduction, *Cold Spring Harbor Symposia on Quantitative Biology* 18 (1953).

79. Thomas D. Brock, *The Emergence of Bacterial Genetics* (Cold Spring Harbor, N.Y.: Cold Spring Harbor Laboratory Press 1990), 144–145.

80. JDW, *Genes, Girls, and Gamow,* 23.

81. JDW and F. H. C. Crick, "The Structure of DNA," *Cold Spring Harbor Symposia on Quantitative Biology* 18 (1953): 123–131.

82. Holmes, *Meselson, Stahl, and the Replication of DNA,* 29.

83. JDW and Crick, "The Structure of DNA."

84. JDW, *Genes Girls, and Gamow,* 25; JDW, 11 March 2002 Sanders Theatre remarks; JDW, Time Inc. Workshop on DNA, Banbury Center, Cold Spring Harbor Laboratory, 3–4 May 1981.

85. JDW, 11 March 2002 Sanders Theatre remarks.

86. JDW, *Talk of the Nation/Science Friday,* NPR, 2 June 2000.

87. Norton D. Zinder, personal communication.

88. JDW, foreword, *Cold Spring Harbor Symposia on Quantitative Biology* 43 (1978); Arthur Kornberg, *For the Love of Enzymes: The Odyssey of a Biochemist* (Cambridge, Mass.: Harvard University Press, 1989), 121–122.

CHAPTER 5 NOW WHAT? THRASHING AROUND

1. JDW, CSHL Annual Report, 1981, 5.

2. JDW, remarks, Sanders Theatre, Harvard University, 11 March 2002 (transcript).

3. Gunther S. Stent, ed., *The Double Helix* (New York: Norton, Critical Edition, 1980), xviii.

4. JDW, *Genes, Girls, and Gamow* (New York: Knopf, 2001), 154–156; Brenda Maddox, *Rosalind Franklin* (New York, HarperCollins, 2002).

5. JDW, 11 March 2002 Sanders Theatre remarks.

6. JDW, *Genes, Girls, and Gamow,* 25–28.

7. Amy Barrett, "Weird Science: Questions for James D. Watson," *New York Times Magazine,* 3 February 2002, 9.

8. Sydney Brenner, "The House That Jim Built," *Nature* (1 June 2000): 511–512; Sydney Brenner, interview by author.

9. Gino Segrè, "The Big Bang and the Genetic Code," *Nature* 404 (30 March 2000): 437.

10. JDW, 11 March 2002 Sanders Theatre remarks.

11. Horace Freeland Judson, *The Eighth Day of Creation: Makers of the Revolution in Biology*, rev. and expanded ed., (Cold Spring Harbor, N.Y.: Cold Spring Harbor Laboratory Press 1996), 256–262.

12. Sydney Brenner, "Think Small, Talk Big: The Heroic Days of Molecular Biology," *Times Literary Supplement*, 19 October 2001, 6–7.

13. Francis Crick to M. Crick, 19 March 53, quoted in Judson, *Eighth Day of Creation*, 153.

14. Judson, *Eighth Day of Creation*, 236–237.

15. JDW, "Early Speculations and Facts about RNA Templates," in Raymond F. Gesteland, Thomas R. Cech, and John F. Atkins, eds., *The RNA World: The Nature of Modern RNA Suggests a Prebiotic RNA World,*" (Cold Spring Harbor, N.Y.: Cold Spring Harbor Laboratory Press, 1999), xvii–xxv.

16. JDW, *Genes, Girls, and Gamow*, 28–29.

17. JDW, 11 March 2002 Sanders Theatre remarks.

18. Judson, *Eighth Day of Creation*, 267.

19. JDW, *Genes, Girls, and Gamow.* 33–36.

20. JDW, 11 March 2002 Sanders Theatre remarks.

21. JDW, *Genes, Girls, and Gamow*, 35.

22. JDW, 11 March 2002 Sanders Theatre remarks.

23. JDW, *Genes, Girls, and Gamow*, 39–40, 254.

24. Ibid., 40.

25. Ibid, 45.

26. JDW, CSHL Annual Report, 1981, 5.

27. Larry Thompson, "The Man Behind the Double Helix," *Washington Post*, 12 September 1989, Z12.

28. JDW, *Genes, Girls, and Gamow*, 48–49.

29. Ibid., 67.

30. Jonathan Weiner, *Time, Love, Memory: A Great Biologist and His Quest for the Origins of Behavior* (New York: Knopf, 1999), 48–52, 55.

31. JDW, "Involvement of RNA in the Synthesis of Proteins: The Ordered Interactions of Three Classes of RNA Controls the Assembly of Amino Acids into Proteins," *Science* 140 (5 April 1963): 17–26.

32. JDW to Delbrück, 25 March 1954, Max Delbrück Papers, California Institute of Technology Archives (hereafter cited as Delbrück Papers), box 23, folder 23.

33. JDW to Delbrück, 1 June 1954, Delbrück Papers, box 23, folder 23.

34. JDW, "Early Speculations and Facts about RNA Templates."

35. Judson, *Eighth Day of Creation*, 269

36. Norton D. Zinder, personal communication.

37. JDW, *Genes, Girls, and Gamow*, 73.

38. Delbrück to Beadle, 2 June 1954, George Beadle Papers, California Institute of Technology Archives (hereafter cited as Beadle Papers), box 8, folder 2.

39. Beadle to Delbrück, 14 June 1954, Beadle Papers, box 8, folder 2.

40. Beadle to JDW, 14 June 1954, Beadle Papers, box 8, folder 2.

41. Beadle to M. H. Trytten, 28 June 1954, Beadle Papers, box 8, folder 2.

42. JDW, *Genes, Girls, and Gamow,* 73.

43. Errol C. Friedberg, *Correcting the Blueprint of Life: An Historical Account of the Discovery of DNA Repair Mechanisms* (Cold Spring Harbor, N.Y.: Cold Spring Harbor Laboratory Press, 1997), 147–150; Walter Gilbert, personal communication.

44. JDW, *Genes, Girls, and Gamow,* 86.

45. JDW quoted in Judson, *Eighth Day of Creation,* 279.

46. Brenda Maddox, *Rosalind Franklin* (New York, HarperCollins, 2002), 246.

47. Friedberg, *Correcting the Blueprint of Life,* 150.

48. JDW, *Genes, Girls, and Gamow,* 85.

49. JDW, sixtieth-birthday remarks, Science Center C, Harvard University, 13 May 1988.

50. JDW, "Early Speculations and Facts about RNA Templates."

51. JDW to Beadle, 5 September 1954, Beadle Papers, box 8, folder 2.

52. JDW, *Genes, Girls, and Gamow,* 67–68.

53. Francis Crick, *What Mad Pursuit: A Personal View of Scientific Discovery* (New York, Basic Books, 1988), 96.

54. Sydney Brenner, interview 21 May 1975, by Charles Weiner, Recombinant DNA Archive, Manuscript Collection 100, MIT Archives, 19.

55. Brenner, "Think Small, Talk Big," 6.

56. JDW, *Genes, Girls, and Gamow,* 139.

57. JDW, "Involvement of RNA in the Synthesis of Proteins," *Science* 140 (5 April 1963): 17–26

58. Crick, *What Mad Pursuit,* 95.

59. Sydney Brenner, interview by author.

60. Judson, *Eighth Day of Creation,* 279.

61. JDW, *Genes, Girls, and Gamow,* 98–103.

62. JDW, "Early Speculations and Facts about RNA Templates," xx.

63. Brenner to Watson, 31 July 1954; JDW, *Genes, Girls, and Gamow,* 89.

64. Brenner, author interview.

65. Ibid.

66. Maddox, *Rosalind Franklin,* 239–240; JDW, *Genes, Girls, and Gamow,* 106–107.

67. Brenner, author interview.

68. Ibid.; JDW, *Genes, Girls, and Gamow,* 102.

69. JDW, *Genes, Girls, and Gamow,* 105.

70. Ibid.

71. JDW, *Genes, Girls, and Gamow,* 131.

72. Paul M. Doty, interviews by author, 9 September 1999, and 29 August 2001.

73. Konrad Bloch, remarks, Watson's seventieth-birthday celebration, Harvard University, 24 May 1998.

74. JDW, *Genes, Girls, and Gamow,* 82.

75. John R. Raper to Max Delbrück, Delbrück papers, box 23, folder 23; John Raper to George Beadle, 3 March 1955, Beadle Papers, box 8, folder 2.

76. Delbrück to Raper, 7 March 55, Delbrück Papers, box 23, folder 23.

77. Salvador Luria, *A Slot Machine, a Broken Test Tube: An Autobiography* (New York: Harper & Row, 1984), 131.

78. Paul Doty, remarks, *Nature* conference, Boston, 19 September 1983 (author's notes).

79. Thompson, "The Man Behind the Double Helix."

80. JDW, "Early Speculations and Facts about RNA Templates," xxi.

81. JDW, *Genes, Girls, and Gamow,* 179.

82. Ibid., 213.

83. Ibid.

84. Maddox, *Rosalind Franklin,* 267–268; JDW, *Genes, Girls, and Gamow,* 224–225 (photo on 224).

85. James Watson and Francis Crick, "Structure of Small Viruses," *Nature* 177 (10 March 1956): 473–476; JDW, "Early Speculations and Facts about RNA Templates," xvii–xxv; JDW, *Genes, Girls, and Gamow,* 217.

86. JDW, *Genes, Girls, and Gamow,* 217.

87. Ibid., 194.

88. Ibid., 233–234.

CHAPTER 6 Harvard: "FEW DARED CALL HIM TO ACCOUNT"

1. JDW, *Genes, Girls, and Gamow: After the Double Helix* (New York: Knopf, 2001), 229.

2. Ibid., 229–230.

3. Ibid., 234–235.

4. Nan Robertson, "Love and Work Now Watson's Double Helix," *New York Times,* 26 December 1980, A24.

5. James Darnell, interview by author.

6. Nancy Hopkins, interview by author.

7. Stephen S. Hall, "James Watson and the Search For Biology's 'Holy Grail,'" *Smithsonian* 20:40–49.

8. Paul Doty, remarks, JDW's seventieth-birthday celebration, Harvard University, 24 May 1998.

9. Paul Doty remarks, *Nature* conference, Boston, 19 September 1983 (author's notes).

10. Victor K. McElheny, "Harvard's Biology Chemistry Depts; Merging for Research," *Boston Globe,* 14 April 1967, 9.

11. JDW, sixtieth-birthday materials.

12. JDW, sixtieth-birthday remarks, Grace Auditorium, Cold Spring Harbor Laboratory, 16 April 1988.

13. JDW, seventieth-birthday remarks, Harvard University, 24 May 1998.

14. Jeremy Knowles, lecture, Science Center B, Harvard University, 30 September 1999 (author's notes).

15. JDW, *Genes, Girls, and Gamow,* 254.

16. Edward O. Wilson, *Naturalist* (Washington, D.C.: Island Press/Shearwater Books, 1994), 222–223.

17. Ibid., 219.

18. David Schlessinger, "The Start at Harvard: Those Golden Years," in "Celebrating Sixty Years" (prepared for JDW's sixtieth-birthday celebration at Harvard University), 5–7.

19. Charles Kurland, interview by author.

20. Wilson, *Naturalist,* 230; Victor K. McElheny, "Harvard Forms Center for Environmental Biology," *Boston Globe,* 17 December 1967, 4.

21. Name TK, personal communication, 1976.

22. Gerald Fink, interview by author.

23. JDW, CSHL Annual Report, 1979, 4–7; reprinted in JDW, *A Passion for DNA* (Cold Spring Harbor, N.Y.: Cold Spring Harbor Laboratory Press 2000), 109–116.

24. Daniel Wulff, interview by author.

25. JDW, CSHL Annual Report, 1998, 1–5.

26. JDW, seventieth-birthday remarks, Harvard.

27. JDW, CSHL Annual Report, 1998, 5.

28. David Schlessinger, "Parallel Lives: My Life in the Lab and Molecular Microbiology," *ASM News,* January 1996, 29.

29. Kurland interview.

30. Doty, remarks, JDW seventieth-birthday celebration.

31. JDW, sixtieth-birthday remarks, Harvard; Larry Thompson, "The Man Behind the Double Helix," *Washington Post,* 12 September 1989, Z12.

32. Julian Fleischman, interview by author.

33. Susan Gottesman, interview by author.

34. Lee Edson, "Says Nobelist James (Double Helix) Watson, 'To Hell with Being Discovered When You're Dead,'" *New York Times Magazine,* 18 August 1968.

35. Horace Judson, *The Eighth Day of Creation: Makers of the Revolution in Biology,* rev. and expanded ed. (Cold Spring Harbor, N.Y.: Cold Spring Harbor Laboratory Press, 1996), 24–26.

36. David Botstein, interview by author.

37. Nancy Hopkins, interview by author.

38. Nancy Hopkins, remarks, JDW seventieth-birthday celebration.

39. Hopkins interview.

40. Hopkins, remarks, JDW seventieth-birthday celebration.

41. Hopkins interview.

42. JDW, CSHL Annual Report, 1998.

43. JDW, sixtieth-birthday, remarks, Harvard.

44. Mario Capecchi, remarks, JDW seventieth-birthday celebration.

45. William Haseltine, interview by author.

46. Fleischman interview.

47. Alfred Goldberg, interview by author.

48. Masayasu Nomura, "Early Days of Ribosome Research," *Trends in Biomedical Sciences* 15 (June 1990): 244–247; Masayasu Nomura, "Bacteriophages, Colicins, and Ribosomes: Some Reflections on My Research Career," in Kornberg et al., eds., *Reflections on Biochemistry* (Oxford: Pergamon Press, 1976), 317–323; Masayasu Nomura, interview by author.

49. Wulff interview.

50. Fleischman interview.

51. Kurland interview.

52. David Zipser, interview by author.

53. Goldberg interview; Darnell interview.

54. Kurland interview.

55. Edson, "Says Nobelist James (Double Helix) Watson."

56. Paul Berg, interview by author.

57. Keith Burridge, interview by author.

58. Edson, "Says Nobelist James (Double Helix) Watson."

59. Haseltine interview.

60. JDW, sixtieth-birthday, remarks, Cold Spring Harbor, 6 April 1988.

61. Botstein interview.

62. Hazeltine interview.

63. JDW, CSHL Annual Report, 1981.

64. JDW, CSHL Annual Report, 1989, 3.

65. JDW, sixtieth-birthday remarks, Cold Spring Harbor.

66. JDW, CSHL Annual Report, 1989, 3–4.

67. JDW, CSHL Annual Report, 1981, 5–6.

68. JDW, *Talk of the Nation/Science Friday,* NPR, 2 June 2000 (transcript).

69. Jerry E. Bishop, "What Causes Life? New Research on How Body Makes Proteins May Yield an Answer," *Wall Street Journal,* 25 April 1960, 1.

70. John Pfeiffer, "The New Biology: The Role Of Nucleic Acids In Medicine," *Medical News,* special supplement, 11 May 1960.

71. Francis Bello, "Great American Scientists: The Biologists," *Fortune,* June 1960.

72. Morris Kaplan, "Williams Seeking Parley on Health," *New York Times,* 21 October 1960, 28.

73. Austin C. Wehrwein, "U. of Chicago Installs 7th Head," *New York Times,* 5 May 1961, 17.

74. *Life* magazine, 2 November 1962.

75. Leonard Engel, "The Race to Create Life," *Harper's Magazine,* October 1962, 9–45.

76. F. H. C. Crick, "The Genetic Code," *Scientific American,* October 1962.

77. Robert K. Plumb, "The Genetic Code Is Held Universal," *New York Times,* 12 October 1962, 33.

78. Goldberg interview.

79. "Biologist Shares Prize: A Nobel for Harvard," *Boston Traveler,* 19 October 1962; Andrew T. Weil, "J. D. Watson Wins Nobel for Medicine," *Harvard Crimson,* 19 October 1962, 1.

80. Werner Wiskari, "Nobel Prize Goes to 3 Biophysicists," *New York Times,* 18 October 1962, 1.

81. Goldberg interview.

82. Ian Menzies, "Harvard Scientist Shares Nobel, *Boston Globe,* 18 October 1962.

83. Herbert Black, "Too Skinny to Date Girls, Dr. Watson Turned to Science," *Boston Globe,* 21 October 1962.

84. Weil, "J. D. Watson Wins Nobel for Medicine."

85. Black, "Too Skinny to Date Girls."

86. "Biologist, 34, Shares Prize."

87. Weil, "J. D. Watson Wins Nobel for Medicine."

88. Wiskari, "Nobel Prize."

89. "Watson, a Quiz Kid and 'Child Prodigy,' Now 'Young Turk.'" *New York Times,* 19 October 1962, 27.

90. Black, "Too Skinny to Date Girls."

91. "Watson, a Quiz Kid and 'Child Prodigy.'"

92. "Biologist, 34, Shares Prize." *Boston Traveler,* 18 October 1962.

93. Noah Gordon, "Colleagues Salute Watson for Nobel," *Boston Herald,* 19 October 1962.

94. John T. Edsall, "Nobel Prize: Two Britons, American Share 1962 Award for Genetic Code Achievement," *Science* 138 (26 October 1962): 498–500.

95. Crick to Delbrück, 6 November 1962, Delbrück Papers, box 5, folder labeled "Francis Crick, 1961–1965."

96. JDW, *Genes, Girls, and Gamow,* 250, Reuters, *Boston Globe,* 5 December 1962.

97. Judson, *Eighth Day of Creation,* 4.

98. Daniel Dennett, personal communication, 25 September 2001.

99. Judson, *Eighth Day of Creation,* 556.

Chapter 7 Manifesto and Marriage

1. JDW, *Molecular Biology of the Gene* (New York: W. A. Benjamin, 1965), x.

2. Sydney Brenner, interview by author.

3. JDW, *Genes, Girls, and Gamow: After The Double Helix* (New York, Knopf, 2001), 252.

4. David Botstein, interview by author.

5. JDW, *Molecular Biology of the Gene,* 267.

6. Ibid., 396.

7. Ibid., 400.

8. Ibid., 401.

9. Ibid.

10. Ibid., 363.

11. Ibid., 468.

12. Ibid., 68.

13. Ibid., 29

14. Ibid., 228.

15. Ibid., 160.

16. Ibid., 102–103.

17. Ibid., 99.

18. Ibid., 277.

19. Salvador Luria, *A Slot Machine, a Broken Test Tube: An Autobiography* (New York, Harper & Row, 1984), 131.

20. Alfred Goldberg, interview by author.

21. Phillip Sharp, interview by author.

22. Nancy Hopkins, interview by author.

23. Mario Capecchi, remarks, JDW seventieth-birthday celebration, Harvard University, 24 May 1998.

24. JDW, sixtieth-birthday, remarks, Harvard Science Center, 13 May 1988.

25. *Molecular Biology of the Gene,* 4th ed. (Menlo Park: Benjamin/Cummings, 1988), v.

26. David Schlessinger, "The Start at Harvard: Those Golden Years," in "Celebrating Sixty Years," prepared for JDW sixtieth-birthday celebration, Cold Spring Harbor Laboratory, 16 April 1988, 5–7.

27. Susan Gottesman, interview by author.

28. Schlessinger, "The Start at Harvard."

29. David Schlessinger, "Parallel Lives: My Life in the Lab and Molecular Microbiology," *ASM News,* January 1996, 29.

30. Ibid., 31.

31. Schlessinger, "The Start at Harvard."

32. JDW, "Involvement of RNA in the Synthesis of Proteins: The Ordered Interaction of Three Classes of RNA Controls the Assembly of Amino Acids Into Proteins," *Science* 140 (5 April 1963): 17–26 (adapted from Nobel lecture, December 1962).

33. Masayasu Nomura, "Early Days of Ribosome Research," *Trends in Biomedical Sciences* 15 (June 1990): 244–247.

34. JDW, *Talk of the Nation/Science Friday,* NPR, 2 June 2000.

35. JDW, sixtieth-birthday, remarks, Grace Auditorium, Cold Spring Harbor, 16 April 1988 (author's notes); Larry Thompson, "The Man Behind the Double Helix," *Washington Post,* 12 September 1989, Z12.

36. JDW, sixtieth-birthday, remarks, Grace Auditorium, Cold Spring Harbor.

37. Gunther Stent, "That Was the Molecular Biology That Was," *Science* 160 (1968): 390–395.

38. JDW, remarks, *Nature* conference, Boston, Mass., 19 September 1983 (author's notes).

39. Francis Crick, *What Mad Pursuit: A Personal View of Scientific Discovery* (New York, Basic Books, 1988), 95–96; JDW, *Genes, Girls, and Gamow,* 240;

John A. Osmundsen, "Biologists Hopeful of Solving Secrets of Heredity This Year," *New York Times*, 2 February 1962, 1, 14; Sydney Brenner, interview by author; Horace Freeland Judson, *The Eighth Day of Creation: Makers of the Revolution in Biology*, rev. and expanded ed. (Cold Spring Harbor, N.Y.: Cold Spring Harbor Laboratory Press, 1996), 271.

40. François Jacob, interview by author, Paris, 1964.

41. Charles Kurland, interview by author.

42. Masayasu Nomura, "Bacteriophages, Colicins, and Ribosomes: Some Reflections on My Research Career," in Arthur Kornberg et al., eds., *Reflections on Biochemistry* (Oxford: Pergamon Press, 1976), 318; Nomura, "Early Days of Ribosome Research."

43. JDW, "Involvement of RNA in the Synthesis of Proteins."

44. JDW, *Nature* conference remarks.

45. JDW, "Involvement of RNA in the Synthesis of Proteins," 21.

46. JDW, *Genes, Girls, and Gamow*, 227–229; Elliot Volkin and Lazarus Astrachan, "Intracellular Distribution of Labeled Ribonucleic Acid After Phage Infection of *Escherichia coli*," *Virology* 2 (August 1956): 433–437; Jerard Hurwitz and J. J. Furth, "Messenger RNA," *Scientific American*, February 1962; JDW, "Early Speculations and Facts about RNA Templates," in Raymond F. Gesteland, Thomas R. Cech, and John F. Atkins, eds., *The RNA World*, 2nd ed. (Cold Spring Harbor, N.Y.: Cold Spring Harbor Laboratory Press, 1999), xxiv.

47. JDW and Alfred Tissières, "Ribonucleoprotein Particles from *E. coli*," *Nature* 182 (1958): 778; Nomura, "Early Days of Ribosome Research."

48. C. G. Kurland, "Molecular Characterization of Ribonucleic Acid from *E. coli* Ribosomes: I. Isolation and Molecular Weights," *Journal of Molecular Biology* 2 (1960): 83.

49. Kurland interview; Walter Gilbert, personal communication.

50. Walter Gilbert, remarks, JDW seventieth-birthday celebration, Harvard; JDW, *Genes, Girls, and Gamow*, 218.

51. JDW, *Genes, Girls, and Gamow*, 222.

52. Celia Gilbert, remarks, seventieth-birthday celebration, Harvard.

53. JDW, *Genes, Girls, and Gamow*, 231.

54. Ibid., 266–268

55. Celia Gilbert, seventieth-birthday celebration remarks.

56. François Gros, interview by author.

57. Victor K. McElheny, "France Considers Significance of Nobel Awards," *Science* 150 (19 November 1965): 1013–1015.

58. JDW, *Molecular Biology of the Gene*, 404, 409, 412.

59. Michel Morange, *A History of Molecular Biology*, trans. Matthew Cobb (Cambridge, Mass.: Harvard University Press, 1998), 143.

60. Jacob interview, 2; A. B. Pardee, F. Jacob, and J. Monod, "The Genetic Control and Cytoplasmic Expression of 'Inducibility' in the Synthesis of Beta-Galactosidase in *E. coli*," *Journal of Molecular Biology* 1 (1959): 165–178; A. B. Pardee, "Experiments on the Transfer of Information from DNA to Enzymes," *Experimental Cell Research Supplement* 6 (1958): 142–151; Arthur Pardee, interview by author.

61. Nomura, "Bacteriophages, Colicins, and Ribosomes," 318.

62. M. Nomura, B. D. Hall, and S. Spiegelman, "Characterization of RNA Synthesized in *Escherichia coli* After Bacteriophage T2 Infection," *Journal of Molecular Biology* 2 (1960): 306–326.

63. Pardee interview.

64. Jacob interview.

65. JDW, *Genes, Girls, and Gamow*, 240–241.

66. François Jacob, *The Statue Within: An Autobiography*, trans. Franklin Phillip (Cold Spring Harbor, N.Y.: Cold Spring Harbor Laboratory Press, 1995), 311.

67. François Gros, "The Messenger," in André Lwoff and Agnes Ullmann, eds., *Origins of Molecular Biology: A Tribute to Jacques Monod* (New York: Academic Press 1979), 120.

68. Crick, *What Mad Pursuit*, 118–112; JDW, *Genes, Girls, and Gamow*, 241–242.

69. Judson, *Eighth Day of Creation*, 418–427; Morange, *History of Molecular Biology*, 144.

70. Morange, *History of Molecular Biology*, 145.

71. JDW, "Early Speculations and Facts about RNA Templates," xxiv; JDW, *Genes, Girls, and Gamow*, 241.

72. Gros, "The Messenger."

73. Kurland interview.

74. Gros, "The Messenger"; Walter Gilbert, personal communication.

75. Walter Gilbert, interview by author; W. Gilbert, remarks, JDW seventieth-birthday celebration, Harvard.

76. W. Gilbert interview; Stephen S. Hall, *Invisible Frontiers: The Race to Synthesize a Human Gene* (New York: Atlantic Monthly Press, 1987), 32–33; Walter Gilbert, remarks, JDW sixtieth-birthday celebration, Cold Spring Harbor Laboratory, 16 April 1988.

77. Morange, *History of Molecular Biology*, 144.

78. Kurland interview; Charles Kurland, e-mail to author, 26 November 2001.

79. W. Gilbert, remarks, seventieth-birthday celebration.

80. Kurland interview; Kurland, e-mail, 26 November 2001; Sydney Brenner, François Jacob, and Matthew Meselson, "An Unstable Intermediate Carrying Information from Genes to Ribosomes for Protein Synthesis," *Nature* 190 (1961): 576–581; François Gros et al., "Unstable Ribonucleic Acid Revealed by Pulse Labeling of *Escherichia coli*," *Nature* 190 (13 May 1961): 581–585; Judson, *Eighth Day of Creation*, 427–428; Walter Gilbert, personal communication.

81. Matthew Meselson, interview by author; JDW, *Genes, Girls, and Gamow*, 245–246; Judson, *Eighth Day of Creation*, 463–464; Morange, *History of Molecular Biology*, 135–136.

82. JDW, *Molecular Biology of the Gene*, 1965, 373.

83. JDW, "The Double Helix Revisited," *Time*, 3 July 2000, 30.

84. JDW, address, American Neurological Society, Marriott Copley Place, Boston, 17 October 2000.

85. Francis Crick, Croonian lecture, Royal Society, London, 5 May 1966 (author notes); Peter Stubbs, "Reading the Genetic Code," *New Scientist*, 12 May 1966, 348.

86. JDW, "Looking Forward," *Gene* 135 (15 December 1993): 315.

87. JDW, *Genes, Girls, and Gamow*, 253.

88. Earl Lane, "Inside James Watson's World Famous Lab," *Newsday*, 4 February 1979, 11.

89. John Cairns, "Through a Magic Casement," review of *What Mad Pursuit*, by Francis Crick, *Nature* 336 (17 November 1988): 269.

90. JDW, *Genes, Girls, and Gamow*, 253.

91. Maxine Singer, interview by author (telephone).

92. JDW, *Genes, Girls, and Gamow*, 254–255.

93. Singer interview.

94. JDW, sixtieth-birthday remarks, Harvard.

95. JDW, CSHL Annual Report, 1989, 4.

96. Walter Gilbert, "Polypeptide Synthesis in *Escherichia coli*; I. Ribosomes and the Active Complex," *Journal of Molecular Biology* 6 (1963): 374–388; Walter Gilbert, "Polypeptide Synthesis in *Escherichia coli*; II. The Polypeptide Chain and S-RNA," *Journal of Molecular Biology* 6 (1963): 389–403; Michael Cannon, Robert Krug, and Walter Gilbert, "The Binding of S-RNA by *Escherichia coli* Ribosomes," *Journal of Molecular Biology* 7 (1963): 360–378; Walter Gilbert, "Protein Synthesis in *Escherichia coli*," *Cold Spring Harbor Symposia on Quantitative Biology* 28 (1963): 287–294; JDW, *Molecular Biology of the Gene*, 332–333; Judson, *Eighth Day of Creation*, 561.

97. Daniel Nathans, "Bacterial Viruses," Review of Norton D. Zinder, ed. *RNA Phages* (Cold Spring Harbor, N.Y.: Cold Spring Harbor Laboratory Press, 1975) *Science* 190 (19 December 1975): 1195.

98. Mario Capecchi, remarks, JDW sixtieth-birthday celebration, Grace Auditorium, Cold Spring Harbor Laboratory, 16 April 1988 (author notes).

99. Norton Zinder, personal communication, 9 May 2002.

100. Susan Sample, "Against the Odds," *University of Utah Health Sciences Report*, Winter 1997.

101. Capecchi, remarks, JDW seventieth-birthday celebration.

102. Mario Capecchi, acceptance speech, Lasker Award ceremony, New York City, 15 September 2001.

103. Sample, "Against the Odds"; Capecchi, Lasker Award acceptance speech.

104. Capecchi, Lasker Award acceptance speech.

105. Mark Ptashne and Walter Gilbert, "Genetic Repressor," *Scientific American*, June 1970, 44.

106. Richard R. Burgess et al., "Factor Stimulating Transcription by RNA Polymerase," *Nature* 221 (4 January 1969): 43–46.

107. JDW, "Transcription of Genetic Material," *Cold Spring Harbor Symposia on Quantitative Biology* 35 (1970); JDW, *Molecular Biology of the Gene*, 2nd ed. (New York: W. A. Benjamin, 1970), 349–353.

108. Richard Burgess, remarks, JDW sixtieth-birthday celebration, Cold Spring Harbor Laboratory, 16 April 1988.

109. Guido Guidotti, remarks, JDW seventieth-birthday celebration, Harvard.

110. Mark Ptashne, personal communication.

111. "The Scientist," *ABC News,* 1969, transcript of videotape, Harvard Department of Molecular Biology and Biochemistry; Ptashne interview.

112. Walter Gilbert and Benno Müller-Hill, "Isolation of the Lac Repressor," *Proceedings of the National Academy of Sciences* 56 (1966): 1891–1898.

113. Ptashne and Gilbert, "Genetic Repressor," 44.

114. Ibid., 40.

115. "The Scientist," *ABC News*; Hall, *Invisible Frontiers,* 34.

116. Mark Ptashne, "Isolation of the Lambda Phage Repressor," *Proceedings of the National Academy of Sciences* 57 (February 1967): 306–313; Gilbert and Müller-Hill, "Isolation of the Lac Repressor"; Victor K. McElheny, "Harvard Scientists Isolate Key Life Control Substance; Repressors Control Genes' Actions," *Boston Globe,* 19 February 1967, 8; Victor McElheny, "Cell-Control Advances Are Claimed," *Washington Post,* 1 May 1967.

117. Ptashne interview; "Turning Genes On and Off," *Mosaic,* July–August 1975, 20–23.

118. William Haseltine, interview by author.

119. Walter Gilbert, seminar, Harvard Medical School, 13 October 1967 (author's notes).

120. Victor K. McElheny, "New Clues Found in Genetic Study," *New York Times,* 11 April 1974, 7.

121. JDW, seventieth-birthday remarks, Harvard.

122. Nan Robertson, "Love and Work Now Watson's Double Helix," *New York Times,* 26 December 1980, A24.

123. Susan Gottesman, interview by author.

124. Susanna Kaysen, *Girl, Interrupted* (New York: Vintage Books, 1994), 25–27.

125. JDW, *Genes, Girl, and Gamow,* 257.

126. Ibid., 255–256.

127. C. Gilbert, remarks, JDW seventieth-birthday celebration.

128. JDW, *Genes, Girls, and Gamow,* 257.

129. Ibid., 259; Marilyn Zinder, personal communication.

Chapter 8 "Fresh, Arrogant, Catty, Bratty, and Funny"

1. JDW, "The Double Helix," parts 1 and 2, *The Atlantic,* January and February 1968.

2. Harold M. Schmeck, Jr., Palo Alto, "Virus Core Produced Artificially; Way Is Opened to Advances in Genetics," *New York Times*, 15 December 1967, section 1, 28; "A Great Achievement," editorial, *New York Times*, 14 December 1967, 28; "'Primitive Form' of Life Duplicated in a Test Tube," 1, 5. Victor K. McElheny, "Simplest DNA Virus Duplicated," *Boston Globe*, 15 December 1967, 3; Walter Sullivan, "Nobel Prizes and Data Multiply in Recent Research on Heredity," *New York Times*, 15 December 1967, 29; "Life Chemistry Expert: Arthur Kornberg," *New York Times*, 15 December 1967, 28; "Uncovering Life's Secrets," editorial, *New York Times*, 15 December 1967, 46; Daniel S. Greenberg, "The Synthesis of DNA: How They Spread the Good News," *Science* 158 (22 December 1967): 1548–1550; Maxine Singer, "In vitro Synthesis of DNA: A Perspective on Research," *Science* 158 (22 December 1967): 1550–1551; Efraim Racker, "The Accidental Talent," *Nature* 340 (13 July 1989): 107–108, Review of *For the Love of Enzymes* by Arthur Kornberg; Victor McElheny, "News analysis: DNA Discovery Kept Quiet for Past 3 Months," *Boston Globe*, 15 December 1967, 2.

3. Victor K. McElheny, "Watson Tells 'DNA Story,' " *Boston Globe*, 29 December 1967, 20.

4. Thomas Wilson, author interview (1968).

5. Joel R. Kramer, "Corporation Vetoed Watson Book, *Harvard Crimson*, 13 February 1968, 1, 8.

6. "The Double Helix," editorial, *Harvard Crimson*, 14 February 1968.

7. Walter Sullivan, "A Book That Couldn't Go to Harvard." *New York Times*, 15 February 1968, 1, reprinted in Gunther S. Stent, ed. *The Double Helix* (New York: Norton, Critical Edition, 1980, xxiv–xxv.

8. JDW, letter to Max Perutz, 19 February 1968.

9. Nan Robertson, "Love and Work Now Watson's Double Helix," *New York Times*, 26 December 1980, A24.

10. Victor McElheny, "Harvard Scholar Defends Book: It Harms No One," *Boston Globe*, 16 February 1968.

11. Lee Edson, "Says Nobelist James (Double Helix) Watson, 'To Hell with Being Discovered When You're Dead,' " *New York Times Magazine*, 26–27.

12. JDW, *The Double Helix: A Personal Account of the Discovery of the Structure of DNA* (New York: Atheneum, 1968), 7.

13. Ibid., 33.

14. Ibid., 17.

15. Ibid., 42–43.

16. Maurice Wilkins, letter to JDW, 25 July 1966.

17. Francis Crick, letter to JDW, 13 April 1967, quoted in Horace Freeland Judson, *The Eighth Day of Creation: Makers of the Revolution in Biology*, rev. and expanded ed. (Cold Spring Harbor, N.Y.: Cold Spring Harbor Laboratory Press, 1996), 157.

18. Francis Crick, *What Mad Pursuit: A Personal View of Scientific Discovery* (New York: Basic Books, 1988), 81.

19. JDW, interview by Anthony Liversidge, *Omni*, May 1984, 120.

20. Sheryl Stolberg, "Chasing the Mysteries of Life; After You Have Helped Crack the Code for DNA, What's Next?," *Los Angeles Times*, 28 February 1994, A1.

21. Peter B. Medawar, "Lucky Jim," review of *The Double Helix*, by James Watson, *New York Review of Books*, 28 March 1968, 3–5; reprinted in Stent, *Double Helix*, 218–224.

22. Stent, *Double Helix*.

23. Peter Farb, "Can a Young Man Whose Mind Is on Girls, Booze, Skiing and High-Class Parties Win the Nobel Prize?" *Washington Post Book World*, 18 February 1968, 3.

24. Walter Sullivan, "The Competition Can Get Personal," *New York Times*, 18 February 1968, 8E.

25. JDW, remarks, Sanders Theatre, Harvard University, 11 March 2002.

26. Erwin Chargaff, "A Quick Climb up Mount Olympus," *Science* 159 (4 April 1968): 1448–1449.

27. Stent, *Double Helix*, 168–171.

28. John Lear, "Heredity Transactions," *Saturday Review*, 16 March 1968, 36, reprinted in Stent, *Double Helix*, 194–198.

29. Richard C. Lewontin, " 'Honest Jim' Watson's Big Thriller about DNA," *Chicago Sunday Sun-Times*, 25 February 1968, 1–2; reprinted in Stent, *Double Helix*, 185–187.

30. Conrad H. Waddington, "Riding on a High Spiral," *The Sunday Times* (London), 1; Stent, *Double Helix*, 204–207.

31. Robert L. Sinsheimer, "The Double Helix," *Science and Engineering*, September 1968, 4, 6; reprinted in Stent, *Double Helix*.

32. André Lwoff, "Truth, Truth, What Is Truth (About How the Structure of DNA was Discovered)?" *Scientific American* 219 (July 1968), 133–138; reprinted in Stent, *Double Helix*, 224–234.

33. "Professor Watson's Memoirs," editorial, *Nature* 217 (25 March 1968).

34. Sydney Brenner, "The House That Jim Built," review of *A Passion for DNA*, by James Watson, *Nature* 405 (1 June 2000): 511–512.

35. Jacob Bronowski, "Honest Jim and the Tinker-Toy Model," *The Nation*, 18 March 1968, 381–382; reprinted in Stent, *Double Helix*, 200–203; JDW, *Genes, Girls, and Gamow: After The Double Helix* (New York: Knopf, 2001), 258–259.

36. Philip Morrison, "The Human Factor in a Science First, *Life*, 1 March 1968, 8; reprinted in Stent, *Double Helix*, 175–177.

37. Medawar, "Lucky Jim."

38. Robert K. Merton, "Making It Scientifically," *New York Times Book Review*, 1; reprinted in Stent, *Double Helix*, 213–218.

39. Robert K. Merton, "Behavior Patterns of Scientists," *American Scholar*, Spring 1969, 197–225.

40. JDW, Liversidge interview, 120.

CHAPTER 9 A PASSION FOR BUILDING: COLD SPRING HARBOR

1. Edward O. Wilson, *Naturalist* (Washington, D.C.: Island Press/Shearwater Books, 1994), chapter 12, "The Molecular Wars," 218–237.

2. Victor K. McElheny, "James Dewey Watson and the Double Helix," *Boston Sunday Globe Magazine,* 19 May 1968, 28.

3. Elizabeth L. Watson, *Houses for Science: A Pictorial History of Cold Spring Harbor Laboratory* (Cold Spring Harbor, N.Y.: Cold Spring Harbor Laboratory Press, 1991), 177–178.

4. JDW, CSHL Annual Report, 1973, 7; JDW, CSHL Annual Report, 1980, 5; Watson, *Houses for Science,* 199; JDW, CSHL Annual Report, 1990, 5; JDW, CSHL Annual Report, 1992, 3; JDW, CSHL Annual Report, 1998.

5. Leslie Roberts, "Cold Spring Harbor Turns 100," *Science* 250 (26 October 1990): 496–498.

6. Carol Strickland, "Watson Relinquishes Major Role at Lab," *New York Times,* 21 March 1993, Long Island ed., section 13, 1.

7. Arthur Kornberg, *For the Love of Enzymes: The Odyssey of a Biochemist* (Cambridge, Mass.: Harvard University Press, 1989 [paperback]), 308.

8. Robert L. Sinsheimer, *The Strands of a Life: The Science of DNA and the Art of Education* (Berkeley: University of California Press), 149.

9. JDW, CSHL Annual Report, 1969, 4–5.

10. Strickland, "Watson Relinquishes Major Role."

11. S. N. Behrman, *Duveen* (Boston: Little, Brown, 1978), 186–201.

12. Joseph Frazier Wall, *Andrew Carnegie* (New York: Oxford University Press, 1970), 866–868.

13. Arthur S. Link, *Wilson: The Road to the White House* (Princeton, N.J.: Princeton University Press, 1968 [paperback]), 59–91.

14. Ron Chernow, *Titan: The Life of John D. Rockefeller, Sr.* (New York: Random House, 1998).

15. James Darnell, interview by author.

16. Charles Stevens, interview by author.

17. Harold Varmus, interview by author.

18. Ibid.

19. Norton D. Zinder, interview by author.

20. Watson, *Houses for Science,* 164–165.

21. Zinder interview; Norton Zinder, personal communication, 9 May 2002.

22. Watson, *Houses for Science,* 166–167.

23. Zinder, personal communication.

24. Harrison Echols, *Operators and Promoters: The Story of Molecular Biology and Its Creators,* ed. Carol A. Gross (Berkeley: University of California Press, 2001), 70, 72.

25. Max Delbrück to Bernard D. Davis, 10 July 1962, Delbrück Papers, California Institute of Technology Archives (hereafter cited as Delbrück Papers), box 5, folder 2.

26. John Cairns, CSHL Annual Report, 1967.

27. Watson, *Houses for Science,* 166–167.

28. Max Delbrück, letter to Terrell L. Hill, 27 October 1966, Delbrück Papers, box 5, folder 2.

29. John Cairns, letter to Max Delbrück, 3 December 1963, Delbrück Papers, box 5, folder 2.

30. JDW, CSHL Annual Report, 1973, 8.

31. Watson, *Houses for Science,* 169–170; JDW, CSHL Annual Report, 1967.

32. Heiner Westphal, interview by author.

33. Robert J. Crouch, interview by author.

34. Zinder, personal communication.

35. Zinder interview.

36. John Cairns, letter to Max Delbrück, 9 January 1968, Delbrück Papers, box 5, folder 2.

37. JDW, CSHL Annual Report, 1973, 8; JDW, CSHL Annual Report, 1977, 13.

38. Alden Whitman, "Cancer Engrosses Nobel Winner," *New York Times,* 18 July 1971, 55.

39. Robert Reinhold, "Watson, Author of 'The Double Helix,' to Direct Laboratory on Long Island," *New York Times,* 29 March 1968, 34,

40. Lee Edson, "Says Nobelist James (Double Helix) Watson, 'To Hell with Being Discovered When You're Dead,'" *New York Times Magazine*, 18 August 1968, 26.

41. Will Bradbury, "Genius on the Prowl: Nobel Prize Winner James Watson of 'Double Helix' Fame Tangles with Another Enigma—Cancer," *Life,* 30 October 1970, 57–66.

42. Horace Judson, *The Eighth Day of Creation: Makers of the Revolution in Biology,* rev. and expanded ed. (Cold Spring Harbor, N.Y.: Cold Spring Harbor Laboratory Press, 1996), 26–27.

43. Reinhold, "Watson."

44. Judson, *Eighth Day of Creation,* 26–27.

45. JDW, "The Cancer Conquerors: Who Should Get All that New Money?" *The New Republic,* 26 February 1972, 19.

46. Bradbury, "Genius on the Prowl," 57–66.

47. National Institutes of Health, press release, 23 May 1969; Watson, *Houses for Science,* 176–177; JDW, CSHL Annual Report," 1972, 5.

48. Zinder, personal communication.

49. JDW, CSHL Annual Report, 1970, 7.

50. Heiner Westphal, interview by author.

51. Crouch interview.

52. Joseph Sambrook, CSHL Annual Report, 1983, 1.

53. JDW, CSHL Annual Report, 1976, 13; JDW, CSHL Annual Report, 1977, 10–11.

54. JDW, CSHL Annual Report, 1972, 8–9; Watson, *Houses for Science*, 188, 190.

55. JDW, CSHL Annual Report, 1971, 10.

56. Watson, *Houses for Science*, 195.

57. JDW, CSHL Annual Report, 1972, 8.

58. JDW, CSHL Annual Report, 1969, 6; JDW CSHL Annual Report, 1972, 5; Watson, *Houses for Science*, 187.

59. JDW, CSHL Annual Report, 1973, 10.

60. JDW CSHL Annual Report, 1975, 11; Watson, *Houses for Science*, 204–206.

61. JDW, CSHL Annual Report, 1969, 8.

62. JDW, CSHL Annual Report, 1973, 8.

63. JDW, CSHL Annual Report, 1987; Watson, *Houses for Science*, 200–201; JDW, CSHL Annual Report, 1993, 4–5; Walter Gilbert, interview by Horace Judson, October 1973, cited in Judson, *Eighth Day of Creation*, 43.

64. Douglas Hanahan, interview by author.

65. Darnell interview; JDW, CSHL Annual Report, 1973, 9, 10–11; Watson, *Houses for Science*, 192–194.

66. Bruce Alberts, interview by author; Zinder interview; Zinder, personal communication.

67. JDW, CSHL Annual Report, 1972, 7–8; JDW, CSHL Annual Report, 1973, 11; Watson, *Houses for Science*, 197–198; Zinder interview; Robert Caro, *The Power Broker: Robert Moses and the Fall of New York* (New York: Knopf, 1979), 277–278, 302, 1152.

68. Tom Maniatis, interview by author.

69. JDW, "Charles Robertson, February 13, 1905–May 2, 1981," obituary, CSHL Annual Report, 1981, 4–5; JDW, "Edward Pulling (1896–1991)," obituary, CSHL Annual Report, 1990; Watson, *Houses for Science*, 198–200.

70. Robert Pollack, interview by author.

71. JDW, CSHL Annual Report, 1973, 7; JDW, CSHL Annual Report, 1974, 8; JDW, CSHL Annual Report, 1975, 13; Bruce Stillman, CSHL Annual Report, 1998, 22.

72. JDW, CSHL Annual Report, 1976, 9; JDW, CSHL Annual Report, 1977, 8.

73. JDW, CSHL Annual Report, 1993, 4.

74. JDW, CSHL Annual Report, 1982, 14.

75. Roberts, "Cold Spring Harbor Turns 100."

76. Watson, *Houses for Science*, 309–321; JDW, CSHL Annual Report, 1987, 7–9; JDW, CSHL Annual Report, 1988, 12–14; JDW, CSHL Annual Report, 1989, 10–11; JDW, CSHL Annual Report, 1990, 8; JDW, CSHL Annual Report, 1991, 8–10.

77. Watson, *Houses for Science*, 182–183.

78. Ibid., 236–238, 259–261.

79. Ibid., 217–221.

80. Ibid., 231–233, 248–254.

81. Ibid., 266–267; JDW, CSHL Annual Report, 1987, 9–10; JDW, CSHL Annual report, 1988, 12; JDW, CSHL Annual Report, 1989, 9–10; JDW, CSHL Annual Report, 1990, 13–14.

82. Strickland, "Watson Relinquishes Major Role"; Harold Varmus, interview by author.

83. JDW, CSHL Annual Report, 1979; JDW, CSHL Annual Report, 1980, 5; JDW, CSHL Annual Report, 1981, 12; author's notes.

84. JDW, CSHL Annual Report, 1979, 11–12; Watson, *Houses for Science,* 222–227; Nan Robertson, "Love and Work Now Watson's Double Helix," *New York Times,* 26 December 1980, A24; Marilyn Zinder, personal communication, 25 August 2002.

85. JDW, in Annual report, 1980, 12; JDW, CSHL Annual Report, 1981, 9; JDW, CSHL Annual Report, 1982, 10; JDW, CSHL Annual Report, 1984, 8; JDW, CSHL Annual Report, 1985, 10–11; JDW, CSHL Annual Report, 1986, 6–7.

86. Stevens interview.

87. Alberts interview.

88. Author notes from 8 January 1979 to 24 February 1981.

89. Author notes from 22 January 1979 to 8 March 1982; JDW, CSHL Annual Report, 1982, 11; JDW, CSHL Annual Report, 1984, 9; JDW, "Walter Hines Page II: 1913–1999," obituary, CSHL Annual Report, 1998, ix-xiii; JDW, CSHL Annual Report, 1985, 5; Watson, *Houses for Science,* 245–248; JDW, CSHL Annual Report, 1986, 7–8; JDW, CSHL Annual Report, 1988, 17.

90. Hanahan interview.

91. Zinder interview; Zinder, personal communication, 9 May 2002.

92. David Botstein, interview by author.

93. Roberts, "Cold Spring Harbor Turns 100."

94. Ibid.

95. Maniatis interview.

CHAPTER 10 HIGHER" CELLS: SCIENCE AT COLD SPRING HARBOR

1. Walter Gilbert, remarks, Sanders Theatre, Harvard University, 11 March 2002.

2. Victor K. McElheny, "Mystery of Heredity Still Eludes Scientists," *Boston Globe,* 16 June 1968, 8A.

3. Victor K. McElheny, "Beer and Barbecue: Biology conference like family party," *Boston Globe,* 16 June 1968, 8A.

4. Will Bradbury, "Genius on the Prowl: Nobel Prize Winner James Watson of 'Double Helix' Fame Tangles with Another Enigma—Cancer," *Life,* 30 October 1970, 57–66; JDW, CSHL Annual Report, 1993, 2.

5. JDW, CSHL Annual Report, 1972, 2-9; Leslie Roberts, "Cold Spring Harbor Turns 100," *Science* 250 (26 October 1990): 496–498.

6. JDW, CSHL Annual Report, 1992; James Darnell, interview by author.

7. Horace Judson, *The Eighth Day of Creation: Makers of the Revolution in Biology*, rev. and expanded ed. (Cold Spring Harbor, N.Y.: Cold Spring Harbor Laboratory Press, 1996), 26–27.

8. JDW, CSHL Annual Report, 1973, 6, 12.

9. David Botstein, interview by author.

10. Author's notes.

11. JDW, CSHL Annual Report, 1977, 7.

12. JDW, foreword, *Cold Spring Harbor Symposia on Quantitative Biology* 39 (1974).

13. JDW, foreword, *Cold Spring Harbor Symposia on Quantitative Biology* 41 (1979).

14. JDW, CSHL Annual Report, 1981, 11.

15. JDW, CSHL Annual Report, 1989.

16. JDW, quoted in Judson, *Eighth Day of Creation,* 168.

17. Alden Whitman, "Cancer Engrosses Nobel Winner," *New York Times,* 18 July 1971, 55.

18. Heiner Westphal, interview by author.

19. JDW, CSHL Annual Report, 1981, 5; JDW, CSHL Annual Report, 1993, 11.

20. JDW, CSHL Annual Report, 1990, 4–5.

21. JDW, CSHL Annual Report, 1981, 6–7.

Harold M. Schmeck, Jr., "Gene Gains Buoy Hopes on Cancer," *New York Times,* 15 April 1983, D19; "Gazing into the crystal ball of research." *The Economist,* 23 April 1983, 97; JDW, CSHL Annual Report, 1985, 12; JDW, CSHL Annual Report, 1988, 19–20; JDW, CSHL Annual Report, 1990, 6; Elizabeth L. Watson, *Houses for Science: A Pictorial History of Cold Spring Harbor Laboratory* (Cold Spring Harbor, N.Y.: Cold Spring Harbor Laboratory Press, 1991), 234, 22; John F. Avedon, " 'Where No One Has Been Before': A Community of Top Scientists, Led By Nobel Laureate James Watson, Is Closing In on the Three Great Riddles of Modern Biology," *Parade,* 16 June 1985, 12–13; JDW, CSHL Annual Report, 1985, 11–12; JDW, CSHL Annual Report, 1988, 19; Leslie Roberts, "Cold Spring Harbor Turns 100," *Science* 250 (26 October 1990): 496–498.

23. John Cairns and Jonathan Logan, "Step by Step into Carcinogenesis," *Nature* 304 (18 August 1983): 582–583 (commenting on papers by Ruley and others); JDW, CSHL Annual Report, 1983, 7; JDW, CSHL Annual Report, 1984, 1984, 3-14; JDW, CSHL Annual Report, 1989, 7.

24. JDW, CSHL Annual Report, 1997, 6.

25. Francis Crick, "Predictions in Biology," *Chemtech,* May 1979, 298.

26. Renato Dulbecco, "From the Molecular Biology of Oncogenic Viruses to Cancer," *Science* 192 (30 April 1976): 437 (adapted from Nobel lecture, December 1975).

27. Robert Reinhold, "Watson, Author of 'The Double Helix,' to Direct Laboratory on Long Island," *New York Times,* 29 March 1968, 34.

28. "Watson Appointment," *Science* 160 (5 April 1968): 50.

29. JDW, CSHL Annual Report, 1968, 5.

30. JDW, CSHL Annual Report, 1997, 5.

31. JDW, CSHL Annual Report, 1989, 4.

32. Westphal interview.

33. JDW, CSHL Annual Report, 1993, 2.

34. Joseph Sambrook, CSHL Annual Report, 1983, 4.

35. Carel Mulder, interview by author (telephone).

36. JDW, CSHL Annual Report, 1993, 3.

37. JDW, CSHL Annual Report, 1990, 4-5.

38. Richard Roberts, interview by author.

39. Keith Burridge, interview by author.

40. JDW, CSHL Annual Report, 1981, 6.

41. Westphal interview.

42. JDW, CSHL Annual Report, 1978; Mulder interview.

43. Joseph Sambrook, CSHL Annual Report, 1983, 4.

44. Westphal interview.

45. Roberts interview.

46. Douglas Hanahan, interview by author.

47. Robert Crouch, interview by author.

48. Tom Maniatis, interview by author.

49. Michael Botchan, interview by author.

50. Ibid.

51. JDW, CSHL Annual Report, 1982, 7-16; JDW, CSHL Annual Report, 1989, 7.

52. Phillip Sharp, interview by author.

53. Botchan interview.

54. Watson, *Houses for Science,* 207; www.cshl.org/History/100years-t39.html.

55. Crouch interview.

56. Ibid.

57. Mulder interview.

58. Richard Roberts, "An Amazing Distortion in DNA Induced by a Methyltranferase," Nobel lecture, Stockholm, 8 December 1993; Tom Maniatis, personal communication.

59. Roberts interview.

60. Watson, *Houses for Science,* 207; JDW, CSHL Annual Report, 1989, 5.

61. JDW, CSHL Annual Report, 1973, 5.

62. Victor K. McElheny, "MIT Biologist Probes Virus-Cancer Link," *Boston Globe,* 12 June 1970, 10.

63. Victor K. McElheny, "In Both Treatment and Prevention, There's Optimism Among the Cancer Experts . . . If Their Funds Last," *Boston Globe,* 6 December 1970, 4A; JDW, foreword, *Cold Spring Harbor Symposia on Quantitative Biology* 35 (1970); David Baltimore, interview, 12 May 1975, Manuscript Col-

lection 100, Recombinant DNA Archive, MIT; JDW, CSHL Annual Report, 1989, 5–6.

64. Westphal interview.
65. Sharp interview.
66. Westphal interview.
67. Sharp interview.
68. Stephen Blose, interview by author.
69. Sharp interview.
70. Westphal interview.
71. Botchan interview.
72. Burridge interview.
73. Robert Pollack, interview by author.
74. Tom Maniatis, interview by author.
75. Roberts interview.
76. Burridge interview.
77. Lan Bo Chen, interview by author.
78. Burridge interview; James Garrels, interview by author.
79. Monte Davis, "The World According to Jim," *Discover,* April 1981, 40–42.
80. Earl Lane, "Inside James Watson's World Famous Lab," *Newsday* (Long Island), 4 February 1979, 10–13, 22, 25; Tom Mariatis, personal communication.
81. David Zipser, interview by author.
82. Blose interview.
83. Mulder interview.
84. Crouch interview.
85. Botchan interview.
86. Westphal interview.
87. Crouch interview.
88. Garrels interview; Tom Mariatis, personal communication.
89. Crouch interview.
90. Robert Tjian, interview by author.
91. Westphal interview.
92. Lane "Inside James Watson's World Famous Lab," 13.
93. JDW, foreword, *Cold Spring Harbor Symposia on Quantitative Biology* 37 (1972).
94. JDW, foreword, *Cold Spring Harbor Symposia on Quantitative Biology* 36 (1971); JDW, foreword, *Cold Spring Harbor Symposia on Quantitative Biology* 38 (1973).
95. Maniatis interview.
96. Judson, *Eighth Day of Creation,* 26-27.
97. Sharp interview.
98. Mulder interview.
99. Westphal interview.
100. David Pendlebury, "Cold Spring Harbor Tops Among Independent Labs, *The Scientist* 4 (19 March 1990): 20; JDW, CSHL Annual Report, 1989, 19.

101. Blose interview.

102. Tjian interview.

103. Crouch interview.

104. Pollack interview.

105. Botchan interview.

106. Robert Reinhold, "Superstars of Culture Twinkle on Florida Beach," *New York Times,* 18 November 1974, 35.

107. Robert Reinhold, "Student's Forgery Perils Key Harvard Research," *New York Times,* 16 December 1974, 1, 56; Norton Zinder, personal communication.

108. Matthew Meselson, interview by author.

109. Robert Horvitz, interview by author.

110. Roberts interview.

111. Maniatis interview.

112. Mark Ptashne, interview by author; Mark Ptashne, personal communication.

113. JDW, remarks, seventieth-birthday celebration, Harvard University, 24 May 1998 (transcript).

114. JDW Remarks, Science Center B, Harvard University, 30 September 1999 (transcript and author notes).

115. Sharp interview.

116. JDW, CSHL Annual Report, 1975, 4-15.

117. JDW, CSHL Annual Report, 1995, 2.

118. Watson, *Houses for Science,* 207-208.

119. Botstein interview. 1

120. Joseph Palca, "National Research Council Endorses Genome Project," *Nature* 331 (11 February 1988): 467 (citing NAS report).

121. Botstein interview.

122. JDW, CSHL Annual Report, 1969, 5-7; JDW, CSHL Annual Report, 1970, 7-8.

123. Botstein interview.

124. JDW, CSHL Annual Report, 1975.

125. David Botstein, "A Phage Geneticist Turns to Yeast," in Michael N. Hall and Patrick Linder, eds., *The Early Days of Yeast Genetics* (Cold Spring Harbor, N.Y.: Cold Spring Harbor Laboratory Press, 1993), 361-373.

126. Gerald Fink, interview by author.

127. Botstein interview.

128. Davis, "The World According to Jim," 40-41.

129. JDW, CSHL Annual Report, 1978, 12-13.

130. CSHL Annual Report, 1980, 115–120.

131. Thomas H. Maugh II, "2 Share Nobel Prize for Finding 'Split Genes,'" *Los Angeles Times,* 12 October 1993, A 1.

132. Roberts interview.

133. Michel Morange, *A History of Molecular Biology,* trans. Matthew Cobb (Cambridge, Mass.: Harvard University Press, 1998), 204–214; Max Perutz,

"Bizarre Behaviour Among the Messengers," *New Scientist,* 5 January 1978, 8-9; W. Ford Doolittle, "Genes in Pieces: Were They Ever Together?" *Nature* 272 (13 April 1978): 581-582.

134. T. R. Broker and L. Chow, CSHL Annual Report, 1977, 43.

135. JDW, CSHL Annual Report, 1977, 7.

136. Jeffrey L. Fox, "New mRNA Segments Intrigue Scientists," *Chemical and Engineering News,* 8 August 1977, 19-21; John Rogers, "Genes in Pieces . . . " *New Scientist,* 5 January 1978, 18-20; Philip Leder, "Discontinuous Genes," editorial, *The New England Journal of Medicine* 298 (11 May 1978): 1079-1081; C. C. F. Blake, "Do Genes-In-Pieces Imply Proteins-In-Pieces?" *Nature* 273 (25 May 1978): 267; James E. Darnell, Jr., "Implications of RNA-RNA Splicing in Evolution of Eukaryotic Cells," *Science* 202 (22 December 1978): 1257-1260; Pierre Chambon, "Split Genes," *Scientific American* 244 (1981): 60-71; James E. Darnell, Jr., "The Processing of RNA," *Scientific American* 248 (1985): 89-99; Jan Witkowski, "Reflections of Biochemistry; the Discovery of 'Split' Genes: A Scientific Revolution," *Trends in Biological Sciences* 13 (March 1988): 110-113; Paul Doty, personal communication.

137. Lawrence K. Altman, "Surprise Discovery About 'Split Genes' Wins Nobel Prize," *New York Times,* 12 October 1993, C3; Walter Gilbert, "Why Genes in Pieces?" *Nature* 271 (9 February 1978): 501; Walter Gilbert, interview by author.

138. Stephen S. Hall, *Invisible Frontiers: The Race to Synthesize a Human Gene* (New York: Atlantic Monthly Press, 1997), 238-239.

139. Francis Crick, "Split Genes and RNA Splicing," *Science* 204 (20 April 1979): 264, 270.

140. Francis Crick, "Predictions in Biology," *Chemtech,* May 1979, 301.

141. James D. Watson, Nancy H. Hopkins, Jeffrey W. Roberts, Joan Argetsinger Steitz, and Alan M. Weiner, *Molecular Biology of the Gene,* 4th ed. (Menlo Park: Benjamin/Cummings, 1988), 621.

142. JDW, CSHL Annual Report, 1993, 5. CSHL Annual Report, 1977, 69.

143. Altman, "Surprise Discovery About 'Split Genes' Wins Nobel Prize."

144. Maugh, "2 Share Nobel Prize for Finding 'Split Genes.' "

145. Sharp interview.

146. Roberts interview.

147. JDW, CSHL Annual Report, 1993, 11-13.

148. Sharp interview.

149. Roberts interview; Lan Bo Chen interview; Anthony Flint, "Behind Nobel, a Struggle for Recognition," *Boston Globe,* 5 November 1993, 1, 20; Jon Cohen, "The Culture of Credit," *Science* 268 (23 June 1995): 1706-1711.

150. Roberts interview.

151. Ibid.

152. Flint, "Behind Nobel, a Struggle for Recognition"; Cohen, "Culture of Credit,"

153. Flint, "Behind Nobel, a Struggle for Recognition."

154. Cohen, "Culture of Credit."

155. JDW, CSHL Annual Report, 1969, 5-7.
156. JDW, CSHL Annual report 1993, 7.
157. Charles Stevens, interview by author.
158. JDW, CSHL Annual Report, 1972.
159. Judson, *Eighth Day of Creation,* 26-27.
160. JDW, CSHL Annual Report, 1980, 5.
161. Watson, *Houses for Science,* 241; JDW, CSHL Annual Report, 1978, 11-12.
162. JDW, CSHL Annual Report, 1975, 9; JDW, foreword, *Cold Spring Harbor Symposia on Quantitative Biology* 40 (1975).
163. JDW, CSHL Annual Report, 1975, 4-15.
164. JDW, CSHL Annual Report, 1980, 9-10; Frances Cerra, "Business Woos Scientists at Cold Spring Harbor Lab," *New York Times,* 22 November 1981, section 11, 1; JDW, CSHL Annual Report, 1984, 5; Watson, *Houses for Science,* 241.
165. JDW, CSHL Annual Report, 1980, 10-11.
166. Stevens interview.
167. JDW, CSHL Annual Report, 1981, 12-13; Ron McKay, interview by author.
168. JDW and Ron McKay, foreword, *Cold Spring Harbor Symposia on Quantitative Biology* 48 (1983).
169. Joseph Sambrook, CSHL Annual Report, 1983, 2-3.
170. Botstein interview.
171. Stevens interview.
172. JDW, CSHL Annual Report, 1991, 8.

CHAPTER 11 "ODD MAN OUT": RECOMBINANT DNA

1. JDW, "The Ethics of Recombinant DNA," public lecture, Cornell University, sponsored by the Division of Biological Sciences and the Program on Science, Technology, and Society, November 29, 1976, typescript of expanded version, 15–18.
2. Lee Edson, "Says Nobelist James (Double Helix) Watson, 'To Hell With Being Discovered When You're Dead,' " *New York Times Magazine,* 18 August 1968, 46; James D. Watson, *Genes, Girls, and Gamow: After The Double Helix* (New York: Knopf, 2002), 248.
3. JDW, interview by Tim Radford, *The Guardian* (London), 25 March 2000, Home Pages, 10; author's notes.
4. Robert M. Krim, "J. D. Watson Advised Government on Chemical-Biological Warfare," *Harvard Crimson,* 4 October 1968.
5. JDW, "Ethics of Recombinant DNA," 15–18.
6. See, for example, Stuart Auerbach, "New Attack Made on Genetic Malady," *Washington Post,* 23 April 1971, C3; Jane E. Brody, "Prenatal Diagnosis Is Reducing Risk of Birth Defects," *New York Times,* 3 June 1971, 41; Ron

Cooper, "Huntington's Chorea, a Genetic Time Bomb, Brings Untold Tragedy," *Wall Street Journal*, 22 June 1971, 1.

7. Christopher Anderson, "In His Own Words: Nobel Laureate James Watson Calls Report of Cloning People 'Science Fiction Silliness,' " interview with JDW, *People*, 17 April 1978, 93–94, 96–97.

8. Ibid.

9. JDW, "Moving Toward the Clonal Man: Is This What We Want?" *The Atlantic*, May 1971, 50–53, adapted from a lecture given at the twelfth meeting of the Panel on Science and Aeronautics, U.S. House of Representatives, January 1971, reprinted in JDW, *A Passion for DNA* (Cold Spring Harbor, N.Y.: Cold Spring Harbor Laboratory Press, 2000); "DNA and the Sorcerer's Apprentice," editorial, *Los Angeles Times*, 4 February 1971, A22; "Man into Superman: The Promise and Peril of the New Genetics," *Time*, 19 April 1971, 51; Daniel J. Kevles, *The Physicists: The History of a Science Community in Modern America* (Cambridge, Mass.: Harvard University Press, 1987 [paperback]), 408; Daniel J. Kevles and Leroy Hood, eds., *The Code of Codes: Scientific and Social Issues in the Human Genome Project* (Cambridge, Mass.: Harvard University Press, 1993 [paperback]), 34–35.

10. Philip Abelson, "Anxiety About Genetic Engineering," *Science* 173 (23 July 1971): 285.

11. JDW, CSHL Annual Report, 1980, 6–7.

12. JDW, CSHL Annual Report, 1981, 5–7.

13. JDW, "The Cancer Conquerors: Who Should Get All That New Money?" *The New Republic* 166 (26 February 1972): 17–21.

14. "The War on Cancer," *The Nation* 217 (21 May 1973): 645; Richard Knox, "Nobel Winner Calls Nixon's Medical Research Policy 'Lunacy,'"*Boston Globe*, 7 March 1973, 67.

15. JDW, CSHL Annual Report, 1973, 4–12; JDW, CSHL Annual Report, 1974, 4–13.

16. "2 Cancer Experts Split on Program," *New York Times*, 9 March 1975, 51; Jane E. Brody, "U.S. Aides Cite Great Progress in Cancer Fight, Disputing Critics," *New York Times*, 22 March 1975, 12; Benno C. Schmidt (chairman, President's Cancer Panel), "Cancer: As the Battle Continues," letter, *New York Times*, 22 April 1975, 34.

17. Harold M. Schmeck, Jr., "War on Cancer Stirs a Political Backlash," *New York Times*, 27 May 1975, 1.

18. Zinder, personal communication, 9 May 2002.

19. Paul Berg, interview by Rae Goodell, 17 May 1975, Manuscript Collection 100, Recombinant DNA Archive, MIT (henceforth cited as MIT Recombinant DNA Archive), box 1, folder 8 (transcript), 40; Paul Berg, interview by author.

20. Nicholas Wade, "Hazardous Profession Faces New Uncertainties," *Science* 182 (9 November 1973): 566–567; Berg, Goodell interview; JDW, "The Ethics of Recombinant DNA," 2; JDW and John Tooze, *The DNA Story: A Documentary History of Gene Cloning* (San Francisco: W. H. Freeman, 1981); JDW, "In Defense of DNA: We're Going Wild over Bugs We've Already Eaten," *The New*

Republic 176 (25 June 1977): 11, reprinted in JDW and Tooze, *DNA Story*, 236; Arthur Lubow, "Playing God with DNA," *New Times*, 7 January 1977, reprinted in JDW and Tooze, *DNA Story*, 118–126; Berg, author interview; T. A. Heppenheimer, "The DNA Dilemma," *American Heritage of Invention and Technology* 16, 4 (Spring 2001): 10; author's notes at Asilomar, 16 February 2002.

21. JDW and John Tooze, *DNA Story*, 2; Berg, Goodell interview; Norton D. Zinder, "The Gene, The Scientists, and The Law," speech, 20 December 1977, reprinted in JDW and Tooze, *DNA Story*, 195–201.

22. Berg, author interview.

23. *Biohazards in Biological Research,* Cold Spring Harbor Laboratory, 1973; CSHL Annual Report, 1973; Wade, "Hazardous Profession Faces New Uncertainties."

24. Author's notes at Asilomar 16 February 2002.

25. Victor K. McElheny, "Animal Gene Shifted to Bacteria; Aid Seen to Medicine and Farm," *New York Times,* 20 May 1974, 1; Stephen S. Hall, *Mapping the Next Millennium: The Discovery of New Geographies* (New York: Random House, 1992), 181.

26. JDW, CSHL Annual Report, 1981, 6.

27. François Gros, *The Gene Civilization*, trans. Lee F. Scanlon (New York: McGraw Hill, 1991), 45.

28. Arthur Kornberg, M.D., "Biochemistry at Stanford, Biotechnology at DNAX," oral history by Sally Smith Hughes, 1997, for Regional Oral History Office, Bancroft Library, University of California, Berkeley (transcript), posted at www.Sunsite. Berkeley.edu /2020/dynaweb/teiproj/science/Kornberg; author's notes.

29. Peter Lobban and A. D. Kaiser, "Enzymatic End-to-End Joining of DNA Molecules," *Journal of Molecular Biology* 79 (1973): 453–471.

30. J. E. Mertz and R. W. Davis, "Cleavage of DNA by RI Restriction Endonuclease Generates Cohesive End," *Proceedings of the National Academy of Sciences* 69 (1972): 3370–3374; Henry Harris, *The Cells of the Body: A History of Somatic Cell Genetics* (Cold Spring Harbor, N.Y.: Cold Spring Harbor Laboratory Press, 1995), 161–163.

31. D. R. Jackson, R. Symons, and P. Berg, "Biochemical Method for Inserting New Genetic Information into DNA of Simian Virus 40," *Proceedings of the National Academy of Sciences* 69 (1972): 2904–2909.

32. Stanley Cohen et al., "Construction of Biologically Functional Bacterial Plasmids in Vitro," *Proceedings of the National Academy of Sciences* 70 (1973): 3240–3244; Mark Ptashne, personal communication.

33. JDW and Tooze, *DNA Story*, 3.

34. Maxine Singer and Dieter Söll, "Guidelines for DNA Hybrid Molecules," *Science* 181 (21 September 1973): 1114; National Academy of Sciences, press conference, 18 July 1974 (typescript), 1–2; JDW, "The Ethics of Recombinant DNA"; Lubow, "Playing God with DNA," 122; JDW, CSHL Annual Report, 1976, 5; Zinder, "The Gene, the Scientists, and the Law"; Zinder, personal communication.

35. Berg, Goodell interview.

36. Stanley Cohen et al., "Construction of Biologically Functional Bacterial Plasmids in Vitro, *Proceedings of the National Academy of Sciences* 70 (1973)," 3240; "Genetic Manipulation," *Nature* 247 (8 February 1974): 336–337. Mark Ptashne,personal communication.

37. John F. Morrow et al., "Replication and Transcription of Eukaryotic DNA in *Escherichia coli*," *Proceedings of the National Academy of Sciences* 71 (May 1974): 1743–1747; Victor McElheny, "Animal Gene Shifted to Bacteria; Aid Seen to Medicine and Farm," *New York Times*, 20 May 1974; Berg, author interview; Mark Ptashne, personal communication.

38. Norton Zinder, interview by Charles Weiner, 2 September 1975, Recombinant DNA Archive, box 16, folder 203 (transcript), 68–77.

39. Berg, Goodell interview, 43.

40. Zinder, Weiner interview, 71.

41. Zinder, personal communication.

42. JDW, "Ethics of Recombinant DNA," 4.

43. JDW, CSHL Annual Report, 1978, 6.

44. JDW, CSHL Annual Report, 1977, 5.

45. Zinder, Weiner interview, 72.

46. Norton D. Zinder, "The Berg Letter: A Statement of Conscience, Not of Conviction," *Hastings Center Report*, October 1980, 14.

47. Zinder, Weiner interview, 73–80.

48. David Baltimore, interview, 12 May 1975, MIT Recombinant DNA Archive, MIT, box 1, folders 4–6 (transcript), 71.

49. Paul Berg et al., "Potential Biohazards of Recombinant DNA Molecules," *Science* 185 (26 July 1974): 303.

50. JDW, "The Ethics of Recombinant DNA," 5.

51. JDW, "A Biologist's Plea for Science: DNA Folly Continues," *The New Republic*, 13 January 1979, 12–15.

52. Author's notes at Asilomar, 16 February 2002.

53. National Academy of Sciences, press conference, 18 July 1974, draft transcript, 18–19.

54. Author's notes on Asilomar.

55. JDW, CSHL Annual Report, 1977, 5–7; JDW, *People* interview.

56. Robert L. Sinsheimer, "Humanism and Science," *Science and Engineering* (Caltech), October–November 1975, 10.

57. Zinder, personal communication.

58. JDW, "The Ethics of Recombinant DNA," 10–11.

59. Ronald Davis, interview by Rae Goodell, 19 May 1975, MIT Recombinant DNA Archive, box 4, folder 39 (transcript), 51.

60. Ibid., 53.

61. David Perlman, interview, MIT Recombinant DNA Archive, box 12 (transcript), 25; JDW, "Ethics of Recombinant DNA," 8.

62. Author's notes at Asilomar, 25 February 1975.

63. Berg, Goodell interview, 97–98.

64. JDW, "The Ethics of Recombinant DNA," 9.

65. Author's notes at Asilomar, 25 February 1975.

66. Ibid.

67. JDW, "Ethics of Recombinant DNA," 13.

68. Lubow, "Playing God With DNA," 122.

69. JDW, "Ethics of Recombinant DNA," 13.

70. JDW, sixtieth-birthday remarks, Grace Auditorium, Cold Spring Harbor, 16 April 1988 (author's notes).

71. Howard Hiatt, James Watson, and Jay Winsten, eds., *Origins of Human Cancer*, Cold Spring Harbor Conferences on Cell Proliferation (Cold Spring Harbor, N.Y.: Cold Spring Harbor Laboratory, 1977).

72. Liebe F. Cavalieri, "New Strains of Life—Or Death,' " *New York Times Magazine*, 22 August 1976.

73. Zinder, personal communication.

74. JDW, "The Ethics of Recombinant DNA," 15–16.

75. Norton D. Zinder, Time Inc. Workshop on DNA, Banbury Center, Cold Spring Harbor Laboratory, 4 May 1981, session 4 (transcript), 22.

76. Nicholas Wade, "Recombinant DNA: A Critic Questions the Right to Free Inquiry," *Science* 194 (15 October 1976): 303, reprinted in JDW and Tooze, *DNA Story*, 146–148.

77. Norton D. Zinder, "The Gene, the Scientists, and the Law," in JDW and Tooze, *DNA Story*, 199.

78. Heppenheimer, "DNA Dilemma," 13.

79. JDW and Tooze, *DNA Story*, 137.

80. JDW, "Ethics of Recombinant DNA," 22.

81. JDW, testimony, New York State, Attorney General's Hearings on Recombinant DNA, World Trade Center, 21 October 1976, MIT Recombinant DNA Archive, folder 450 (transcript), 90–91.

82. JDW, "The Ethics of Recombinant DNA," 22.

83. JDW, Recombinant-DNA testimony, 21 October 1976, 98.

84. JDW, "The Ethics of Recombinant DNA," 24.

85. JDW, "Recombinant DNA Research: A Debate on the Benefits and Risks," *Chemical and Engineering News*, 30 May 1977, 26.

86. JDW, "Trying to Bury Asilomar," *Clinical Research*, April 1978.

87. JDW, "A Biologist's Plea for Science: DNA Folly Continues," *The New Republic*, 13 January 1979.

88. JDW, "The DNA Biohazard Canard," in JDW, *A Passion for DNA* (Cold Spring Harbor, N.Y.: Cold Spring Harbor Laboratory Press, 2000), 75–79.

89. JDW, "Let Us Stop Regulating DNA Research," *Nature* 278 (8 March 1979), 113.

90. JDW, *People* interview.

91. Lubow, "Playing God with DNA."

92. JDW and Tooze, *DNA Story*, 186; "Concentrates: Science," *Chemical and Engineering News*, 2 May 1977, 23.

93. Lubow, "Playing God with DNA."

94. Stanley N. Cohen, "Recombinant DNA: Fact and Fiction," *Science* 195 (18 February 1977): 654–657.

95. Nicholas Wade, "Gene Splicing: Senate Bill Draws Charges of Lysenkoism," *Science* 197 (22 July 1977): 348, 350.

96. "Genetic Research Ban Extended by Cambridge," *New York Times*, 7 January 1977, 19.

97. "California Weighing Curbs on Gene Study," *New York Times*, 7 February 1977, 13; "California to Consider DNA Limits," *New York Times*, 4 March 1977, A10.

98. Zinder, "The Genes, the Scientists, and the Law."

99. JDW and Tooze, *DNA Story*, 140.

100. JDW, CSHL Annual Report, 1978, 8.

101. Kornberg, "Biochemistry at Stanford, Biotechnology at DNAX," oral history, Berkeley.

102. Nicholas Wade, "NIH Seeks Law on Gene-Splice Research," *Science* 195 (25 February 1977): 762; Gail Bronson, "Control of All Genetic Research Facilities by HEW Proposed by Interagency Panel," *Wall Street Journal*, 14 March 1977, 20; Roger Lewin, "U.S. Genetic Engineering in a Tangled Web," *New Scientist*, 17 March 1977, 640–641; "Gene Splicing Ties Scientists in Knots," *Chemical Week*, 23 March 1977, 41.

103. Donald S. Frederickson, "A History of the Recombinant DNA Guidelines in the United States," prepared for the Conference on Recombinant DNA and Genetic Experimentation, Wye College, Kent, U.K., 1–4 April 1979 (presented by W. J. Gartland), reprinted in JDW and Tooze, *DNA Story*, 398.

104. Harold M. Schmeck, Jr., "Science Group Urges New Delays On Experiments in Gene Splicing," *New York Times*, 8 March 1977, 12; Harold M. Schmeck, Jr., "Gene Engineering Causes Sharp Split at Scientist Forum," *New York Times*, 10 March 1977, 19; Roger Lewin, "U.S. Genetic Engineering in a Tangled Web," *New Scientist*, 17 March 1977, 640–641; Jeff Fox, "Concern Widens over Recombinant DNA Research," *Chemical and Engineering News*, 21 March 1977, 23, 26; Norton Zinder, letter to Paul Berg et al., 6 September 1977, reprinted in JDW and Tooze, *DNA Story*, 258; Raymond A. Zilinskas and Burke Zimmerman, eds., *The Gene Splicing Wars: Reflections on the Recombinant DNA Controversy* (New York: Macmillan, 1986), 38.

105. Robert L. Sinsheimer, *The Strands of a Life: The Science of DNA and the Art of Education* (Berkeley: University of California Press, 1994), 140–141; Tom Maniatis, interview by author.

106. Philip Handler, "Editor's Page," *Chemical and Engineering News*, 9 May 1977, 3.

107. Roger Lewin, "Scientists' Backlash Against US Legislation on DNA," *New Scientist*, 23 June 1977, 692; Walter Gilbert, "Recombinant DNA Re-

search: Government Regulation," *Science* 197 (15 July 1977): 208; Wade, "Gene Splicing: Senate Bill Draws Charges of Lysenkoism."

108. Burke K. Zimmerman, "DNA comes to Washington," in Zilinskas and Zimmerman, *Gene Splicing Wars,* 43; Stanley Cohen, letter to Harley Staggers, 7 October 1977, in JDW and Tooze, *DNA Story,* 187–188.

109. Author's notes at Asilomar, 16 February 2002.

110. State of New York, Executive Chamber, Hugh L. Carey, Governor, "An Act to Amend the Public Health Law," press release, 12 August 1977; Heppenheimer, "DNA Dilemma"; JDW, letter to Stanley Cohen, Stanford University, 17 August 1977, in JDW and Tooze, *DNA Story,* 254; Zinder, "The Genes, the Scientists, and the Law," 200.

111. Zinder, letter to Berg et al., 6 September 1977, in *DNA Story* 255–259.

112. Edward M. Kennedy, address to medical writers, New York, 27 September 1977, reprinted in JDW and Tooze, *DNA Story,* 173–174; Barbara Culliton, "Recombinant DNA Bills Derailed: Congress Still Trying to Pass a Law," *Science* 199 (20 January 1978), 274–277.

113. JDW, CSHL Annual Report, 1978.

114. Harold M. Schmeck Jr., "Rules on DNA Studies Viewed as Too Strict," *New York Times,* 18 December 1977, 19; Frederickson, "History of the Recombinant DNA Guidelines," 399.

115. Tony Durham, "The Race to Map Human Life," interview of James Watson," *The Times Higher Education Supplement,* 2 October 2000.

116. Natalie Angier, "Great 15-Year Project to Decipher Genes Stirs Opposition," *New York Times,* C1.

117. Todd Ackerman, "The New Millennium; A Brave New World of Designer Babies," *Houston Chronicle,* 27 June 1999, A1; Gina Kolata, "Scientists Brace for Changes in Path of Human Evolution," *New York Times,* 21 March 1998, A1; Joannie Fischer, "Passing on Perfection," *U.S. News & World Report,* 2 October 2000.

118. JDW, Radford interview, 10.

119. Fischer, "Passing on Perfection"; JDW, speech at ARCO Forum, John K. Kennedy School of Government, Harvard University.

120. JDW, speaking on *Donahue,* broadcast 25 September 1981; author's notes.

Chapter 12 Genome: "It Is So Obvious"

1. Leslie Roberts, "Academy Backs Genome Project," *Science* 239 (12 February 1988): 725–726; Harold M. Schmeck, Jr., "Scientists Urge Huge Project to Chart All Human Genes," *New York Times,* 12 February 1988, A1; Larry Thompson and Susan Okie, "$3 Billion Effort Urged to Map Human Genes," *Washington Post,* 12 February 1988, A3; Larry Thompson, "Mapping the Human Genes; Is the Mega-Project Politically a 'Go'?" *Washington Post,* 16 February 1988, Z8; Joseph Palca, "National Research Council Endorses Genome Project," *Nature* 331 (11 February 1988): 467; Will Lepkowski, "Program to

Map Entire Human Genome Urged," *Chemical and Engineering News,* 15 February 1988, 5; "Top U.S. Bio-scientists Urge $3-billion, 15-year Expedition to Map Genome," *Biotechnology Newswatch,* 22 February 1988, 7; Reginald Rhein, "Genetics: Making War on Genetic Diseases; a $3 Billion Effort Is Proposed," *Business Week,* 29 February 1988, 74F.

2. Robert C. Cowen, "Reading the Human Rosetta Stone," *Christian Science Monitor,* 9 September 1987, 16; Roger Lewin, "In the Beginning Was the Genome," *New Scientist,* 21 July 1990, 34–38.

3. Thompson, "Mapping the Human Genes"; Larry Thompson, "Science Under Fire: Behind the Clash Between Congress and Nobel Laureate David Baltimore," *Washington Post,* 9 May 1989, Health section, 13.

4. Harold Varmus, interview by author.

5. Roberts, "Academy Backs Genome Project."

6. Robert Locke, "$3 billion Gene-Mapping Project Debated," *San Diego Union-Tribune,* 16 February 1988, A3.

7. Leslie Roberts, "Human Genome: Questions of Cost," *Science* 237 (18 September 1987): 1411–1412.

8. JDW, CSHL Annual Report, 1988, 4–5.

9. Lewin, "In the Beginning Was the Genome"; Robert Cook-Deegan, *The Gene Wars: Science, Politics, and the Human Genome* (New York: Norton, 1994), 166 (citing 1991 interviews with James Wyngaarden, Charles Cantor, and Rachel Levinson); Leslie Roberts, "Watson May Head Genome Office," *Science* 240 (24 June 1988): 878–879.

10. JDW, "The Human Genome Project: Past, Present, and Future."

11. JDW, CSHL Annual Report, 1988, 5.

12. Leslie Roberts, "Controversial from the Start," *Science* 291 (16 February 2001): 1181–1188.

13. Heiner Westphal, interview by author; Renato Dulbecco, "Perspective: A Turning Point in Cancer Research: Sequencing the Human Genome," *Science* 231 (7 March 1986): 1055–1056.

14. Louis Kunkel, personal communication.

15. Jerry E. Bishop and Michael Waldholz, *Genome* (New York: Simon & Schuster, 1999 [paperback]), 49–68; Cook-Deegan, *Gene Wars,* 36–41; "Gazing into the Crystal Ball of Genetic Research," *The Economist,* 23 April 1983, U.S. ed., 97; Matt Clark and Mia Gosnell, "Medicine: A Brave New World," *Newsweek,* 5 March 1984, 65; Harold M. Schmeck, "Potent Tool Fashioned to Probe Inherited Ills," *New York Times,* 11 August 1987, C1.

16. Natalie Angier, "Great 15-Year Project to Decipher Genes Stirs Opposition," *New York Times,* 5 June 1990, C1.

17. Christopher Wills, *Exons, Introns, and Talking Genes: The Science Behind the Human Genome Project* (New York: Basic Books, 1991), 71–75; B. D. Davis, "Sequencing the Human Genome: A Faded Goal," *Bulletin of the New York Academy of Medicine* 68, 1 (January–February 1992): 115–125.

18. Harold M. Schmeck, "Gene Gains Buoy Hopes on Cancer," *New York Times,* 15 April 1983, D19; Harold M. Schmeck, "Cancer Gene Linked to Nat-

ural Human Substance," *New York Times,* 30 June 1983, B11; Harold M. Schmeck, "Burst of Discoveries Reveals Genetic Basis for Many Diseases," *New York Times,* 31 March 1987, C1; Bishop and Waldholz, *Genome.*

19. JDW, CHSL Annual Reports, 1980, 1981, 1982, 1985, 1988, 1989, 1997; Watson, *Houses for Science,* 234–235; Victor Cohn, "Discoveries Buttress Theory That Viruses Cause Human Cancer," *Washington Post,* 17 October 1981; Mark J. Murray et al., "Three Different Human Tumor Cell Lines Contain Different Oncogenes," *Cell* 25 (August 1981): 355–361; Harold M. Schmeck, "Some Genes Linked to Cancer Process," *New York Times,* 19 September 1981, 1; Sharon and Kathleen McAuliffe, "The Genetic Assault on Cancer," *New York Times Magazine,* 24 October 1982, 39; Schmeck, "Gene Research Gives New Hope on Cancer Causes," 16; Harold M. Schmeck, "Gene Gains," *New York Times,* 15 April 1983; Harold M. Schmeck, "Cancer Gene Linked to Natural Human Substance," *New York Times,* 30 June 1983, B11.

20. "Gazing into the Crystal Ball of Genetic Research"; Harold M. Schmeck, "Treatment Is Nearing for Genetic Defects," *New York Times,* 10 April 1984, C1,

21. "Computers: Machines That Splice Genes Automatically," *Business Week,* 12 January 1981, 29–30; Barnaby Feder, "Technology," *New York Times,* 15 January 1981, D2; "Automated DNA/RNA Synthesizers Debut," *Chemical and Engineering News,* 26 January 1981, 12; "Gene Machines Will Add Players to the DNA Game," *Chemical Week,* 4 February 1981, 52; David Fishlock, "Genetics' Pick and Shovel; How Applied Biosystems Produces Biotechnology Tools," *Financial Times* (London), 22 August 1983, 5; Marilyn Chase, "After Slow Start, Gene Machines Approach a Period of Fast Growth and Steady Profits," *Wall Street Journal,* 13 December 1983, 33, 51.

22. Stephen S. Hall, "Botstein's Caveat," *Technology Review,* September–October 2000, 115; Walter Gilbert, personal communication.

23. Robert L. Sinsheimer, "Historical Sketch: The Santa Cruz Workshop—May 1985," *Genomics* 5 (1989): 954–956; JDW, CHSL Annual Report, 1988; JDW, "The Human Genome Project: Past, Present, and Future"; Lewin, "In the Beginning Was the Genome"; S. S. Hall, "Genesis: The Sequel," *California,* July 1988, 62–69; Cook-Deegan, *Gene Wars,* 1994; Robert L. Sinsheimer, *The Strands of a Life: The Science of DNA and the Art of Education* (BerkeleyL University of California Press, 1994), 257–270; Hall, *Technology Review,* September–October 2000, 115; Roberts, "Controversial from the Start"; Walter Gilbert personal communication.

24. Roger Lewin, "Proposal to Sequence the Human Genome Stirs Debate," *Science* 232 (27 June 1986): 198–1600; Roger Lewin, "The Man with Credibility and Charisma," *New Scientist,* 21 July 1990, 36–37; JDW, CSHL Annual Report, 1988; Wills, *Exons, Introns, and Talking Genes,* 76.

25. Roberts, "Human Genome: Questions of Cost."

26. Roberts, "Agencies Vie Over Human Genome Project," *Science* 237 (31 July 1987): 488.

27. Tony Durham, "The Race to Map Human Life," interview of James Watson," *Times Higher Education Supplement,* 2 October 1998.

28. JDW, "Foreword," *Cold Spring Harbor Symposia on Quantitative Biology* 51 (1986): xv–xvi.

29. Cook-Deegan, *Gene Wars*, 110.

30. Ibid.

31. Thomas Maugh II, "Caltech Scientists Develop Superfast DNA Analyzer," *Los Angeles Times*, 12 June 1986, Part 1, 3; Jay Mathews, "Caltech: New DNA Analysis Machine Expected to Speed Cancer Research," *Washington Post*, 12 June 1986, A26; Keith Schneider, "Gene Mapping Is Improved," *New York Times*, 26 June 1986, D2; David Fishlock, *Financial Times*, 10 July 1986, 24; Jeannine Stein, "Superscientist Balances Home Life and Lab Life," *Los Angeles Times*, 14 July 1986, part 5, 1; Ricki Lewis, "Computerized Gene Analysis," *High Technology*, December 1986, 46–50.

32. William Allman, "The Amazing Gene Machine," *Business Week*, 16 July 1990, 53–54; Norman Arnheim and Corey Levenson, "Polymerase Chain Reaction," *Chemical and Engineering News*, 1 October 1990, 36–47; Henry A. Erlich, David Gelfand, and J. Sninsky, "Recent Advances in the Polymerase Chain Reaction," *Science* 252 (21 June 1991): 1649–1650; JDW et al., *Recombinant DNA*, 2d ed. (New York: Scientific American Books, 1992), xiii, 79–98; Cook-Deegan, *Gene Wars*, 72–77; Nicholas Wade, "Title TK," *New York Times*, 15 September 1998, F1.

33. Cook-Deegan, *Gene Wars*, 110.

34. David Botstein, author interview.

35. JDW, address, American Neurological Association, Marriott Copley Place, Boston, 17 October 2000 (transcript), 1430; Roger Lewin, "Proposal to Sequence the Human Genome Stirs Debate," *Science* 232 (27 June 1986): 1598.

36. Ibid., 1599.

37. Ibid., 1600.

38. Roberts, "Human Genome: Questions of Cost."

39. Philip M. Boffey, "Rapid Advances Point to the Mapping of All Human Genes," *New York Times*, 15 July 1986, Science Times, C1.

40. JDW, CSHL Annual Report, 1996, 1.

41. JDW, CSHL Annual Report, 1988, 2.

42. Walter Gilbert, "Genome Sequencing: Creating a New Biology for the Twenty-first Century," *Issues in Science and Technology*, Spring 1987, 26–35; Larry Thompson, "In Gene Mapping, an Opening Gambit," *Washington Post*, 21 July 1987, Z9; Sharon Begley, "The Genome Initiative: Will It Cure Disease or Rook the Taxpayer?" *Newsweek*, 31 August 1987, 60.

43. JDW, "The Double Helix Revisited," *Time*, 3 July 2000, 30.

44. Bruce Alberts, interview by author.

45. Cook-Deegan, *Gene Wars*, 128, 354; JDW, CSHL Annual Report, 1988, 1–11; JDW, "The Human Genome Project," *Science* 248 (6 April 1990): 45.

46. Cook-Deegan, *Gene Wars*, 132.

47. JDW, CSHL Annual Report, 1988, 3; Lewin, "In the Beginning Was the Genome."

48. JDW and John Tooze, *The DNA Story: A Documentary History of Gene Cloning* (San Francisco: W. H. Freeman, 1981).

49. Cook-Deegan, *Gene Wars*, 127–134.

50. National Research Council, *Mapping and Sequencing the Human Genome* (Washington, D.C.: National Academy Press, 1988), 108–110; Roberts, "Controversial from the Start."

51. John Burris, Robert Cook-Deegan, and Bruce Alberts, "The Human Genome Project After a Decade: Policy Issues," *Nature Genetics* 20 (20 December 1998): 333–335; David Botstein, interview by author; Alberts interview.

52. Lewin, "In the Beginning Was the Genome."

53. Botstein interview.

54. Gayle Golden, "Cataloguing the Genetic Code," *Toronto Star*, 31 January 1987, M13.

55. Roberts, "Controversial from the Start."

56. Charles Murray and Catherine Bly Cox, *Apollo: The Race to the Moon* (New York: Simon & Schuster, 1989), 113–120, 124–128; Walter A. McDougall, . . . *the Heavens and the Earth: A Political History of the Space Age* (New York: Basic Books, 1985), 289, 378–80.

57. JDW, CSHL Annual Report, 1996, 15.

58. Cook-Deegan, *Gene Wars*, 129; JDW, American Neurological Association address.

59. Harold Varmus, interview by author.

60. JDW, CSHL Annual Report, 1988, 4.

61. JDW, interview by Charlie Rose, 21 June 2000.

62. JDW, "The Double Helix Revisited," 30.

63. Daniel J. Kevles, *The Baltimore Case: A Trial of Politics, Science, and Character* (New York: Norton, 1998); David Baltimore, "Dear colleague," open letter, 17 May 1988 (author's collection); Barbara J. Culliton, "A Bitter Battle over Error," part 1, *Science* 240 (24 June 1988): 1720–1723; ibid., part 2, *Science* 241 (1 July 1988): 18–21; Walter W. Stewart and Ned Feder, "Battle over Error," letter, *Science* 242 (14 October 1988): 167; Barbara J. Culliton, letter (reply to preceding letter), *Science* 242 (14 October 1988): 167–168; Barbara Culliton, "NIH Panel Finds No Fraud in *Cell* Paper but Cites Errors," *Science* 242 (16 December 1988): 1499; Barbara Culliton, "Baltimore Cleared of All Fraud Charges," *Science* 243 (10 February 1989): 727–728; Phillip Sharp, "Dear colleague," letter, 18 April 1989, and David G. Nathan, letter to John D. Dingell, 26 April 1989 (both in author's collection); Barbara Culliton, "Dingell v. Baltimore," *Science* 244 (28 April 1989, 412–414; Robert E. Pollack, "In Science, Error Isn't Fraud; Dingell's Inquiry Is a Witch Hunt," *New York Times*, 2 May 1989, A25; Kathy A. Fackelman, "Trouble in the Laboratory," *Science News* 137 (31 March 1990): 200–203, 205; Peter G. Gosselin, "US Panel Orders Criminal Probe of Ex-MIT Researcher," *Boston Globe*, 15 May 1990, 1; "When Science Turns Nasty," *The Economist*, 9 June 1990, 87–88.

64. Kevles, *Baltimore Case*, 140–143.

65. William Booth, "A Clash of Cultures at Meeting on Misconduct," *Science* 243 (3 February 1989): 598; Joseph Palca, "Research, Misconduct, and Congress," *Nature* 337 (9 February 1989): 503.

66. Roberts, "Genome Project Under Way, at Last," *Science* 243 (13 January 1989): 167–168; "Genome Mapping Efforts Turning from Men and Mice to Microbes, Molds," *Biotechnology Newswatch*, 16 January 1989, 6; ibid., February 20, 1989, 6; Cook-Deegan, *Gene Wars*, 168; JDW, "The Human Genome Project"; Stu Borman, "Human Genome Project: Five-Year Plan Taking Shape," *Chemical and Engineering News*, 11 December 1989, 4.

67. Larry Thompson, "Science Under Fire: Behind the Clash Between Congress and Nobel Laureate David Baltimore," *Washington Post*, 9 May 1989, Health section, 12–16.

68. William K. Stevens, "Nobel Prize Winner Asked to Head Rockefeller U," *New York Times*, 4 October 1989, A18; William K. Stevens, "Dispute on New President Shatters Tranquil Study at Rockefeller U," *New York Times*, 10 October 1989, A1, C6.

69. Kevles, *Baltimore Case*, 283–288.

70. Robert Lee Hotz, "Biomedicine's Bionic Man," *Los Angeles Times Magazine*, 27 September 1997, 10.

71. Kevles, *Baltimore Case*, 1998, 256, 258–260, 264, 454–455; Paul Berg, interview by author; Paul Doty, "Responsibility and Weaver et al.," *Nature* 352 (1991): 183–184; John Edsall, "Something About Science and Ethics," *Journal of NIH Research* 3 (August 1991): 31–32; Richard Saltus, "Baltimore, Citing Furor, Quits as Head of University," *Boston Globe*, 3 December 1991, 1.

72. Robert Bazell, "A Pillar of Molecular Biology," review of *Ahead of the Curve: David Baltimore's Life in Science*, by Shane Crotty, *Science* 410 (12 April 2001): 746–747.

73. JDW, "No Campaign to Strip Baltimore of His Nobel," letter, *Science* 411 (10 May 2001): 131–132.

74. Cook-Deegan, *Gene Wars*, 166–167; JDW, CSHL Annual Report, 1988, 5.

75. Elaine Blume, "NIH Launches Human Genome Effort," *Journal of the National Cancer Institute* 80 (November 1988): 1356–1357.

76. Nicholas Wade, "Double Landmarks for Watson: Helix and Genome," *New York Times*, 27 June 2000, D5.

77. Pamela Zurer, "Watson to Head NIH's Human Genome Effort," *Chemical and Engineering News*, 3 October 1988, 7.

78. JDW, CSHL Annual Report, 1991, 4–6.

79. Blume, "NIH Launches Human Genome Effort."

80. Harold M. Schmeck Jr., "DNA Pioneer to Tackle Biggest Gene Project Ever," *New York Times*, 4 October 1988, C1.

81. Norton D. Zinder, personal communication, 9 May 2002.

82. Robert Cook-Deegan, *Gene Wars*, 163; Schmeck, "DNA Pioneer"; Wade, "Double Landmarks for Watson"; JDW, CSHL Annual Report, 1991, 4; Daniel J. Kevles, "Out of Eugenics: The Historical Politics of the Human

Genome," in Daniel J. Kevles and Leroy Hood, eds., *The Code of Codes: Scientific and Social Issues in the Human Genome Project* (Cambridge, Mass.: Harvard University Press, 1993 [paperback]), 35.

83. Schmeck, "DNA Pioneer."

84. Roberts, "Genome Project Under Way, at Last."

85. Richard Saltus, "World Interest Grows in a US Gene Project," *Boston Globe*, 13 June 1989, 7.

86. Catherine Arnst, "Ethics Issues Stand Out in Relief As Gene Mapping Advances," *Los Angeles Times,* 25 June 1989, part 1, 3.

87. Daniel S. Greenberg, "Q & A with James Watson, Genome Project Chief; Deplores Japan for 'Lack of Gratitude,' " *Science and Government Report* 20 (number 5): 1–5.

88. "Genetics Must Not Be Abused Again, Says DNA Pioneer," *Daily Telegraph* (London) 7 April 1990, 6.

89. Steve Connor, "DNA Pioneers Urge Tougher Controls on Genetic Data," *The Independent* (London), 8 April 1990, 3.

90. Nigel Williams, "The Monday Profile," *The Guardian* (London), 9 April 1990.

91. Robert Wright, "Mad Scientist: James D. Watson and the Human Genome Project," *The New Republic,* 9–16 July 1990, 21–5.

92. Bob Davis, "Watson Doesn't Use Gentle Persuasion to Enlist Japanese and German Support for Genome Effort," *Wall Street Journal,* 18 June 1990, A12.

93. "Japan's Science Council Urges Sequencing Genome in 10 Years," *Biotechnology Newswatch*, 20 February 1989, 7.

94. Thompson, "Science Under Fire."

95. Davis, "Watson Doesn't Use Gentle Persuasion"; Tom Maniatis, personal communication.

96. Davis, "Watson Doesn't Use Gentle Persuasion"; Norton Zinder, author interview.

97. Leslie Roberts, "Genome Center Grants Chosen," *Science* 249 (28 September 1990): 1497; Zinder interview.

98. JDW, CSHL Annual Report, 1995, 4; "European Firms Support Sequencing Yeast Genome," *Biotechnology Newswatch* 10 (19 March 1990): 1; Leslie Roberts, "The Worm Project," *Science* 248 (15 June 1990): 1310–1313; Leslie Roberts, "Controversial from the Start," *Science* 291 (16 February 2001): 1197; J. E. Ferrell, "Genetic Mapping Charts New World," *Los Angeles Times,* 17 December 1990, B3.

99. Angier, "Great 15-Year Project"; Scott Jaschik, "Many Scientists Charge Genome-Mapping Project Threatens Other Research," *Chronicle of Higher Education,* 18 July 1990, A1, A24; B. D. Davis et al., "The Human Genome and Other Initiatives," *Science* 249 (27 July 1990): 342–343.

100. Zinder, personal communication, 9 May 2002.

101. Jeffrey Mervis, "One Day in the Hard Life of the Genome Project," *The Scientist* 4, 16: 1, 4, 14.

102. Stephen S. Hall, "James Watson and the Search for Biology's 'Holy Grail,' " *Smithsonian,* February 1990, 40–49.

103. Mervis, "One Day in the Life of the Genome Project"; "Genome Critics Charge: Feds Spend Too Much, Achieve Little," *Biotechnology Newswatch* 10 (5 November 1990): 1.

104. James Watson and Norton Zinder, "Genome Project Maps Paths of Diseases and Drugs," letter, *New York Times,* 13 October 1990, 24.

105. Ronald Davis, author interview.

106. "Genome Critics Charge"; Wills, *Exons, Introns, and Talking Genes,* 82–83.

107. Natalie Angier, "Life's Machinery, Seen in a Translucent Worm," *New York Times,* 8 January 1991, C1; Roger Highfield, "Decode the Worm, Then Decode the Man; Breakthrough," *Daily Telegraph* (London), 19 August 1991, 11; Matthew Schofield, "Rejection Gets Credit for Researchers' Later Success," *Kansas City Star,* 1 February 1998, A20; Elizabeth Pennisi, "Worming Secrets from the *C. elegans* Genome," *Science* 282 (11 December 1998): 1972–1974; Nicholas Wade, "Animal's Genetic Program Decoded, in a Science First," *New York Times,* 11 December 1998, A1; Nicholas Wade, *Life Script: How the Human Genome Discoveries Will Transform Medicine and Enhance Your Health* (New York: Simon & Schuster, 2001), 31–32.

108. JDW, CSHL Annual Report, 1988, 23–24.

109. Lewin, "In the Beginning Was the Genome"; Schofield, "Rejection Gets Credit."

110. Robert J. Crouch, interview by author.

111. "Parasitologists Alter Protozoan Genome to Thwart Tsetse-Fly Pathogen," *Biotechnology Newswatch* 11 (18 February 1991): 4.

112. "Children's Hospital Is 7th Genome Center," *Biotechnology Newswatch* 11 (20 May 1991): 2.

113. Joseph Palca, "The Genome Project: Life After Watson," *Science* 256 (15 May 1992), 956.

114. "Berg to Head NIH Genome Committee," *Science* 252 (31 May 1991): 1249.

115. Leslie Roberts, "Report Card on the Genome Project," *Science* 253 (18 July 1991): 376.

116. Roger Highfield, "Science: Dr. Watson and a Worm That Turned: The Controversy," *Daily Telegraph,* 19 August 1991, 11.

117. Burris, Cook-Deegan, and Alberts, "Human Genome Project After a Decade."

118. Deborah Shapley, "Gene Genie—A look at One Scientist's Potentially Revolutionary Effect on Genetic-Related Business," *Financial Times* (London), 14 June 14 1994, 17; Keith Davies, *Cracking the Code: Inside the Race to Unlock Human DNA* (New York: Free Press, 2001), 59.

119. M. D. Adams et al., "Complementary DNA Sequencing: Expressed Sequence Tags and Human Genome Project," *Science* 251 (21 June 1991):

1651–1656; Richard Saltus, "Gene Patents: Weighing Protection vs. Secrecy," *Boston Globe*, 2 December 1991, Science and Technology, 25.

120. William Haseltine, interview by author.

121. JDW, *Talk of the Nation/Science Friday,* NPR, 2 June 2000.

122. JDW, interview by Charlie Rose, *Charlie Rose,* NPR.

123. JDW, CSHL Annual Report, 1995, 5–7.

124. Roberts, "Controversial from the Start"; Alex Barnum, "Biotech Labs Enraged by Bid to Patent Human Genes," *San Francisco Chronicle*, B1.

125. Cook-Deegan, *Gene Wars*, 326.

126. Durham, "Race to Map Human Life."

127. Richard Preston, "The Genome Warrior," *The New Yorker*, 12 June 2000, 71.

128. Keith Davies, *Cracking the Code: Inside the Race to Unlock Human DNA* (New York: Free Press, 2001), 62; Larry Thompson, "NIH's Rush to Patent Human Genes," *Washington Post*, 28 October 1991; Roberts, "Controversial from the Start."

129. "Gene Scientists Ponder: To Patent or Not to Patent cDNA?" *Biotechnology Newswatch* 11 (4 November 1991): 1.

130. Alex Barnum, "Biotech Labs Enraged by Bid to Patent Human Genes," *San Francisco Chronicle*, B1.

131. Christopher Anderson and Peter Aldhous, "Genome Project: Secrecy and the Bottom Line," *Nature* 354 (14 November 1991): 96; "Free Trade in Human Sequence Data?" *Nature* 354: 171–172; Christopher Anderson, "More Questions Than Answers," *Nature* 354: 174; David L. Wheeler, "Britain and Congress Respond to Controversy Sparked by NIH Plan to Patent Genes," *The Chronicle of Higher Education*, 27 November 1991, A29; Leslie Roberts, "MRC Denies Blocking Access to Genome Data," *Science* 254 (13 December 1991): 1583.

132. "NIH Patent Fight 'Fascinates'; Fair Price Dominates House Subcommittee," *Biotechnology Newswatch* 11 (2 December 1991): 4.

133. Thompson, "NIH's Rush to Patent Human Genes."

134. Warren E. Leary, "U.S. Scientists to Seek Patent on 2,375 More Genes," *New York Times,* 13 February 1992, B16; "U.S. Pursuit of Gene Patents Riles Industry," *Wall Street Journal,* 13 February 1992, B1, B3; Malcolm Gladwell, "NIH Seeks Patent Protection for Human Genes," *Washington Post,* 13 February 1992, A16; Richard Saltus, "Scientists Criticize NIH Bid for Patent on Gene Fragments," *Boston Globe,* 13 February 1992, 26; Alex Barnum, "NIH Plan to Patent Genes Continues to Draw Fire," *San Francisco Chronicle,* 13 February 1992, B1.

135. Jerry Bishop, "Opposition to Businessman's Worm Genome Project Led to Conflict Charges," *Wall Street Journal,* 13 April 1992, B9.

136. David Brown and Malcolm Gladwell, "Nobel Prize Biologist Watson Plans to Resign U.S. Position," *Washington Post,* 9 April 1992, A3.

137. Cook-Deegan, *Gene Wars*, 338.

138. Brown and Gladwell, "Nobel Prize Biologist Watson Plans to Resign U.S. Position."

139. Cook-Deegan, *Gene Wars,* 339.

140. "DNA Pioneer Quits Gene Map Project: Watson Resigns After Federal Review of His Holdings in Biology Companies," *New York Times,* 11 April 1992, 12.

141. JDW, March 2002 Sanders Theatre remarks.

142. JDW, American Neurological Association address.

143. Burris, Cook-Deegan, and Alberts, "Human Genome Project After a Decade," 335.

144. Haseltine interview.

145. Preston, "The Genome Warrior"; Nicholas Wade, "Gains are Reported in Decoding Genome," *New York Times,* 22 May 1999, A11; Nicholas Wade, "Scientist's Plan: Map All DNA Within 3 Years," *New York Times,* 10 May 1998, 1; Bill Richards, "Perkin-Elmer Jumps into Race to Decode Genes," *Wall Street Journal,* 11 May 1998, B6; "Team Says It Can Map Human DNA Within 3 Years," *Boston Globe,* 11 May 1998, A3; Nicholas Wade, "Beyond Sequencing of Human DNA," *New York Times,* 12 May 1998, F3; Carl T. Hall, "New Fight in Gene Race," *San Francisco Chronicle,* 13 May 1998, D1; Nicholas Wade, "International Gene Project Gets Lift," *New York Times,* 17 May 1998, 20; Wade, *Life Script,* 45–48.

146. JDW, Charlie Rose interview.

147. White House, "Remarks . . . On the Completion of the First Survey of the Entire Human Genome Project"; Walter Gilbert, personal communication.

148. Andrew Pollack, "The Microsoft (and Gates) of the Genome Industry," *New York Times,* 23 July 2000, section 3, 1.

149. Julia Boguslavsky, "Genomic Race Reaches Milestone," *Drug Discovery & Development,* June 2000, 11; Walter Gratzer, "Glitz Steals Glory Off Gene," *The Times Higher Education Supplement,* 14 September 2001, 28.

150. White House, "Remark. . . on the Completion of the First Survey of the Entire Human Genome Project."

Epilogue: "I'm an Optimist"

1. JDW, CSHL Annual Report 1993, 28–29.

2. JDW, Time 100 Scientist and Thinker, 24 March 1999, Time Room, online question and answer session, 23 February 2001.

3. Brenda Maddox, "The Dark Lady of DNA?" *The Observer,* 5 March 2000, 1.

4. Felicity Barringer, "Nobel Laureate Warns, Don't Learn Too Much," *New York Times,* 10 March 1993, B8.

5. JDW, "Values from a Chicago Upbringing," *Annals of the New York Academy of Sciences* 758 (1995): 171–173.

6. JDW, "Looking Forward," *Gene* 135 (15 December 1993): 309–315.

7. Dennis L. Breo, "The Double Helix—Watson and Crick's 'Freak Find of How Like Begets Like,'" *Journal of the American Medical Association* 269 (24 February 1993).

8. Lea Wee, "Gene Genius Continues up the Spiral Staircase," *The Straits Times* (Singapore), 29 November 1996, 2.

9. Carolyn Hong, "How Beautiful It Was, This Thing Called DNA," *New Straits Times* (Malaysia), 1 December 1996, 15.

10. Bryan Christie, *The Scotsman,* 1 October 1993.

11. Clive Cookson and Daniel Green, "Gene Is Out of the Bottle," *Financial Times* (London), 15 October 30, 1997, 15.

12. FDCH Political Transcripts, Federal Document Clearing House (transcript), 16 December 1997.

13. "Evening Hours," *New York Times,* 8 March 1998, Style section, 6.

14. Daniel Dennett, personal communication, 25 September 2001.

15. Gina Kolata, "Hope in the Lab: A Special Report—A Cautious Awe Greets Drugs That Eradicate Tumors in Mice," *New York Times,* 3 May 1998, 1; JDW, "High Hopes on Cancer," *New York Times,* 7 May 1998, A30; "EntreMed Stock Surge Wanes on New Wave of Information," *Medical Industry Today,* 8 May 1998; "More Cautions Issued on Cancer Advances," *Star Tribune* (Minneapolis), 9 May 1998, 21A; Atul Gawande, "Mouse Hunt; Forget Cancer—Is There a Cure for Hype?" *The New Yorker,* 18 May 1998, 5–6; Michael Shapiro, "Pushing the 'Cure': Where a Big Cancer Story Went Wrong," *Columbia Journalism Review,* July–August 1998, 15; Howard J. Lewis, "The Kolata Story. . . When No News Made Big News Over and Over," *Science Writers,* Spring–Summer 1998, 1–10; Matthew Miller, "Signs of the Times," *U.S. News and World Report,* 13 July 1998, 23; JDW, *Talk of the Nation/Science Friday,* NPR, 2 June 2000.

16. "Random Samples," *Science* 283 (5 January 1999): 323.

17. JDW, "Five Days in Berlin," In JDW, *A Passion for DNA: Genes, Genomes, and Society* (Cold Spring Harbor, N.Y.: Cold Spring Harbor Press, 2000), 209–222.

18. JDW, "All for the Good: Why Genetic Engineering Must Soldier On," *Time,* 11 January 1999, 91.

19. George Sweeney, review of *A Passion for DNA,* by James Watson, *Clinical and Investigative Medicine* (Canadian Medical Assn.) 242 (April 2001): 118–119.

20. Tom Shakespeare, "No Hope of Reality Modifying Brilliance," *Times Higher Education Supplement,* 19 January 2001, Books, 27.

21. Tom Abate, "All-Day Party to Honor the Discovery That Launched Biotech Revolution," *San Francisco Chronicle,* 15 March 1999; Tom Abate, "Nobel Winner's Theories Raise Uproar in Berkeley," *San Francisco Chronicle,* 13 November 2000, 1; Tom Abate, "Readers Defend Nobel Laureate's Opinions," *San Francisco Chronicle,* 11 December 2000; Jonathan Leake, "U.S. Nobel Laureate Sparks Sex-Race FLAP," *Toronto Star,* 7 January 2001, 1.

22. "DNA Pioneer Moves on to Bigger Things," *Daily Telegraph* (London), 21 June 2000, 29; Tom Peterkin, "Why Big Really Is Beautiful," *The Scotsman,* 25 June 2000, 6; Tracy McVeigh, "DNA pioneer: Fat Is Key to Good Sex: Top Geneticist Identifies Pleasures of the Flesh," *The Observer,* 2 July 2000, 7; "Fat and Sexy, Slim and Grim," editorial, *Toronto Star,* 7 July 2000; "Heavier Women Are Happier and Have Better Sex, Nobel Prize Scientist Says," *Jet,* 24 July 2000, 46.

23. Clare Thompson and Abi Berger, "Agent Provocateur Pursues Happiness," *British Medical Journal,* 1 July 2000.

24. Michael Botchan, interview by author.

25. Abate, "Nobel Winner's Theories Raise Uproar."

26. JDW, seventieth-birthday remarks, Harvard University, 24 May 1998 (transcript).

27. JDW, CSHL Annual Report, 1999.

28. JDW, interview by Kam Patel, "God's Little Helper," *The Times Higher Education Supplement,* 14 March 1997, 15.

29. Susanna Sirefman, "Jencks Spirals in Long Island," *Architectural Record,* October 2000, 48; Matt Ridley, personal communication.

30. JDW, interview by Charlie Rose, *Charlie Rose,* NPR, 21 June 2000 (transcript).

INDEX

ABOUT THE AUTHOR

Victor McElheny has been covering the explosive advances in post–double-helix biology since the beginning of his career as a science reporter in the 1950s. Biology was a dominant theme in his work in Europe for *Science* magazine in the mid-1960s, at the *Boston Globe* in the late 1960s and early 1970s, and at the *New York Times* in the 1970s; and also in organizing numerous seminars over 15 years for the Knight Science Journalism Fellowships at MIT, of which he was the inaugural director. In June 1970, while at the *Globe,* he wrote the first newspaper report of the discovery of the reverse transcriptase enzyme by David Baltimore and Howard Temin. While at the *Times,* he wrote the first newspaper story about the recombinant-DNA techniques for transferring genes from one species to another in May 1974 and covered the February 1975 international conference at Asilomar in California, which debated the potential risks of recombinant DNA. His first book, *Insisting on the Impossible,* a biography of Edwin H. Land, founder of Polaroid and the field of instant photography, was published in 1998 by Perseus Books.